LAURENCE E. JACOBSON
401 BROADWAY, ROOM 1001
NEW YORK CITY, N. Y. 10013

Vertical Transportation:

Elevators and Escalators

Vertical Transportation:

Elevators and Escalators

Second Edition

George R. Strakosch
Jaros, Baum & Bolles

A Wiley-Interscience Publication

JOHN WILEY & SONS

New York • Chichester • Brisbane • Toronto • Singapore

Library of Congress Cataloging in Publication Data:

Strakosch, George R.
  Vertical transportation.

  "A Wiley-Interscience publication."
  Bibliography: p.
  Includes index.
  1. Elevators.   2. Escalators.   I. Title.

TJ1370.S77 1982   621.8′77   82-10866
ISBN 0-471-86733-0

Printed in the United States of America

10 9 8 7 6

Preface

This second edition of *Vertical Transportation* is a tribute to the many people who have commented to me about the use and benefit of the first edition. My discussions with them revealed the need for an expanded and more detailed information source.

The realm of vertical transportation has also changed dramatically in the past decade. The people who are engaged in all the aspects of the application of vertical transportation to buildings—the initial design, installation, and ultimate use—are becoming increasingly aware of these changes and are finding it difficult to keep up with them. In recent years, numerous fires in high-rise buildings, at least one major earthquake, and many power failures have stranded people in elevators. In addition, a rapidly developing building technology has taken place with the use of lighter high-rise structures and a greater emphasis on safety. Vertical transportation has also been affected by social and urban development. Elevators and escalators are frequently being used in many diverse types of structures rather than the traditional multistory building. There has been an expansion of mass transit with a critical need for pedestrian-level change, and the needs of the mobility-limited or handicapped have been emphasized by both public awareness and legislation.

Automated material handling systems are coming to the forefront to relieve the increasing cost of mail and package delivery in both tall buildings and the newer, campus-type office complexes. Newer methods of automated horizontal transportation, utilizing the passenger-controlled techniques developed from elevator experience, are being encountered in many applications. The list can go on and on.

This edition of *Vertical Transportation* attempts to recognize and comment upon these various changes and trends. The person familiar with the first edition will quickly recognize what has happened. The size of the volume is almost doubled and closer scrutiny will reveal that the text and the techniques suggested for calculating elevator and escalator requirements demand more discipline from the user. Hopefully, these disciplined steps are not onerous since additional charts and tables are provided which can easily fit into the memory storage of a home-type computer.

The reader will also note a distinct emphasis on the suggested transportation design requirements of handicapped people as well as an emphasis on pedestrian circulation. Design requirements for the handicapped are increasingly being imposed by law and good pedestrian circulation can ensure the success of any project. In addition, various legislative bodies have imposed greater emphasis on safety requirements for elevators and escalators during building emergencies, such as earthquake, fire, power failure, and severe wind conditions, and these aspects are discussed.

Much additional information is given about unique and unusual vertical transportation approaches. There is also a discussion of the possible application of not so vertical (inclined) elevators and of horizontal transportation aspects utilizing the pedestrian-handling ability to which an elevator or escalator approach can be readily adapted.

The revision also reflects the changing status of the writer. After more than 30 years with the Otis Elevator Company and the opportunities afforded me at Otis to view and become involved in practically all aspects of vertical transportation, I am now an associate in one of the most progressive consulting engineering firms in the world specializing in commercial, institutional, and residential buildings. My experience with Jaros, Baum & Bolles has allowed me the opportunity to become involved in the additional aspects of applying and integrating vertical transportation in a building in conjunction with various other disciplines such as HVAC, power distribution, and life safety systems. The expeditious establishment of vertical transportation space into the building design and the constant change as the building design becomes final is an exciting aspect of the engineering field. It is thoroughly amazing how many different elevator and escalator schemes can be proposed for a given building of a specified gross area while building height, gross area per floor, zoning restrictions, building use, architectural requirements, and owners' desires are being resolved.

Once the vertical transportation system is decided on, the continuing demands on the elevator engineer include the allocation of space, the electrical requirements, the ventilation of machinery space, life safety systems, and, primarily, pedestrian access and circulation. A beautiful elevator or escalator scheme can be developed but if people cannot be comfortably and expeditiously accomodated the system can be a source of persistent aggravation. Hopefully, this book will be useful in providing sound and useful advice and aid the reader to consider all the aspects of good pedestrian circulation and transportation plus the essential ancillary functions of proper material handling.

I wish to acknowledge the help, guidance, and encouragement of my friends in the elevator industry while undertaking this revision. The job is far from finished since day by day new ideas, changes, and approaches are developing, technology is expanding and the future will be more exciting than the past.

I also want to express my appreciation to William S. Lewis for his review and critique which, I feel, has improved this volume and to Robert S. Caporale for his special help both of whom are my associates at Jaros, Baum and Bolles. My special thanks to my wife and family for their patience and understanding during the many hours of homework this extensive revision required.

George Strakosch

Albertson, New York
December 1982

Contents

Contents

Contents   xiii

Vertical Transportation:

Elevators and Escalators

CHAPTER ONE

The Essentials of Elevatoring

EARLY BEGINNINGS

Since the time man has occupied more than one floor of a building, he has given consideration to some form of vertical movement. The earliest forms were, of course, ladders, stairways, animal-powered hoists, and manually driven windlasses. Ancient Roman ruins show signs of shaftways where some guided movable platform type of hoist was installed. Guides or vertical rails are a characteristic of every modern elevator. In Tibet, people are transported up mountains in baskets drawn up by pulley and rope and driven by a windlass and manpower. An ingenious form of elevator, vintage about the eighteenth century, is shown in Figure 1.1. (note the guides for the one "manpower"). In the early part of the nineteenth century, steam-driven hoists made their appearance, primarily for the vertical transportation of material but occasionally for people. Results often were disastrous because the rope was of fiber and there was no means to stop the conveyance if the rope broke.

In the modern sense, an elevator* is defined as a conveyance designed to lift people and/or material vertically. The conveyance should include a device to prevent it from falling in the event the lifting means or linkage fails. Elevators with such safety devices did not exist until 1853 when Elisha Graves Otis invented the elevator safety device. This device was designed to prevent the free fall of the lifting platform if the hoisting rope parted. Guided hoisting platforms were common at that time, and Otis equipped one with a safety device that operated by causing a pair of spring-loaded dogs to engage the cog design of the guide rails when the tension of the hoisting rope was released (see Figure 1.2).

ELEVATOR SAFETY DEVICES

Although Otis' invention of the safety device improved the safety of elevators, it was not until 1857 that public acceptance of the elevator began. In that

---

* In England and other parts of the world the word "lift" is used. The legally recognized definition of an elevator can be found in ANSI/ASME A17.1, Safety Code for Elevators and Escalators.

Figure 1.1. A very early type of vertical transportation.

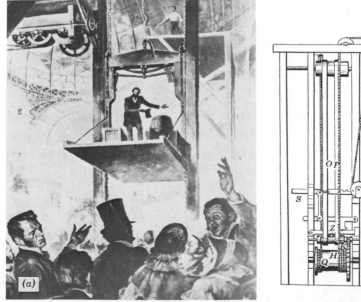

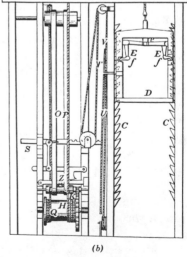

Figure 1.2. (a) Otis' demonstration, Crystal Palace, New York, 1853. (b) Otis' patent sketch for a safety device (Courtesy Otis Elevator).

year the first passenger elevator was installed in the store of E. V. Haughwout & Company in New York. This elevator traveled five floors at the then breathtaking speed of 40 fpm (0.20 mps).* Public and architectural approval followed this introduction of the passenger elevator. Aiding the technical development of the elevator was the availability of improved wire rope and the rapid advances in steam motive power for hoisting. Spurring architectural development was an unprecedented demand for "downtown" space. The elevator, however, remained a slow vertical "cog" railway for quite a few years. The hydraulic elevator became the spur that made the upper floors of buildings more valuable through ease of access and egress. Taller buildings permitted the concentration of people with multiple disciplines to bind together and caused the cities to grow in their present form during the 1870s and 1880s.

HYDRAULIC ELEVATORS

The hydraulic elevator provided a technological plateau for quite a few years; it was capable of higher rises and higher speeds than the steam-driven hoist-type elevator with its limits of winding drums (Figure 1.3). The hydraulic elevator also evolved from the direct ram-driven elevator to the so-called geared or roped hydraulic (Figure 1.4) capable of speeds of up to 700 fpm (3.5 mps) and rises of 30 or more stories. The cylinder and sheave arrangement was developed to use multiple sheaves and was mounted vertically for the higher rises. The 30-story building did not appear until after 1900, well after steel-frame construction was introduced, but the hydraulic elevator served practically all of the 10- to 12-story buildings of the 1880 to 1900 era.

It was in this era that many of the aspects of elevators as we know them today were introduced. Hoistways became completely enclosed and doors were installed at landings. Before that time many hoistways were simply holes cut in the floor—occasionally protected by railings or grillage. Simple signaling was introduced using bells and buzzers with annunciators to register a call which was manually canceled. Groups of elevators were installed, the first recorded group of four elevators being in the Boreel Building in New York City and the "majordomo" of "elevator buildings"—the starter—entered the scene and was assigned to direct the elevator operators to serve the riding public.

The first electric elevator quietly made its appearance in 1889 at the Demarest Building in New York. This elevator was a modification of a steam-driven drum elevator, the electric motor simply replacing the steam engine. It continued in service until 1920 when the building was torn down. Electric power was here to stay and the Otis Elevator Company installed the first automatic electric or push-button elevator in 1894.

With the tremendous building activity of the early 1900s and the increased size and height of buildings at that time, the questions of quantity, size, speed, and location of elevators began to arise. With these questions began the applied

---

* Elevator speed is traditionally stated in fpm (feet per minute) or mps (meters per second).

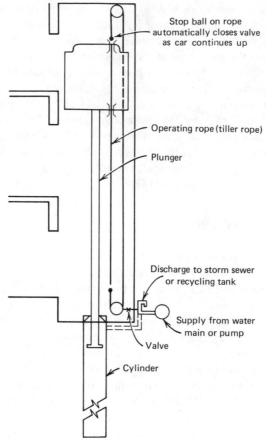

Figure 1.3. Hydraulic elevator with handrope operation.

technology of elevatoring. A typical but wrong logic pattern of the time was, "Joe Doe has two elevators in his building and seems to be getting by all right. Since my building is twice as big give me two twice the size." It rapidly became evident that people in the latter building had to wait twice as long for service as in Joe Doe's building and complaints and building vacancies reflected their dissatisfaction. The example is typical, and soon elevatoring emerged as a special design discipline.

ELEVATORING

Elevatoring is the technique of applying the available elevator technology to satisfy the traffic demands in multiple- and single-purpose multifloor buildings. It involves careful judgment in making assumptions as to the total population expected to occupy the upper floors and their traffic patterns, the appropriate

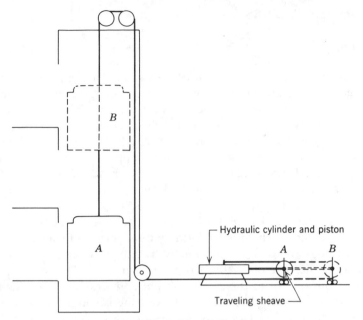

Figure 1.4. Roped hydraulic elevator.

calculation of the passenger elevator system performance, and a value judgment of the results so as to recommend the most cost effective solution or solutions.

A major part of elevatoring is the understanding of pedestrian flow, pedestrian queing, and the associated human engineering factors that will provide a nonirritating "lobby to lobby" experience. The traffic demands of passengers, service functions, and materials must be evaluated and simultaneously satisfied for an optimal solution.

Elevatoring, in the modern sense, is the process of applying elevators and the building interfaces necessary for the vertical transportation of personnel and material within buildings. Service should be provided in the minimum practical time and equipment should occupy a minimum of the building's space. The need for refinement in this process became apparent in the early 1900s as the height and cost of buildings increased.

Elevators changed radically in the early 1900s. As electricity became common, and with the introduction of the traction elevator, the water hydraulic was rapidly superseded. Helping its demise was the rapid rise of building heights—the Singer Building, 612 ft (185 m); the Metropolitan Life Tower, 700 ft (212 m); the Woolworth Building, 780 ft (236 m); all in New York City and built by 1912. The roped hydraulic could not be stretched to compete with such rises, and the direct-plunger-driven elevator required a hole as high as the rise. Telescoping rams were tried and proved unsatisfactory. These buildings were made possible by the invention of the traction elevator which was introduced in 1903.

TRACTION ELEVATORS

Description

Up until about 1903, either drum-type elevator machines wherein the rope was wound on a cylindrical drum, or the hydraulic-type elevator, either direct plunger or the roped hydraulic machine were the principal means of hoisting force. Both had severe rise limitations, the drum type, the size of the drum and the hydraulic type, the length of the cylinder. The drum type elevator had the further disadvantage of requiring mechanical stopping devices to shut off power to prevent the car from being drawn into the overhead if the machine failed to stop by normal electrical means. On a hydraulic machine this is prevented by a stop ring on the plunger.

The traction machine had none of the rise disadvantage of either the hydraulic or drum machines. The traction principle is a means of transmitting lifting force to the hoist ropes of an elevator by friction between the grooves in the machine drive sheave and the hoist ropes (Figure 1.5). The ropes are simply connected from the car to the counterweight and wrapped over the machine drive sheave in grooves. The weight of both the car and the counterweight ensure the seating of the ropes in the groove or, for higher-speed elevators, the ropes are double wrapped, i.e., pass over the sheave twice.

The safety advantages of the traction-type elevator are manyfold: multiple ropes are used, each capable of supporting the weight of the elevator, which increases the suspension safety factor as well as improving traction. The drive sheave loses traction if the car or counterweight bottoms on the buffers in the pit. Thus, the possibility of the car or counterweight being drawn into the overhead in the event of electrical stopping switch failure is reduced.

Traction elevators are capable of exceedingly high rises, the highest (or lowest) being in a mine application in South Africa for a depth of 2000 ft (600 m). The critical factors become the weight of the ropes themselves and the load imposed on the sheave shaft and its bearings. It was the traction elevator in addition to other advances in building technology which made today's tall buildings of 100 or more stories practical.

The traction principle has been available for centuries. The capstan on a ship is an example. The first known elevator application was the "Teagle" hoist which was present in England about 1845 as shown in Figure 1.6. This old print shows the traction drive and the counterweight. Motive force was provided by means of belts to the line shafting in the building where the lift was installed. The operation was by handrope as described for the hydraulic elevator as shown in Figure 1.3. The handrope acted to engage the belt to the drive pulley usually to the right or left of an idler pulley to move the lift up or down.

Performance

With the application of electrical drives to elevators, the versatility of electrical versus mechanical controls allowed certain standards of elevator operation

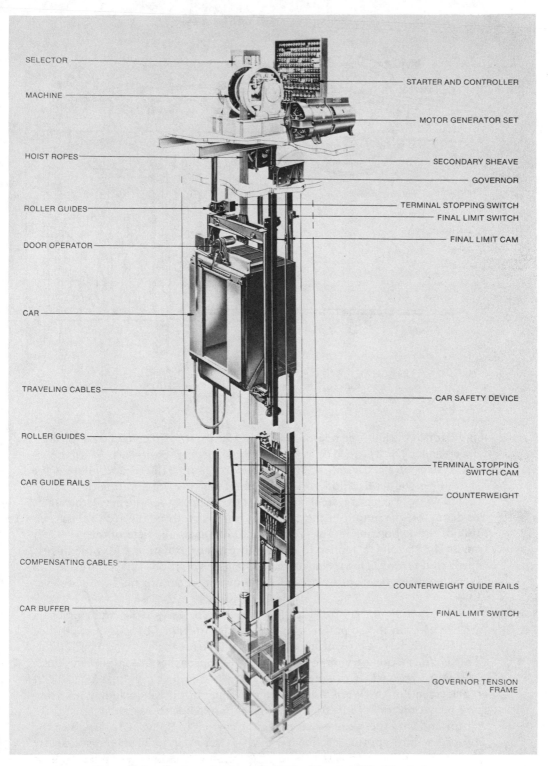

SELECTOR

MACHINE

HOIST ROPES

ROLLER GUIDES

DOOR OPERATOR

CAR

TRAVELING CABLES

ROLLER GUIDES

CAR GUIDE RAILS

COMPENSATING CABLES

CAR BUFFER

STARTER AND CONTROLLER

MOTOR GENERATOR SET

SECONDARY SHEAVE

GOVERNOR

TERMINAL STOPPING SWITCH

FINAL LIMIT SWITCH

FINAL LIMIT CAM

CAR SAFETY DEVICE

TERMINAL STOPPING
SWITCH CAM

COUNTERWEIGHT

COUNTERWEIGHT GUIDE RAILS

FINAL LIMIT SWITCH

GOVERNOR TENSION
FRAME

Figure 1.5. Gearless elevator installation (Courtesy Otis Elevator).

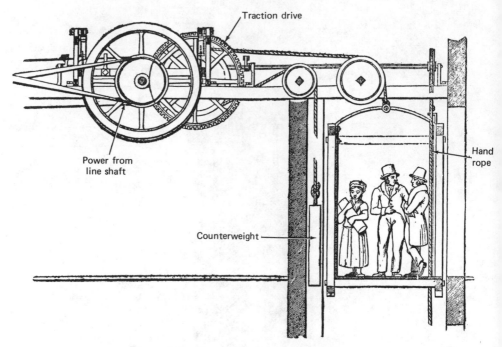

Figure 1.6.   Teagle elevator (circa 1845).

and control so that time related factors in an elevator trip could be established. Speed no longer depended on varying water or steam pressure. The Ward–Leonard system of electric motor speed control was introduced and allowed the smoothness of acceleration and deceleration common in elevators of today.

The Ward–Leonard system employs a motor generator driven by either an ac or dc motor, the output of the generator being directly connected to the armature of the dc hoisting motor. Varying the voltage on the field of the generator varies the dc voltage applied to the hoisting motor armature and consequently speed and torque. This speed control system or variations thereof are used on high-quality elevators of today and are known as generator field control elevators.

Solid-state motor control was introduced in the early 1970s and employs a system wherein the ac power from the line is controlled by silicon controlled rectifiers (SCRs) to provide a varying dc voltage for the elevator drive motor. This is an emerging technology and various approaches are in present (1982) use. One approach is to convert line ac into varying dc voltage as a direct replacement for the Ward–Leonard motor generator drive, including reversing and provision for regeneration. Another approach is to use an ac hoist drive motor and vary the speed by reversing the current to "plug" or slow down the motor. A third approach is the use of an ac servo drive wherein a solid-state feedback system controls the speed of the motor. Still another approach is to use a varying dc voltage from SCRs without provisions for reversing armature current and reversing the dc hoist motor field to cause either up or down travel.

At present, the use of solid-state motor drives creates interface concerns such as structural borne electrical noise and harmonic vibrations. The operation of the elevators on emergency or standby power needs to be resolved, especially when such standby is required to operate sensitive computer equipment. The ultimate solid-state motor drive system has yet to be established and may take years as did the introduction of the Ward–Leonard system in the period between 1910 and 1930.

In the course of this book the operating characteristics of electric elevators are described and a basis for time study calculations of elevator trips are established. These time factors will become the basic tools of establishing the number of elevators necessary for any type of building and will be related to the speed at which people can be moved from place to place vertically. As a preliminary, familiarity with modern elevator types is necessary.

GEARLESS TRACTION ELEVATORS

Description

The preceding brief discussion of early elevator history introduced the traction-type elevator. The first high-rise application of this type of elevator was in the Beaver Building in New York City in 1903, followed by such notable installations as the Singer Building (demolished in 1972) and the Woolworth Building. These elevators were of the gearless traction type that are at present the accepted standard for the high-rise, high-speed [over 400 fpm (2.0 mps)], and high-quality elevator installation.

The gearless traction elevator consists of a large, slow-speed (50 to 200 rpm) dc motor of four to eight poles directly connected to a drive sheave of about 30 to 48 in. (750 to 1200 mm) in diameter. An electrically released, spring-applied brake is arranged to apply to the drive sheave. Slow-speed dc motors and in the future, ac motors, though expensive and massive, are necessary to maintain the necessary torque to directly drive large diameter sheaves. The larger diameters sheaves also conform to the bending radius of elevator steel ropes. A limitation is imposed by safety codes as good practice for long rope life and is generally established at a minimum of 40 times the diameter of the wire rope used. For example, a ½-in. (13-mm) wire rope would require a minimum sheave size of 20 in. (500 mm).

The slow speed of the direct drive gearless traction machine is necessitated by the speed of the elevator it serves. For example, for a 500 fpm (2.5 mps) elevator and sheave diameter of 30 in. (750 mm), a top speed of 86 rpm is required. To level this elevator to a landing at a maximum speed of 25 fpm (0.125 mps), 4.3 rpm is necessary. Gearing with higher-speed motors has been tried with moderate success to gain these higher speeds. The continous operation of elevators [up to 25,000 mi (40,000 km) per year] and the relative ease of maintenance of the gearless machines, as well as their dependability, makes them the preferred type for higher speeds.

On higher-speed gearless traction machines of 800 fpm (4.0 mps) or more, the double wrap principle is generally applied to obtain traction and to minimize rope wear. The ropes from the car are wrapped around the drive sheave, around a secondary or idler sheave, around the drive sheave, and down to the counterweight (Figure 1.7). The groove seats are round, providing support on the full half of the rope thus eliminating pinching action and minimizing wear. Traction is obtained by the pressure of the ropes on the sheave. As may be noted, increasing the weight on the car or counterweight increases the force so that friction between the ropes and the sheave increases traction.

Polyurethane groove liners have been introduced which have the effect of increasing the traction between the groove and the rope. In addition, the poly-

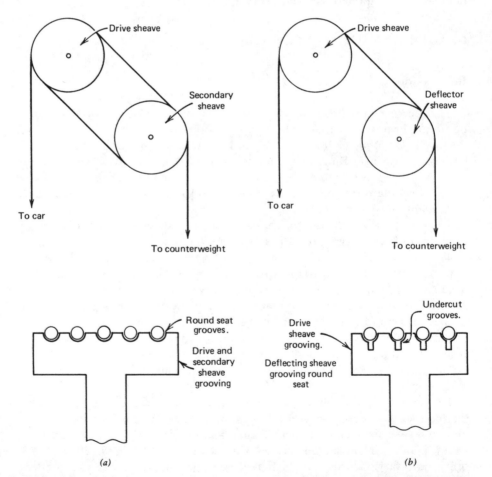

Figure 1.7. (*a*) Double wrap traction arrangment. (*b*) Single wrap traction arrangment.

urethane cushions the rope and prolongs rope life. The design of the groove is such that, when the polyurethane wears and is not replaced, traction is ensured by pinching action of metal sheave groove. As this development progresses, the double wrap traction (DWT) principle may be superseded by single wrap using this groove liner principle.

Elevator machines are also roped with a single wrap arrangement which is applied to both gearless and geared machines. The single wrap arrangement provides traction by the use of grooves which will pinch the ropes with varying degrees of pressure depending upon the shape of the groove, and its undercutting (see Figure 1.7 and later discussion). The most effective single wrap arrangement provides 180 degrees of rope contact with the sheave without a deflecting sheave as shown in Figure 1.8 [for single wrap traction (SWT), 2:1 roping].

Conventional elevators are roped either 1:1 or 2:1 (Figure 1.8) for both car and counterweight. In some unusual installations and special applications, 1:1 car and 2:1 counterweight roping has been used. In that event the counterweight must be at least twice as heavy as the weight of the car. The 1:1 arrangement is the most popular for higher speeds and has been used for a load and speed of 10,000 lb (4500 kg) at 1600 fpm (8 mps). The 2:1 arrangement allows the use of a higher-speed and therefore a smaller but faster elevator motor. The mechanical advantage of 2:1 roping requires that only half the weight be lifted, so 2:1 is generally used whenever loads in excess of 4000 lb (1600 kg) must be lifted. The economy of the faster motor, which can be built smaller and lighter

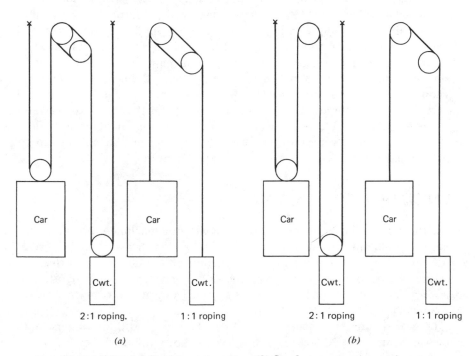

Figure 1.8. (*a*) Double wrap roping. (*b*) Single wrap traction roping.

than lower-speed dc motors, also makes 2:1 roping attractive for a full range of speed requirements from 100 to 700 fpm (0.5 to 3.5 mps) or more and for any lifting capacity.

Any of the above 1:1 and 2:1 roping arrangements can be provided with the elevator machine in the basement or at a lower level. The appropriate sheaves are installed in the overhead space to direct the ropes from the machine to the car and counterweight. The preferred arrangement is the single wrap traction types. A foundation must be provided for the machine which will overcome the uplift and solidly anchor the machine under all conditions of operation and safety application.

The long life, smoothness, and high horsepower of gearless traction elevators provide a durable elevator service that can outlive the building itself. The original gearless machines in the Woolworth Building were reused when that building's elevators were modernized in 1950, and again in 1970. The gearless machine not only provides speed if necessary but also is capable of performance essential to any well-elevatored building.

Gearless Machines—Performances

Essential to elevatoring considerations is the requirement that a gearless traction machine, no matter what its lifting capacity or speed, must be capable of optimum floor-to-floor operating time commensurate with passenger comfort. Stated another way, the machine must be capable of starting a filled elevator car, accelerating to a maximum speed for the distance traveled, and slowing to a stop in a minimum time of about 4.5 to 5.0 sec. This must be performed under all conditions of loading, either up or down. The elevator system must be so arranged that such acceleration and deceleration takes place without discomfort to the passenger from too rapid change in the rate of acceleration or deceleration (with optimum jerk). Furthermore, the elevator must be capable of releveling, while passenger load is changing at a floor (correcting for rope stretch) with almost imperceptible movement. The aspects of performance are discussed further in a later chapter.

GEARED TRACTION MACHINES

As the name implies, the geared traction elevator machine utilizes a reduction gear with a high-speed motor to drive the traction sheave. A high-speed ac or dc motor drives a worm and gear reduction unit which in turn drives the hoisting sheave, the net result being the slow sheave speed and high torque necessary for elevator work. A brake is applied by spring to stop the elevator and/or hold the car at a floor level.

The geared traction machine is used for elevators and dumbwaiters of all capacities from 25 to 30,000 lb (10 to 14,000 kg) or more, and speeds from 25 to 450 fpm (0.125 to 2.3 mps). The complete flexibility of worm gear ratios and

Figure 1.9. (*a*) Early steam-driven hoisting machine. (*b*) Early electric-driven hoisting machine.

13

motor speeds and horsepowers as well as drive sheave diameters and roping arrangements (1:1, 2:1, and, sometimes, 3:1) makes this vast range of application practical. In some materials handling applications, geared machines are used for speeds of 600 fpm or more (3.0 mps) with excellent results.

The geared traction elevator is an outgrowth of the earlier drum-type elevators. The steam engine gave way to the electric motor and gear (Figure 1.9) and the drum gave way to the drive sheave (Figure 1.10). The grooved drive sheave was an outgrowth of the traction principle applied to gearless elevators; instead of ropes being wrapped around the sheave, grooves were cut into the sheave and the necessary friction was created by the pinching action of the grooves on the rope (Figure 1.11). Various types of grooving are used for different loads and traction requirements. Generally, the sharper the undercut angle, the greater the traction (and usually, greater rope and sheave wear).

Again, with the introduction of polyurethane groove liners, greater traction and less rope wear will become available on geared machines as well as on gearless machines.

Geared machines are driven by either one-speed or two-speed ac motors, by dc motors utilizing the Ward–Leonard means of control, or by ac or dc motors with SCR or other solid-state control. The ac motor machines are generally used for speeds from 25 to 150 fpm (0.125 to 0.75 mps) and with solid-state

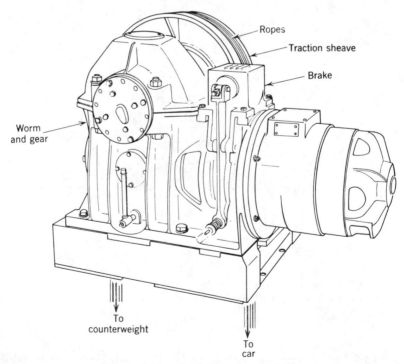

Figure 1.10. Geared traction machine.

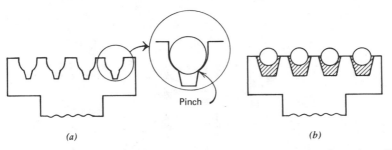

Figure 1.11. (*a*) Undercut sheave groove. (*b*) Sheave groove with polyurethane liner.

drives to 350 fpm (1.75 mps). Stopping with single-speed motors is accomplished by disconnecting the power from the motor and stopping the car by a combination of slide and brake action. Two-speed ac operation employs a double-wound motor, a fast-speed winding for full-speed running, and a slow-speed winding (which can be any ratio as high as 6:1, i.e., the slow speed being ⅙ full speed) for stopping, leveling, and, if required, releveling . Operation is generally to start at full speed, run, switch to low speed at a measured distance from the stop, and accomplish the final stop by combination of brake and slide. The floor-level accuracy of plus or minus ½ to 1 in. (13 to 24 mm) can be obtained under all conditions of load as contrasted with one-speed accuracy of 1 to 3 in. (24 to 75 mm), which will vary with load. Much greater accuracy can be obtained when solid-state ac motor drives are employed. In contrast, the dc Ward–Leonard drive or a solid-state motor drive allows the car to be stopped electrically before the brake is applied, resulting in leveling accuracy from ¼ to ½ in. (6 to 13 mm) under all conditions of load, and much softer stops than the ac machine. Geared machines with dc motors are used for speeds from 50 to 450 fpm (0.25 to 2.3 mps).

With either ac or dc geared elevators, the performance can be established, which is essential in the calculations when estimating the numbers of elevators for a particular building.

HYDRAULIC ELEVATORS

A third major type of elevator in use today is a modern version of the hydraulic elevator. Modern hydraulics are direct-plunger-driven from below (the cylinder extending into the ground as high as the elevator rises) and the operating fluid is oil moved by high-speed pumps rather than water under pressure (Figure 1.12*a*). Hydraulic elevators today are used for both passenger and freight service in buildings from two to six stories high and for speeds from 25 to 200 fpm (0.125 to 1.0 mps). Single-ram capacities will range from 2000 to 20,000 lbs (1000 to 10,000 kg) or more. Multiple rams are used for high capacities of 20,000 to 100,000 lb (10,000 to 50,000 kg). Varied speeds and high capacities are obtained through multiple pumps. Elevatoring considerations of perform-

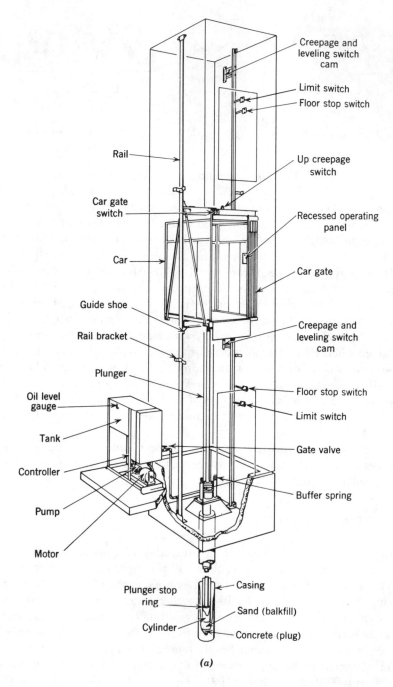

Figure 1.12. (a) Hydraulic freight elevator;

16

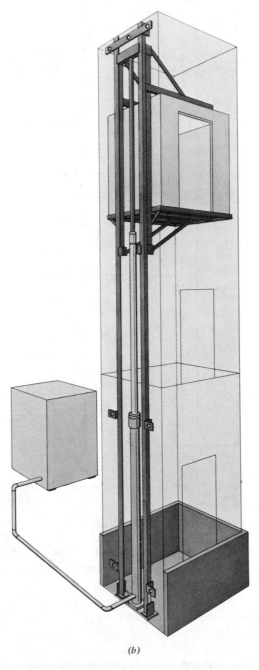

Figure 1.12. (b) "Holeless" hydraulic elevator with telescoping plunger (Courtesy Otis Elevator);

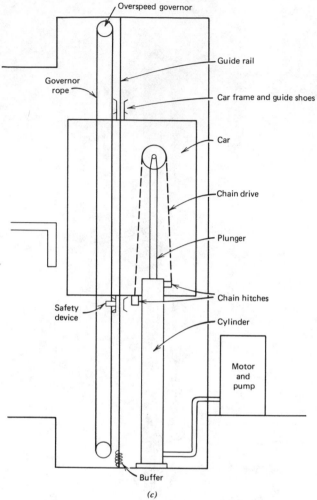

Figure 1.12. (*c*) Chain-driven hydraulic elevator.

ance time of hydraulic elevators are approximately the same as for geared elevators.

A variation of the hydraulic elevator, the "holeless," is becoming prominent for low-rise applications. The operating cylinder and plunger are located adjacent to the elevator platform and lift by applying force to the upper members of the car frame. Multiple plungers are used for heavy duty applications and telescoping plungers are used for higher rises (Figure 1.12*b*).

A holeless hydraulic elevator utilizing roller chains actuated by a sheave on a vertical hydraulic cylinder similar to the roped hydraulic in Figure 1.4 is becoming quite prominent in Europe. Such an elevator must have a safety on the car to prevent falling if the hoisting chain should part (Figure 1.12*c*).

ESCALATORS AND MOVING WALKS

A very important factor in vertical transportation is the escalator (moving stairway) and moving inclined walks. Before the 1950s, escalators were mainly found in stores and transportation terminals. Today their use has expanded to office buildings, schools, hospitals, banks, and other places where either large flows of people are expected or it is desired to direct people vertically in a certain path. Escalators and elevators are often used in combination either to provide necessary traffic-handling capacity or to improve elevator operation by directing people to one elevator loading level. They are essential in the lobby of a building with double-deck elevators. Many office buildings and some factories have found escalators ideal for rapid shift changes or rapid floor-to-floor communication.

Inclined, flat, or contoured moving walkways are closely related to the escalator form of vertical transportation (Figure 1.13). The passenger-handling ability of such conveyances is based on speed and density of passenger loading per step. Nominal ratings are in passengers per 5 mins. Qualification of capacities and application of the moving stairway or walk are discussed in later chapters.

As a historical note, the flat step escalator was first introduced by Otis Elevator Company at the Paris Exposition in 1900. Preceding the flat step escalator, the Reno type was an endless series of inclined "indentations" that were boarded at the same angle of rise (Figure 1.14). Major development took place starting in 1920, the initial features being flat steps with cleats, followed by flat boarding and debarking areas, narrower step cleats and combs, extended newels, and glass balustrading (Figure 1.15).

The value of escalators in vertical transportation is the providing of a continuous flow of people as contrasted to the batch approach of elevators. This continuous flow principle proves valuable where the required movement of large numbers of people occurs such as in an airport when an airplane unloads or at a sports event both prior to its starting and at the end. A detailed discussion of the application of escalators takes place in a later chapter.

DUMBWAITERS AND MATERIALS-HANDLING SYSTEMS

Other forms of vertical transportation discussed in the later chapters in this book are dumbwaiters and materials-handling systems. Modern buildings use these devices for a variety of purposes: delivery of books in libraries, distribution of mail in office buildings, delivery of food and supplies in hospitals, and so on. The dumbwaiter (Figure 1.16) is actually a small elevator that can have all the performance characteristics of an elevator. Loading and unloading can be either at counter or floor level, and either manual or automatic. Size can vary from letter size to car sizes consisting of any arrangement of 9 ft² (0.9 m²) or less of platform area, and a car with an effective height of no more than 4 ft (1200

Figure 1.13. Moving walkways: (a) inclined; (b) flat (Courtesy Montgomery Elevator).

Figure 1.14.   "Reno" escalators (Courtesy Otis Elevator).

Figure 1.15.   Glass balustrade escalator.

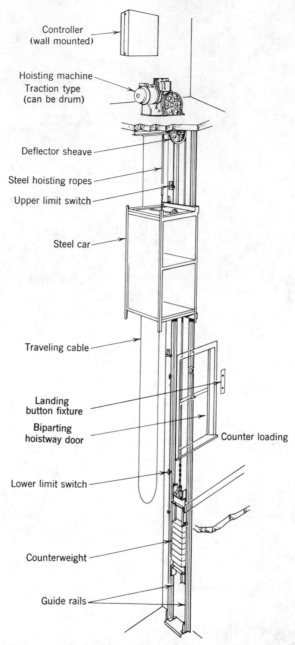

Controller
(wall mounted)

Hoisting machine
Traction type
(can be drum)

Deflector sheave

Steel hoisting ropes

Upper limit switch

Steel car

Traveling cable

Landing
button fixture

Biparting
hoistway door

Counter loading

Lower limit switch

Counterweight

Guide rails

Figure 1.16. Electric dumbwaiter with cantilever car.

mm). This is a limitation imposed by elevator safety codes (covered in detail in a later chapter), and anything over that size must be classified as an elevator. Dumbwaiters need not have safeties and are strictly for material handling. They are always operated from the landing, not from in the cab as in an elevator.

Other forms of vertical materials-handling systems include tote box conveyors, automatic loading and unloading cart lift systems, and self-propelled vehicle systems either with a dedicated rail system or following a guide path on the floor. These will be discussed in detail in a later chapter.

The foregoing represents the principal forms of vertical transportation and correctly applying these forms is the major thrust of elevatoring. Our earlier simplified definition of elevatoring can be restated as follows.

STUDY OF ELEVATORING

Elevatoring is the analysis of the requirements of vertical transportation of people and materials in a building, under all operating conditions. Such transportation requirements may be studied from a compatible aspect, as in an office building, or a functional aspect, as in a hospital or merchandising in a department store.

The first essential step in elevatoring is pedestrian planning, i.e., how many people will require transportation, what will be the peak traffic, and how will it occur—all up, up with partially down traffic, or equally up and down simultaneously. Some well-established guidelines may be available as will be demonstrated in the various chapters on commercial, institutional, or residential buildings. For some projects, extensive study of the expected pedestrian movements must be made before the process of determining elevators or escalators can start.

Once the critical pedestrian traffic is established or estimated, the next steps can take place.

Elevatoring requires consideration of all the time factors and movements that take place during the operations providing transportation for people and/ or materials. These time factors must be related to a total time required for the service based on the actual or estimated demands. Efficient elevatoring requires minimizing the time factors to maximize service.

The time components of an elevator round trip that will be studied and evaluated are as follows:

Loading Time. The time required for a number of people to board an elevator car, moving stairway or walk, or the time required to load material or a vehicle onto an elevator or lift. Loading time must be considered under many conditions of operation consisting of narrow or wide elevator cars, wide doors, narrow doors, arrangement of elevators, and partially filled or empty elevators.

Transfer Time. The time to unload (or reload) an elevator at a local stop above the main landing. Transfer time is based on all the considerations of loading time plus, essentially, the density of the passenger or other load remaining on the elevator, and the direction of the transfer either entering or leaving.

These two elements, loading and transfer time, are the most difficult to quantify because in general these times are based on the interaction of people. Estimates have been made based on hundreds of field traffic studies of human behavior and the conclusions are reluctantly (because of the doubt that such a person exists) based on "the average person."

The other factors in an elevator or escalator trip are the mechanical times, which can be established accurately and assured by specifications that can be developed before installing an elevator or escalator. These time factors are as follows:

Powered Door-Closing Time. A function of door weight (mass). Width of opening and type of opening for horizontally sliding doors—center opening (Figure 1.17*a*), single-slide (1.17*b*), two-speed (1.17*c*), or the height of the opening for vertical biparting doors (Figure 1.17*d*) for freight application—involve different masses that affect closing speed. The kinetic energy of closing doors is limited by elevator safety codes and is usually established at no more than 7 ft poundal (0.29 Joules). In practical terms, this means that the familiar 48-in (1200-mm) center-opening sliding door will require about 3 sec to close. Closing and opening time is a vital consideration in elevatoring, for the door operation on a typical elevator occurs hundreds of times a day. An elevator cannot leave a floor until the doors are closed and locked, and passengers do not transfer until the doors are essentially fully opened.

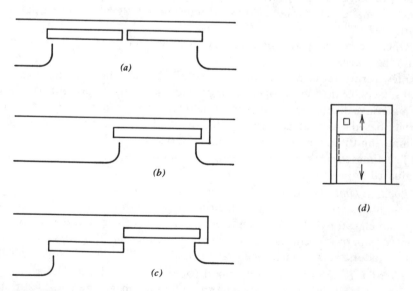

Figure 1.17 Door arrangements for elevator entrances: (*a*) center-opening; (*b*) single-slide; (*c*) two-speed slide; (*d*) vertical biparting.

Powered Door-Opening Time. This can be minimized by proper arrangement. The door-opening time can and should be much faster than the door closing time. Doors can start to open while an elevator is leveling under certain conditions. This preopening must be limited so that the opening is not wide enough to allow passenger transfer before the elevator is level enough with the landing to avoid a tripping hazard. The time necessary to open the doors will vary with the width and type of doors. For example, center-opening doors take less time than single-slide or other types of the same width, and wide openings require more time than narrower openings.

Operating Time. A function of the speed control arrangement of the elevator and the number of stops the car will make in a round trip. The considerations necessary are the times required for a one-floor run, a two-floor run, and full-speed operation. The operating speed of escalators or moving ramps is constant and maximum movement capacity is fixed.

Because the floor-to-floor operations of elevators are repeated over and over again, an estimate of the probable number of stops an elevator will make in the course of a single trip is required. Knowing or estimating the number of stops provides a means to calculate the total time for functions and leads to the cycle time or round-trip time of a single elevator. The number of stops can be established by applying a statistical formula, by inspecting the various attractions of each floor at which an elevator is required to stop, or a combination of both statistical and logical determination. The various approaches are discussed in the chapter on calculating elevator performance.

On a single elevator or escalator, once the time required to serve a given number of people is determined, the number and size of elevators or escalators that will serve the critical pedestrian traffic can be established. Although this is the essence of our studies, elevatoring cannot end here.

The grouping and operation as well as location of those elevators (or escalators) must be also established so that the installation will provide the expected service. The following chapters establish principles of arrangement and location of elevators and escalators. In addition, a discussion of elevator grouping, stops served, lobby arrangements, skip-stopping, operation, and all of the aspects of elevatoring are presented.

CHAPTER TWO

The Basis of Elevatoring a Building

EARLY POPULATION

Before any thought is given to the elevators in a building, a thorough and detailed study must be made of how people will arrive at the building, occupy that building, and move about the building. Occupancy is an obvious prerequisite to the design and size of the building itself. It must be expected to perform its function to provide a living or working environment and to have an economical, physical, and functional life span.

Basic factors in elevatoring a building include the number of occupants and visitors, their distribution by floors, and the times and rates of arrival, departure, and movements. We can determine the population in an existing building by census and the average population in a new building based on so many square feet of space per person. We can also determine that people in arriving at or departing from an existing building relate to existing vertical transportation systems and will require an estimated quantity of vertical transportation in a new building. Surveys of existing buildings of similar nature can help confirm those estimates.

The population factors and the intensity of pedestrian traffic for all types of buildings are suggested in later sections of this book. You will note the term "suggested"; the actual use of the building is beyond the control of the architect and the elevator engineer. The basic population and usage estimates should be a consensus of the assumptions of the entire building team in order to properly elevator a building. When the level of uncertainty is high, the basis of elevatoring must be conservative, since the expense of adding additional equipment is often prohibitive.

ELEVATOR TRAFFIC

In every type of building there is a critical elevator traffic period. The type, direction, and intensity of elevator traffic during this period determines the quantity of elevator service for the building. If the elevators serve traffic well during the critical time, they should be capable of satisfying traffic at all other times. The quality of elevator service during this critical period will be set by the class of occupancy.

Critical traffic periods vary with building types and in various areas of the United States and Canada as well as the world. For example, in office buildings in downtown areas served by mass transit, the critical traffic period is often the morning in-rush complicated by persons who have arrived early and are traveling down. If elevators are sufficient to serve the peak of that in-rush period, the rest of the day usually does not present a problem. In other cities the critical traffic in an office building may be the noontime period when lunch hours may be standardized and the entire building may go to and return from lunch at designated times.

In recent years there has been an introduction of staggered work hours and "flextime" wherein employees set their own arrival and departure times. This practice has reduced the intensity of the morning up peak arrival rate, but has added an opposing down traffic to the predominant up traffic.

The critical traffic period in some hospitals may be in the forenoon when doctors are visiting patients, transfers for treatment are being made, operations are performed, and essential hospital traffic reaches a peak. In other hospitals, critical traffic may occur during visiting hours or the afternoon shift change period.

In apartment houses the critical traffic is usually the late afternoon or early evening period when tenants are returning from work, children are coming home from school, shoppers are returning from stores, and other are leaving for evening entertainment. A downtown apartment with a predominant business-person tenancy may find that the critical traffic period is in the morning when practically everyone is leaving for work.

Once the critical traffic period for a building is determined and evaluated in terms of required elevator handling capacity, the choice of the proper number, speed, size, and location of elevators may proceed.

Critical traffic periods for various types of building have been determined by observations, traffic studies and tests, discussions with building managers and owners, and by research in occupancy and population use requirements. The estimations of these traffic factors are reviewed in the sections of this book dealing with various building types.

PEDESTRIAN PLANNING CONSIDERATIONS OF ELEVATORING

The pedestrian planning considerations in elevatoring cover the process of locating elevators in a building, providing proper access and queuing space to passenger elevators, designing and shaping them to best accommodate people, and determining door sizes and lobby arrangements, to make sure the optimum use and benefit is gained from the total elevator plant in a building.

The direct capital cost of elevators and the indirect cost of the space dedicated to their function is a major cost component of any building. Elevators must be placed, arranged, and designed to provide the most cost effective performance. Unsatisfactory elevator service can damage a building's reputation and cause loss in the productivity of its occupants.

Elevator Platform Shape

The elevator platform—the area on which passengers ride—must be large enough to accommodate a passenger (or freight) load without undue crowding and allow each passenger ready access to the elevator doors.

An average person will require about 3 ft^2 (0.28 m^2) of floor area to feel comfortable (touch zone) (Figure 2.1). Passengers can be crowded, however, to a minimum of about 2 ft^2 (0.19 m^2) for the average person. If no one is pushing or forcing people to crowd, each person will take close to the three square feet (0.28 m^2). There are exceptions: at office building quitting time, and if passengers know each other, densities of 1.5 ft^2 (0.14 m^2) per person have been observed.

The average space per passenger in elevator cars means that the elevator capacity—expressed in pounds (kilograms) and translated to square feet (square meters) so that the car will not exceed its rated load if packed full—must be arranged in the best dimensions to accommodate the shape of people. The arrangement of ranks and files has been found to be best, and inside car dimensions shown in Table 2.1 for common-size passenger elevators have been adopted partially as industry standards.

As an example, the average loading of a 3500-lb (1600-kg) elevator is 16 passengers. You will note from Figure 2.2 how these 16 people may arrange themselves inside the elevator car.

Of course, the neat rank and file is idealistic. In actual practice, the initial passengers generally arrange themselves with their backs to the walls and subsequent passengers fill up the center.

In Figure 2.3 note how the same area, with a different width and depth, may lead to awkward unloading situations. More ranks of passengers now make access to the door difficult and generally require someone to step out of the car to let others out. These complications add a time delay to each elevator stop, which accumulates during the total trip and seriously reduces efficiency. The deep, narrow arrangement also leads to loss in passenger capacity—15 passengers versus 16 passengers shown in Figure 2.2. Part of this loss is from the extra space required for the car doors, but most is from platform shape.

Study of the two illustrations suggests the conclusion that the most efficient elevator car is only one person deep! This is true but not practical because efficient door arrangement must also be considered.

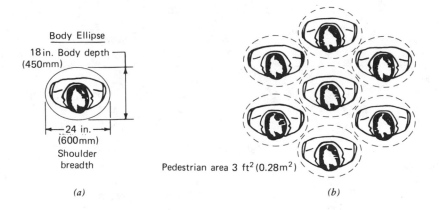

18in. Body depth
(450mm)

Body Ellipse

24 in.
(600mm)
Shoulder
breadth

(a)

12in. (300mm) radius—touch zone

Pedestrian area 3 ft^2(0.28m^2)

(b)

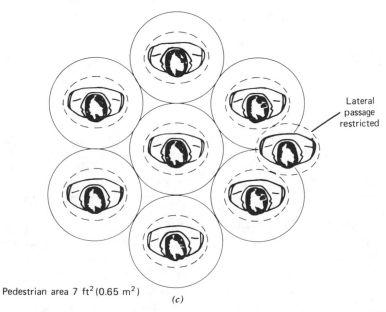

18in. (450mm) no touch zone

Lateral
passage
restricted

Pedestrian area 7 ft^2(0.65 m^2)

(c)

Figure 2.1. Pedestrian queuing: (a) body ellipse; (b) crowded; (c) nominal (Courtesy *Pedestrians*, by J. Fruin).

Door Arrangement

The most efficient door is one that opens and closes in minimum time and allows two persons to enter or leave an elevator simultaneously. The doors must also be reasonably economical and adaptable to efficient platform sizes. The 48-in. (1200-mm) center-opening door meets most of these requirements

Table. 2.1.   Area Versus Capacity—Suggested Standards.

1. Inch/Pound Units

Capacity (lb)	Car Inside (in.)		Area (ft²)	A17.1 Area Allowed (ft²)	Observed Loading (people)
	Wide	Deep			
2000	68	51	24.0	25.4	8
2500	82	51	29.0	30.5	10
3000	82	57	32.5	35.4	12
3500	82	66	37.5	38.0	16
3500 (alt.)	92	57	36.4	38.0	16
4000	82	73	41.6	42.2	19
4000 (alt.)	92	66	42.2	42.2	19
4500	92	73	46.6	46.2	21
5000	92	77	48.2	50.0	23
6000	92	90	57.5	57.7	27

2. Metric Units

Capacity, kg (lb)	Car Inside (mm)		Area (m²)	Code Area Allowed (m²)	Observed Loading (people)
	Wide	Deep			
1200 (2640)	2100	1300	2.7	2.8	10
1400 (3080)	2100	1450	3.05	3.24	12
1600 (3520)	2100	1650	3.5	3.56	16
1600 (alt.)	2350	1450	3.4	3.56	16
1800 (3960)	2100	1800	3.8	3.88	18/19
1800 (alt.)	2350	1650	3.9	3.88	18/19
2000 (4400)	2350	1800	4.2	4.2	20
2250 (4950)	2350	1950	4.6	4.6	22
2700 (5940)	2350	2150	5.1	5.32	25

and is recommended for high-quality elevators when optimum performance is required (Figure 2.4). It can fit the average 86-in. (2200-mm) wide platform and can be opened in slightly less than 2 sec. Closing speed, as we mentioned previously, must be within the 7 ft poundal (0.29 joules) kinetic energy limitation. Because each panel of the door is half the weight of the entire door [no more than 100 or so lb (45 kg) per panel] and the distance traveled is only half the opening width, the 48-in. (1200-mm) center-opening door can be closed in 2.9 sec within this kinetic energy limitation.

Doors 42 in. (1100 mm) and less can be considered one-person doors. Note how awkward it becomes for two people to pass each other (Figure 2.4); the natural tendency is to allow one person to leave while the other holds up elevator service until he or she can enter.

Wider doors are often necessary for special purposes, such as the 60-in. (1550-mm) door on a hospital elevator that must accommodate a hospital bed

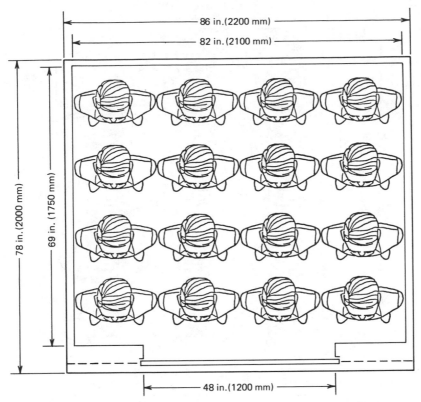

Figure 2.2.   Nominal loading, 3500-lb (1600-kg) "passenger-shaped" elevator.

with an attendant or on a service elevator that must accommodate wide containers or carts. In these cases the efficiency of the door is secondary to the function the elevator must perform.

In apartment houses the economy of the single-slide door prevails. Because passengers are expected to move at a somewhat leisurely pace, some efficiency may be justifiably sacrificed for economy. To effectively allow a wheelchair or ambulance stretcher to enter or leave an elevator, door width should be a minimum of 42 in. (1100 mm).

Lighting and Signals

The interior of an elevator car should be well lighted and the lights arranged so that they cannot be turned off by unauthorized persons. To this end there should be a key operated light switch. The car threshold should have a minimum of 5 ft-candles (50 lux) of light without the benefit of outside lighting as prescribed by elevator safety codes since inaccuracy in floor-level stops can create a possible tripping hazard.

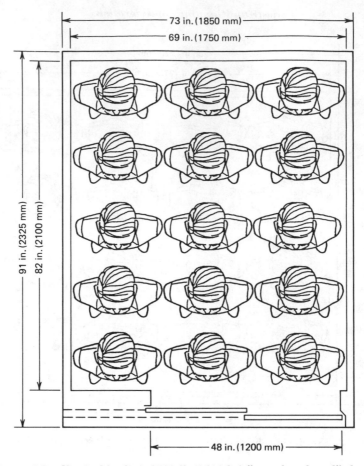

Figure 2.3. Nominal loading, 3500-lb (1600-kg) "stretcher-shaped" elevator.

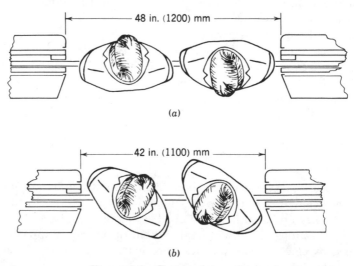

(a)

(b)

Figure 2.4. Passenger transfer.

A person should be able to look at an elevator and tell whether it will travel up or down when it leaves the floor. With a single-car installation in a normally quiet building this is not serious, for if the person is the only one on the floor, the car will generally travel in the direction chosen. Considerations for the handicapped require a lantern to sound a gong or chime when illuminated, one stroke for an up traveling car and two strokes for down.

With more than one car serving a floor, some form of directional indication becomes essential, for a car going either way may stop. Lighted directional arrows in both door jambs of the arriving elevator offer a simple, relatively inexpensive solution to this dilemma which may not be totally desirable when efficient elevator performance is wanted. The car direction is not visible until the doors are open which may be too late for the eager passenger, resulting in unnecessary passenger movement (Figure 2.5).

The most effective waiting passenger indication is to provide lanterns over or next to each hoistway entrance which will inform the prospective passenger of the next car to arrive at the floor and the direction it will travel when it leaves the floor (Figure 2.6). Such lanterns should be arranged to light up sufficiently in advance (about 4 sec) of a car's arrival at each floor to give passengers time to walk to the entrance of the arriving car and be ready to board it. This is not always accomplished, but as will be noted later each second thus saved in

Figure 2.5.   "In-car" directional lantern.

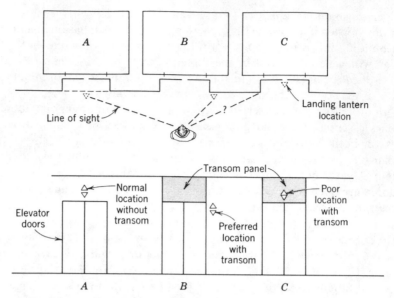

Figure 2.6. Landing lanterns should be located to be seen clearly.

passenger boarding time is worth extra elevator capacity. The lantern must also be prominent enough so that it may be seen from any point where passengers are likely to stand. Note, in Figure 2.6, arrangement *C*, that a lantern on a transom or in a depression cannot be clearly seen. Arrangement *B* is the most acceptable.

Landing call buttons that light when touched or pressed inform the prospective passenger that the call for service has been acknowledged, and arranged to extinguish when the car arrives. For efficient service, the button should not hold the elevator, lest a stuck button cause an elevator to stay at a floor, or people keep an elevator at a floor until their friends arrive, thus delaying service.

Once passengers are in the elevator car they should be readily able to indicate where they are headed and be promptly informed when they arrive. The car call buttons should be conveniently located so passengers can register their call as they enter an elevator, the side locations proving to be the most acceptable (Figure 2.7). The car call button should also light to acknowledge the call which will help improve service. Entering passengers can see that their floor stop will be made if someone has registered it. Numerals or symbols at least ⅝ in. (16 mm) high and raised or engraved 0.030 in. (0.8 mm) adjacent to each car button are being required as an aid to the visually handicapped.

Some form of position indicator in the elevator car is required and should be plainly visible to all passengers. Car position indicators over the car door head jamb, the usual location, are often combined with directories to show what is on each floor of an office building, department store, or hotel (Figure 2.8). Other messages have also been placed in the same location.

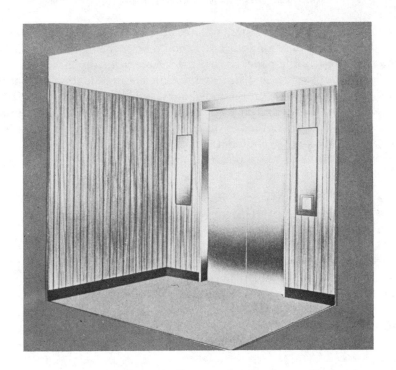

Figure 2.7.   Car operating panels.

Figure 2.8.   Car position indicator and directory. (Courtesy Westinghouse Electric Co.)

In many buildings, it is desired to have doors 8 ft (2400 mm) or higher. With the high doors, car position indicators should be of the digital type and placed over each car operating panel so they are visible from any location within the car. The readout should be distinct with sufficient illumination so that the car lighting does not overwhelm the indications.

In many buildings, especially hotels and hospitals, in which the usual passenger may not be accustomed to riding in elevators, the floor designation may be helpfully placed on the jamb of the hoistway door so it may be seen when the door opens. For the handicapped, the jamb mounted markings using 2-in (50-mm) high numerals raised 0.030 in. (0.8 mm) are a necessity.

POSITIONING OF A BUILDING'S VERTICAL TRANSPORTATION SYSTEM

Elevators and escalators should be accessible and centrally located. All entrances should lead to the vertical transportation nodes, which should be near the main entrance of the building. If a parking lot or subway entrance is near the building it is reasonable to expect that a significant portion of the passenger traffic will come from that direction. If the flow is expected to be heavy, escalators or moving walkways should be employed to deliver passengers to the elevator lobby.

Locating elevators in the geometric center of the population on each floor in a building allows all parts of each floor to be equally accessible to the elevator core. Elevators located on one end of a building detract from the desirability of the other end and adds considerable extra time for passengers to cover the horizontal distance (Figure 2.9).

It is often proposed to provide two elevator cores serving the same floors at separate locations to avoid a long horizontal distance to a central elevator core. This is practical; however, additional elevator capacity must be provided in each core since the imbalance of traffic can be totally unpredictable. In addition, the creation of an attraction, such as a coffee shop or shopping center, near one core can totally change the nature of traffic and may cause excessive loading on one core.

Experience has shown that the walking distance from the elevators to the farthest office or suite should not exceed 200 ft (60 m) with a preferred maximum distance of about 150 ft (45 m) For example, in a building 300 ft (90 m) long, the elevators should be located at the center point (Figure 2.10). Buildings with X, Y, or T floor layouts have a natural center point.

In some buildings, where the busiest entrance is near one end rather than the middle, a central location for the elevators may still be desirable. If people use the main entrance only when entering or leaving the building but use elevators repeatedly during the day, locating them near the center of the floor population may achieve greater net savings of time and energy for all users. This saving of time accrues from the fact that persons usually walk 2 to 4 fps (0.6 to 1.2 mps).

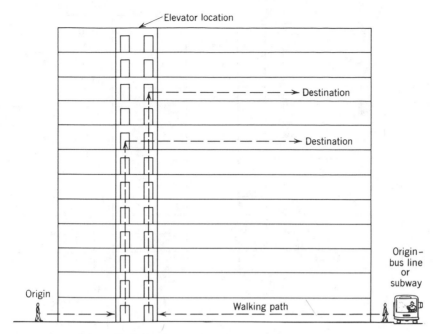

Figure 2.9. Noncentralized elevators mean additional walking time.

Grouping of Elevators

If a building requires more than one passenger elevator, all the passenger elevators should be grouped. Single elevators in various parts of a building have serious disadvantages and are generally unsatisfactory.

With a single-car installation, a passenger who just misses an elevator must wait for that car to return based on the round trip time. With two or three elevators in the same group and with the help of a good operating system, the wait is reduced by one-half or one-third. This reduction can be up to one-eighth depending on the number of elevators in the group which should not exceed eight (Figures 2.11a and 2.11b).

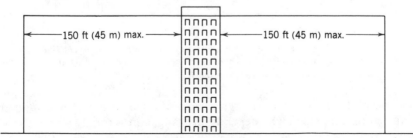

Figure 2.10. A distance of 150 ft (45 m) should be the maximum distance people have to walk to an elevator on any floor.

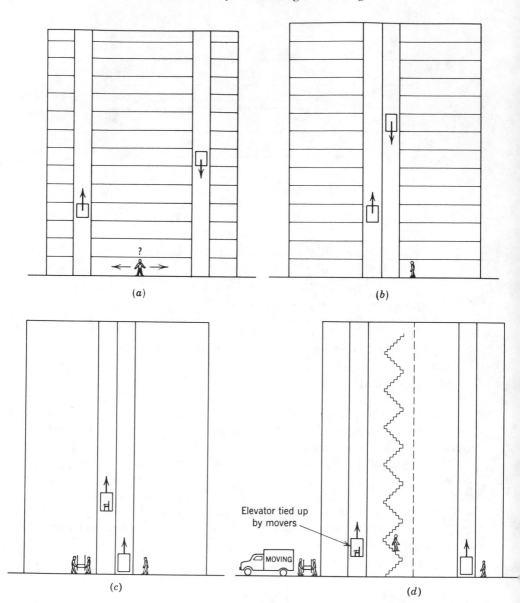

Figure 2.11. (*a*) Separated elevators reduce transportation efficiency. (*b*) Grouped elevators avoid passenger indecision and improve service. (*c*) Grouped elevators afford continuity of service. (*d*) Separated elevators can deny service.

Any elevator requires periodic servicing and replacement of wearing parts. During that time a person who depends on a single car is totally without service and must walk. Making repairs at night is extremely costly and not favored. A group of elevators minimizes this problem and permits periodic repairs without total passenger inconvenience. A well-designed elevator plant will always have excess capacity so that vital traffic is maintained even though one car is out of service.

If a tenant is moving into or out of a building, or if building renovations are scheduled, one car of a group of elevators can be assigned to these tasks. With a single car such moving or renovation is a weekend or nighttime proposition or the passenger again must walk (Figures 2.11*c* and 2.11*d*).

Notable exceptions where elevators are not grouped (if there is more than one elevator in a building) are service elevators designed for specialized use in apartment houses, motels, nursing homes, or office buildings. These service elevators are mainly for building maintenance purposes, moving operations, or food service, and, as such, are generally subject to marring, scuffing, and other damages. They should not be considered for general passenger use and the car interior should be of durable material. If passengers must be carried, removable pads can be used during service operations and building services made to wait until off peak passenger travel times. Food service elevators for upper-floor dining facilities are in the category of possible acceptable single-car installations. A garage shuttle elevator serving the basement garage in an office building is also a possible exception—provided no more than a few stops are served, handling capacity is sufficient, and there is a convenient stairway. The safe-deposit elevator in a bank is a logical single-car installation. Providing access for the handicapped from grade level to a raised or lower main elevator lobby or where escalators are the main access to a building lobby is an essential building service and more than one elevator should be considered.

As a general rule, if elevator service is essential to the building operation two elevators should be considered as the minimum equipment in a vertical transportation node.*

Serving Floors.   All elevators in a group should serve the same floors. This is a common-sense rule that is often violated for false economy. If, for example, only one car out of a group of three serves the basement, people wishing to go to the basement from an upper floor have only one chance out of three that the next elevator that comes along will take them to the basement. Conversely, people in the basement must wait three times as long for elevator service than upper-floor passengers. Ideally, all cars should serve the basement but if not, a special shuttle elevator must be considered, to run only between the main floor and the basement. This later scheme is favored if multiple basements exist and a large group of main passenger elevators is required. The difference in the cost of providing entrances on all the main elevators at the basement levels and the cost of the shuttle elevator or elevators may be negligible when general construction costs are considered in addition to elevator costs.

The expedient of providing a separate call button at an upper floor to call the single car that serves the basement has often been tried and never proved satisfactory. The average person will operate both the normal call button and the basement call button, take the first car that comes along, and cause the

---

* William S. Lewis: "Two elevators are a system, one is a toy."

basement car to make a false stop. These false stops will add up in lost elevator efficiency over the years to more than pay for the cost of the extra entrances on all elevators.

Skip Stops.   Many schemes have been tried to reduce the number of stops an elevator will make, thereby attempting to improve operating efficiency and to reduce equipment costs. Each has disadvantages with respect to floor access and is not legal where legislation for the handicapped has been enacted. The most notable are schemes involving skipped stops on either a single elevator or groups of elevators.

Single-car skip-stop schemes are elementary; the elevator serves alternate or every third floor (Figures 2.12*a* and 2.12*b*). The disadvantage is obvious: with the single-skip scheme persons on the floor skipped must walk either up one floor or down two. In apartment buildings elderly or handicapped persons cannot be asked to walk up stairs, baby carriages become difficult for people on the nonelevator floors to manipulate, and moving operations are extremely difficult. The passenger load remains but is concentrated on fewer floors, and the intended increase in elevator capacity owing to the limited stops becomes dubious because more persons move in and out at each stop, thus delaying the elevator. The main saving is in the lower cost of fewer total elevator entrances, which may be from 5 to 10% of the total installation cost.

In two-car installations, one car may serve odd floors and the other, even floors, thereby saving one entrance per floor (Figure 2.12*a*). The total installation is equivalent to two single elevators side by side with most of the disadvantages of a single-car and the sole advantage of only having to walk one floor if an elevator is shut down. Service is not improved as it would be in a conventional, all-stop, two-car installation.

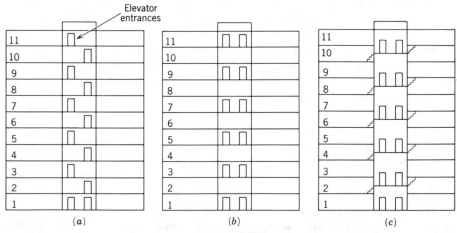

Figure 2.12.   (*a*) **Odd-even skip-stopping—single elevator performance on each floor. (*b*) Alternate floor skip-stopping—dual elevator performance on half the floors, no elevator service on the other half. (*c*) Intermediate landings—dual elevator performance to intermediate landings; everybody must use the stairs.**

Another skip-stop variation is a two-car installation with both cars serving every other floor (Figure 2.12*b*) where waiting time is reduced but half the passengers must walk. Still another variation is the intermediate elevator landing scheme with the cars serving every other floor (Figure 2.12*c*); passengers must walk up or down only half a floor.

Skip-stop elevators are used advantageously in school dormitory buildings where furniture moving is minimal and used to create duplex apartments in apartment buildings where the feeling of luxury offsets the inconvenience of the alternate-floor elevator stopping.

Arrangements

As essential as grouping elevators and the proper location of the group in a building is the arrangement of the elevators in the group. As a rule elevators should be arranged to minimize walking distance among cars.

With any group of elevators, one spot in the lobby is usually favored by waiting passengers, often near the location of the elevator call button. As an elevator arrives at that floor waiting passengers must react and walk to the elevator while it waits for them. Because an average passenger walks at a rate between 2 and 4 fps (0.6 to 1.2 mps), the time spent by the elevator at each stop must be adjusted to this rate of walking.

Advance landing lantern operation will help minimize that waiting time. Excessive distances between the cars in a group results in longer time delays at each stop, which cannot be overcome by lantern operation. Excessive delay from this cause, coupled with other inefficiencies resulting from poor group arrangement, may require the installation of additional elevators.

This is especially true for the physically and visually handicapped. Added time at each elevator stop must be considered to provide ample time for a handicapped person to move to an elevator after the landing lantern operates. Handicapped codes and requirements are defining the minimum time between a landing lantern actuation of an elevator and the closing of the elevator doors to allow sufficient time for a handicapped person standing near the landing call button to travel to the stopped elevator.

Experience has demonstrated the desirability of the elevator arrangements illustrated and discussed in the following pages.

Two-car Groupings. For a two-car group, side-by-side arrangement is best. Passengers face both cars and react immediately to a direction lantern or arriving car. Two cars facing each other constitute an acceptable alternative as the passenger need only turn around to be facing an elevator.

Separation of the elevators should be avoided. The greater the separation, the longer each elevator must be held until a passenger can arrive at that car. Excessive separation tends to destroy the advantages of group operation; passengers will wait at the call button, and, rather than run for the second car if it is too far away, will let it go and reregister a call. Adding a second call button will not relieve this situation but will only result in the effect of two individual elevators serving the same floors, each providing only half the service of which it is capable.

The lobby in front of the elevators on upper floors should, as a minimum, be as wide as the elevators are deep if the elevators are side by side, usually from 4 to 6 ft (1.2 to 1.6 m), and from one and one-half to two times as deep, 8 to 10 ft (2.4 to 3 m) if the cars are opposite each other (Figure 2.13). The side-by-side arrangement requires more space in the main floor lobby because more people are expected to wait in this area. With the cars opposite each other, an assembly area should be provided and the elevator lobby maybe dead-ended (alcove arrangement).

In hospitals or other buildings where elevators carry vehicles, the lobby must be wide enough to accommodate the vehicles as they are turned; hence, the diagonal of the vehicle must be considered as the major vehicle dimension. Vehicles must be pushed straight in or out of the elevator to avoid hitting the protective edge on the car door. About two additional feet (0.6 m) should be allowed to accommodate an attendant.

Three-car Groupings.   The arrangement of three cars in a row is preferable or two cars opposite one is acceptable, the main problem being the location of the elevator call button. The type of elevator door may influence the choice. With center-opening doors, three cars in a row may be preferable as this will give the elevators a balanced appearance (Figure 2.14). With two-speed or single-slide doors, unequal space between elevator entrances (Figure 2.14) may make the two-opposite-one arrangement desirable.

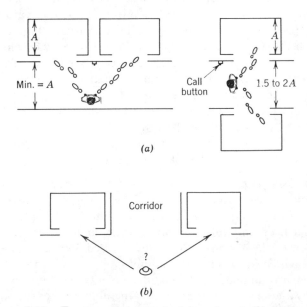

Figure 2.13.   Two-car arrangements: (*a*) preferred; (*b*) wrong.

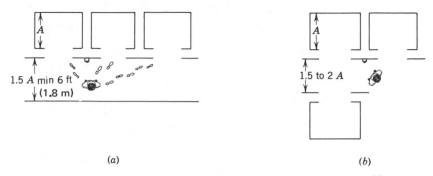

(a)   (b)

Figure 2.14. Three-car arrangements: (a) preferred; (b) acceptable.

There is very little difference in walking time with either of the three-car arrangements, as the turn-around reaction time offsets the shorter distance. Lobby widths must be slightly greater than for two-car arrangements. With the row arrangement, a person should be able to stand back far enough to see the entrances of all three cars; thus the lobby width should be about one and one-half times the car depth or 6 ft (1.6 m) minimum. With the one-opposite-two arrangement, lobby width should be from one and one-half to two times the car depth, or a minimum of about 8 ft (2.4 m).

Four-car Groupings.   Four elevators in a group are common in larger, busier buildings. Experience has shown that a two-opposite-two arrangement is the most efficient.

The alternative arrangement, four cars in a row, has some disadvantage because of the increased distance between the landing call button and the last car in the row. With average-sized elevators, this walking distance is about 12 ft (3.7 m).

The preferred two-opposite-two arrangement (Figure 2.15) should have a lobby from one and one-half to two times the depth of an individual elevator but no less than 10 ft (3 m) for those times when all cars arrive at the main lobby filled and passengers entering or leaving must pass each other. In addition, the 10 ft (3 m) is about the minimum space required for elevator machinery at the machine room level. The closed-end alcove arrangement is acceptable, for even if people wait at the end of the alcove they are only a car length away from the next car to arrive or depart.

If architectural factors necessitate four cars in a row, the lobby should be at least one and one-half times the depth of an individual elevator but no less than 8 ft (2.4 m). This width will allow a person to stand back far enough to see the directional lantern of any elevator. At the main floor the lobby should be wider and longer to accommodate the assembly of people waiting to board the elevators. This is especially necessary during the incoming rush, and to provide quick exit for outgoing passengers.

Six-car Groupings.   Groups of six elevators are often found in large office buildings, public buildings, and large hospitals. Six elevators frequently provide

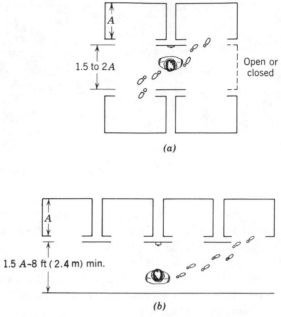

(a)

(b)

Figure 2.15.   Four-car arrangements: (a) preferred; (b) acceptable.

the combination of quantity and quality of elevator service required in these busy buildings. The arrangement of six cars, three opposite three is the preferred architectural core scheme.

The waiting passenger can see all six elevators simply by turning around. The distance to the next arriving elevator is a minimum and the car need be held at a stop for a minimum time. The main floor lobby should be opened at both ends but may be alcoved if necessary.

The lobby width for a six-car group of elevators should be from one and three-fourths to two times the depth of an individual elevator but no less than 10 ft (3 m). If the lobby is to be used as a passage for other than elevator passengers (never recommended at the main floor), its width should be no less than 12 ft (3.6 m).

An acceptable arrangement other than three cars opposite three, is two cars opposite four (Figure 2.16). Passenger response time now begins to become appreciable however, and finally, as with six cars in a row, becomes totally unacceptable. It was formerly believed necessary in a department store to have all the elevators visible to passengers. This may have been true so that attendants of manually operated elevators could see the approaching passengers, but in-line arrangement is hard to justify in view of the universality of automatic operation today and since the time per stop is allowed for the slowest passenger, the random rotation of elevator arrivals at a floor may result in the two end cars arriving simultaneously. If one fills quickly the passengers who could not board must run to the other end and risk missing that car also (Figure 2.17).

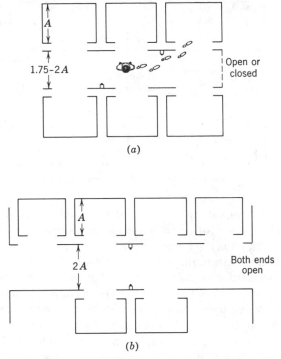

(a)

(b)

Figure 2.16. Six-car arrangements: (a) preferred; (b) acceptable.

Eight-car Groupings. The largest practical group of elevators in a building is eight cars—four opposite four. The main lobby is required to be open at both ends, each of which should be equally accessible to a main entrance to the building, for the handling capacity of eight elevators will require passengers to assemble both in the elevator lobby and in the space beyond both ends of the lobby (Figure 2.18). Equally important is that departing passengers must be able to leave the lobby without having to pass other elevators, which will soon be arriving with capacity passenger loads.

The lobby width of an eight-car group should be about two times the depth of an individual elevator and never less than 10 ft (3 m). The maximum width of the lobby should never exceed 14 ft (4.3 m) and even then the passenger response time to an elevator will be long enough to require extra waiting time at landing calls for some of the cars.

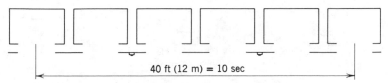

40 ft (12 m) = 10 sec

Figure 2.17. Unacceptable six-car arrangement.

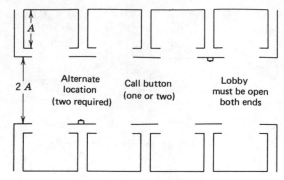

Figure 2.18.   Eight-car arrangement.

Landing call buttons should be located at the center of the lobby, on opposite sides. The best place for persons to wait and, where they usually do wait, is dead center so they may see each elevator. Passengers can then respond as quickly as possible, with minimum walking distance.

Using the main floor elevator lobby as a passage is not recommended, as accommodating any kind of traffic other than elevator traffic would require too wide a lobby. At the upper floors such an arrangement is possible because traffic is expected to be minimal. A main floor lobby, which is open-ended, and upper-floor lobbies in an alcove arrangement is completely acceptable.

Additional considerations to the arrangement of elevators and especially to lobby arrangements are the concern of the structural engineer as well as the architect and elevator engineer. The building columns should be located at the back of the elevator shafts. If they are placed at the front, the columns can interfere with the elevator door size or, if placed in the elevator lobby space, can obscure the landing lanterns and inhibit free passage (Figure 2.19).

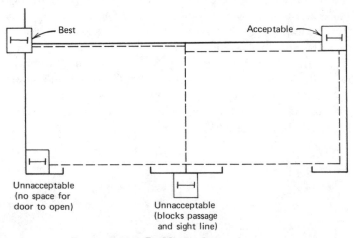

Figure 2.19.   Building column placement.

Unique Arrangements. Certain distinctive elevator arrangements have either unique merit or serious disadvantages. The architecture of particular buildings, rather than economic considerations, is usually responsible for their use.

The angular arrangement of elevators is feasible with a single group of cars serving a building, but has certain inherent disadvantages that must be compensated for by additional elevator service. As can be noted in Figure 2.20*a*, cars at the narrow end of the lobby must be a minimum distance apart, which tends to make the wide-end cars too far apart. This leads to extra time losses per upper-floor landing call stop.

The cornered arrangement in Figure 2.20*b* consolidates the elevator lobby but leads to interference between passengers entering or leaving the corner cars. This is serious if the full capacity of the elevators is required, and is not recommended for that reason.

The circular arrangement (Figure 2.20*c*) is a variation of the angular and cornered, with a premium in cost because of the shape and special mechanical features of the elevator car and doors. The closeness of the elevator entrances requires either special door arrangements or substantial extra hoistway space to accommodate conventional elevator doors. Traffic congestion possible with such an arrangement of elevators is an equally important discouraging factor.

Odd-shaped elevator cars, glass elevators, atrium building elevators, and other unique arrangements of elevators are discussed in the chapter on elevators for special applications.

Front and Rear Entrances. One of the most important of the special arrangements of elevators is the use of both front and rear entrances on one or more elevators in a group. This is one of the areas of elevatoring in which the operation of the elevators must be closely coordinated with their arrangement.

As a general rule, when two sets of entrances are furnished on one elevator in a group but are not on all of the elevators, one set or another but never both should be in operation at one time. The reason is the same as if only one car in a group serves an odd floor: special controls must be furnished to intercept that

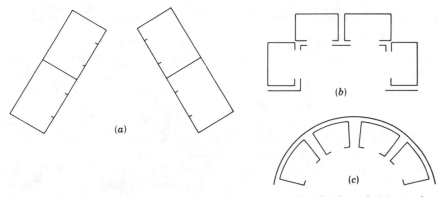

Figure 2.20. Unique elevator core arrangements: (*a*) angle; (*b*) alcoved; (*c*) circular.

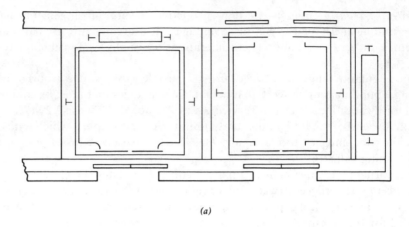

(a)

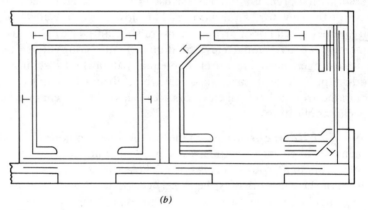

(b)

Figure 2.21. Two entrances on an elevator: (*a*) preferred arrangement—front and rear entrance; (*b*) awkward and costly arrangement—side and front entrances require corner-post type elevator.

car for the rear entrance and a person's wait at a rear-entrance landing may become intolerable.

A suitable arrangement for rear entrances is the use of one of the cars in a group as a part-time service elevator. The elevator is then disconnected from group operation and made to serve a separate landing call button for rear service (Figure 2.21). The rear entrance arrangement is the only acceptable arrangement for new buildings. In the past, side entrances were offered and the elevator car frame and rails had to be arranged "corner-post." At present this arrangement has become prohibitively expensive and should not be considered except under extremely special circumstances.

Another acceptable application of front and rear entrances is the use of a group of elevators between old and new structures to bridge uneven floors. Here all elevators in a group should serve both the front and rear entrances at

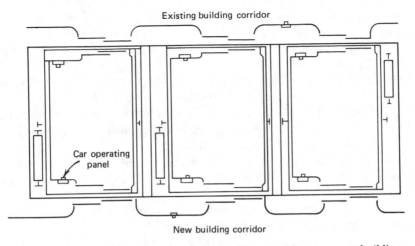

Figure 2.22. Three cars—connecting an existing building to a new building.

all times and car call buttons should be provided for each opening. For example, if the third floor rear is a different level from the third floor front, a person should be able to choose the level required. Separate buttons should be provided even if both front and rear are on the same level. Front and rear openings served in this way make for a building with elevator requirements almost equivalent to one with as many floors as the total number of openings. Because

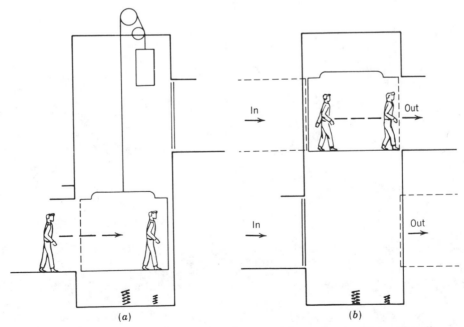

Figure 2.23. Pedestrian movement on elevators: (*a*) transfer on; (*b*) transfer off.

most of an elevator's trip time in a low building is spent in stopping, the increased number of possible stops tends to increase total trip time. Serving both front and rear entrances generally requires additional elevator capacity (Figure 2.22).

When a passenger elevator serves only two stops and heavy pedestrian traffic is expected, front and rear openings allow more prompt loading or unloading of the car. Passengers can walk directly into the elevator and face the entrance from which they will leave, and a loading group at the opposite entrance can follow them in as they are leaving (Figure 2.23).

Front and rear entrance on automobile and freight elevators may expedite loading and unloading. An auto that can be driven off without backing can often be unloaded in half the time.

Other special elevator arrangements are discussed in the sections dealing with specific building types and special applications.

CHAPTER THREE

Passenger Traffic Requirements

If we know how many people will require elevator service within a given period of time, the task of providing vertical transportation is one of time and motion study. It embraces many variables, the most important being the human engineering factors and the human reactions to multiple elevators.

THE ELEVATOR TRIP

In modern buildings people are accustomed to operating a call button to summon an elevator and will move to an elevator that opens its doors to offer them service. Usually, but not universally, passengers will note a lighted lantern showing diretion of car travel and will choose the car corresponding to the desired direction. In a building with light traffic and a tendency for only one elevator to stop at a floor at one time, passengers may ignore the lantern and get into the next elevator that arrives. In busier buildings there is a strong possibility of two cars stopping at a floor at one time. If passengers have taken the wrong car once or twice they become conditioned to be cognizant of the directional signal.

Once passengers board the elevator car they are expected to operate car buttons for the destination floors. Failure to do so may take them where they do not want to go, which quickly teaches them to register car calls each time. The passengers must do one more thing before their trip is complete: to get off at their floor they must note where the elevator is stopping as shown by the car position indicator mounted on the car front and leave the car at their stop.

REQUIREMENTS OF GOOD ELEVATOR SERVICE

The elevator engineer is interested in the passengers' impatience while waiting as well as during their trip. While they are waiting at some intermediate

51

floor their impatience is growing. In a commercial atmosphere they are less tolerant of waiting than in a residential or recreational atmosphere. Frequent studies have indicated that passengers become impatient after waiting about 30 sec in a commercial building, and about 60 sec in a residential building.

From these observations come the first requirement for good elevator service—the elevator system must be designed to provide average waiting time of less than 30 sec in commercial buildings and less than 60 sec in residential buildings.

The second requirement is to provide sufficient quantity of elevator service for the maximum passenger arrival or departure rate expected within a peak traffic period. This can be accomplished by either a platform of sufficient area to accommodate all persons waiting to ride or, alternatively, a sufficient number of smaller platforms. The alternative of more platforms is usually preferred because it reduces the waiting time.

A good analogy is to compare vertical transportation to a batch and continuous-flow conveyor system. The continous-flow system transports from a reservoir into a stream or hose in which the material is moved to its destination. A batch system moves measured quantities to the destination from the reservoir, where they accumulate until another batch is moved, usually in a bucket. An elevator can be compared to a batch conveyor. The arrival of people at a building is in a continual flow and the elevator system is the batch conveyor moving these people from the reservoir (lobby) to their destination. The ideal elevator arrangement is to have the multiplicity of elevators approximate the continuous flow so that the lobby (reservoir) is never filled in excess of the quantity of people a single elevator will transport.

A good example of continuous flow in vertical transportation is the escalator. Platforms (steps) are provided with minimum waiting time (usually 0 sec waiting because the steps are constantly moving) so that a person has immediate access to vertical transportation. Because the platforms are large enough to accommodate only one or two persons at a time, if more than one or two require service at the same instant, someone must wait. The wait is short since the prospective passengers can see the escalator is in service and the extent of their wait. On the other hand, a person waiting for an elevator at an upper floor may not be able to see if the cars are in service and therefore becomes impatient while waiting.

Once people board an escalator they know they will be delivered to the next floor in a relatively short period of time, and except for the extreme high-rise escalators in some subway stations, they can see the top landing. Elevator passengers often do not know how long they will be in the car. If it serves many floors in a busy building and the number of elevators is limited, a person may be on an elevator for a considerable period of time.

Studies have found that about 100 sec becomes the limit of tolerance for people in an elevator making several stops, each for one person. Tolerance will lengthen to about 150 sec if a few people are being served at each stop; the "average person" feels more tolerant if two people are being served. Finally, if monotony is relieved by a changing scene, our passenger may tolerate a ride as long as 180 sec. These time factors are necessarily approximate since an individual's tolerance varies with the urgency of mission or other factors affecting feelings or atmosphere.

The third requirement of good elevator service, therefore, is to design the system so that a person will not be required to ride a car longer than a "reasonable" time. If the first two requirements are met, the third is usually satisfied as a natural consequence.

Three more considerations are necessary to develop a "quality" elevator installation as opposed to an ordinary or utilitarian installation. One, the platform areas as indicated by the capacity of the elevator should be large enough to allow a comfortable area of about 2.5 to 3 ft^2 (0.19 to 0.28 m^2) per person. Two, the door width should be wide enough, 48 in. (1200 mm) recommended, to allow ease of transfer on and off the elevator. And three, a study should be made of the effect of one elevator out of service, and, if critical, an additional elevator or larger platform areas and capacities should be recommended.

CALCULATING THE TIME FACTORS

Two-stop Elevators

To calculate the total time for an elevator trip, a practical procedure is to break the trip down into its components. A simple example of a two-stop elevator will be followed by analysis of more complex and multiple-stop trips.

Suppose we have an elevator that makes two stops about 10 ft (3 m) apart and wish to calculate how long it will take a person to ride to the higher or lower landing.

When the passengers arrive at the landing and operate the elevator call button the trip is, in effect, starting. When they leave the elevator at the other landing they have completed their trip. Once the call is registered the elevator is serving the passenger and the time factors will be as follows.

Referring to Figure 3.1, if a car is at the lower landing when the passenger arrives and operates the call button, the elevator doors need only open (*a*). A typical door requires about 2 to 3 sec to open depending on the width and type of door. About 2 sec must be allowed for the passenger to enter the car and operate the car button (*b*). The doors must close again (about 3 sec) and the car must travel the 10 ft to the next landing (about 7.5 sec) (*c*). The doors must again open and will take 2 sec, with another 2 sec for the passenger to leave (*d*). The total time consumed by that passenger is 19.5 sec.

Before another person can get service more time must necessarily elapse. The doors must close again (3 sec), and the car return to the opposite landing (7.5 sec). At this point the cycle can be repeated. The elevator's total cycle time or round-trip time has been 30 sec. Thus 30 sec is the approximate time a person who just missed the elevator at the first floor will have to wait for it to return

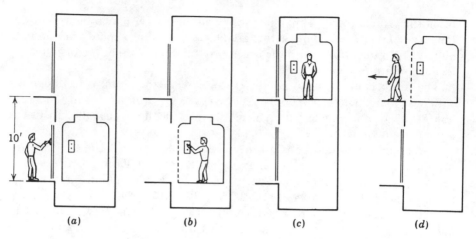

Figure 3.1. Pedestrian–elevator interaction.

and give service. This is called the "interval" between elevator service at a floor. If we view this as a continuous process with a stream of passengers moving in one direction, the average passenger can expect to wait an average of one-half the round-trip time of the elevator. Some will arrive just before the elevator leaves and will have to wait 0 sec, whereas others will just miss the elevator and will have a 30-sec wait, so the average wait is 15 sec.

If two elevators are side by side, each serving two stops and operating as described (each with a round trip time of 30 sec), the interval will be one-half the round trip time or 15 sec and the average wait will be about 7.5 sec. This is established by the expected operation of the elevators; when one is at the upper floor the other is at the lower floor and vice-versa.

The foregoing example serves to show two important considerations in elevatoring: the total time required to serve a person and its relation to elevator handling capacity, and the wait for an elevator as indicated by interval and its direct relationship to round-trip time.

Handling capacity—the number of people served in a given period of time— is calculated from the round-trip time of an elevator. The basic time period is generally established as 5 min for the following reasons.

Five-minute Peaks.   Peak requirements in an office building usually occur in the morning when people are trying to be at their desks by a certain starting time. Human nature being what it is, many employees arrive at the building within a few minutes of the deadline.

Figure 3.2 shows a typical arrival rate at an office building lobby where prospective elevator passengers peak in the 5 min preceding starting time. If there are not sufficient elevators to serve this peak, lobby congestion persists past the deadline. Some conscientious people arrive earlier to avoid being late but, in general, office workers consider themselves on the job if they arrive at the building just prior to the normal starting time. It is the responsibility of

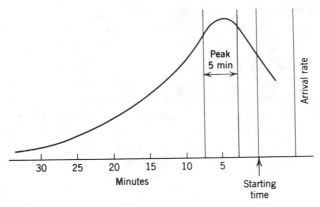

Figure 3.2. Typical arrival rate at an office building.

their employer to transport them quickly up from the lobby to their office floors.

Because elevators for office buildings received the most attention when formal study of elevatoring began, the standard of a 5-min peak has persisted. It has been found that 5 min is a convenient time period to measure peak traffic on elevators in any type of building. For that reason our calculations are concerned with critical 5-min traffic peaks and 5-min elevator handling capacity.

Handling Capacity. Translating our first example into 5-min handling capacity by means of the formula, we find that the elevator has a 5-min handling capacity of 10 people:

$$5\text{-min handling capacity} = \frac{\text{number of passengers per trip} \times 300 \text{ sec}}{\text{round-trip time in seconds}} \quad (3.1)$$

$$\frac{(1) \text{ passenger per trip} \times 300}{30} = 10 \text{ passengers/5 min}$$

If only one passenger is served per trip in one direction, our 5-min handling capacity is only 10 people; with two passengers per trip (or one person up and one person down per trip) in the same time, handling capacity would be 20, and so on. If more passengers are served each trip, however, the time factors, especially entering and exiting transfer time, would increase.

Interval. Interval or the time between elevators is determined directly from round-trip time and is inversely proportional to the number of elevators in a group. It is calculated by the formula:

$$\text{Interval} = \frac{\text{round-trip time of an elevator}}{\text{number of elevators in the group}} \quad (3.2)$$

In our example, with only one elevator, interval is equal to round-trip time. Obviously the interval measures the theoretical longest time a person should have to wait for an elevator. In practice the actual interval varies from trip to trip because of passenger delays or random traffic on an elevator, and should be stated as an average for a given period of time.

Waiting Time. Here we introduce the concept of waiting time related to interval. The theoretical average wait of all persons is one-half the interval, the interval being the theoretical longest wait of any person. The term "theoretical" is used for two reasons: (a) intervals must be considered as averages for a period of time and (b) the operating system must maintain uniform spacing of elevators. With the best available operating systems, average waiting times should be about 55 to 60% of the interval depending on the refinement of the operating system.

For example, if four elevators are provided and the average round trip over a period of time is 120 sec, the average interval for that period is 30 sec. Some people will get immediate service, 0 sec wait, and others will wait the full 30 sec. The average waiting time is therefore 0 sec plus 30 sec divided by 2 for an average of 15 sec. The maximum waiting time should be 30 sec. However, if during the time period an elevator is filled and arranged to bypass waiting landing calls for that trip only three elevators are available and the interval is 40 sec. In the same period, someone may delay an elevator and the average round-trip time for the four elevators can increase to 140 sec so the interval becomes 35 sec. If all the incremental trips are calculated and an average taken, the average interval may still be 30 sec (some trips will be faster than the 120-sec round-trip time) but the average waiting time will be longer than 15 sec, probably in the range of 18 to 20 sec.

Multistop Elevators

A two-stop elevator is the simplest and most efficient elevator system. Everyone that gets on at one landing is expected to get off at the other landing. Transfer time is minimized and no question of probable stops occurs since there is only one possible stop.

Even groups of two-stop elevators have very little complexity. Scheduling can be simple: if traffic is two-way, one car should be at the top and another car at the bottom; with one-way traffic, elevators should be concentrated top or bottom.

Elevatoring calculations become more complex with three or more stops. With three stops, for example, a number of elevator trips are possible: a person at the lower landing may wish to go either to the middle or top landing; a person at the second landing may wish to go either to the top or bottom landing; or a person at the top may wish to go either to the second or the lower landing.

The time required for a typical trip may be calculated on either of two bases. The more conservative basis assumes that the elevator will make every stop up and down. A second more typical and more complex method determines the probable stops the elevator will make related to the traffic it is expected to handle. We will investigate the latter approach in detail.

Probable Stops

The first step in calculating the time required for a typical trip is to determine the number of stops an elevator will make in an average round trip. This depends on the following factors:

1. The number of people entering the elevator at the lobby floor greatly influences the number of stops the elevator makes. If the elevator serves more floors than the number of passengers on the car during that particular trip, assuming passengers enter at the ground floor only, the elevator is not expected to make more stops than passengers carried. The probability is it will make fewer stops than the number of passengers carried. Viewed differently, the number of floors the elevator serves influences the number of stops an elevator is expected to make with a given passenger load. For example, an elevator that serves 20 floors and carries 10 passengers per trip is more likely to make 10 stops each trip, than an elevator that serves 10 floors and carries 10 passengers per trip which will probably make fewer than 10 stops per trip.

2. The normal population of each floor the elevator serves also influences the number of stops an elevator makes. For example, if some floor served by the elevator is a storage or mechanical floor, the elevator's tendency to stop there is zero. If one floor has 100 persons and the other floors have only 10, the tendency to stop at the 100-person floor is greatly increased.

3. The expected direction of traffic imposed on an elevator influences the number of stops each trip. For example:

   a. When everyone is coming into the building, the elevator makes predominantly up car stops and returns to the lobby floor with very few down landing stops.

   b. When persons are traveling between upper floors in a building each person causes the elevator to make two stops: an up or down landing call for that person to board the elevator and a subsequent car call for that person to leave.

   c. When visitors are coming to and going from the building or when the occupants are leaving and returning during lunch or other times, up car stops are generated by people entering the elevator at the lobby and down landing stops by persons on upper floors. Under extremely busy situations and with inadequate elevators, it is possible for an elevator to make every stop up and every stop down, resulting in intolerably long round-trip times in a building of any height. This is especially true if elevators have large capacities and carry many people which can cause many stops.

d. When many people wish to leave the building within a short period of time, as at quitting time in an office building, a different pattern of stops is likely. The elevators are expected to make predominantly down landing stops and to fill quickly at each stop. Because at least a carload of people is eagerly awaiting the elevator and each person is concentrating on leaving the building as quickly as possible, the number of down stops in a typical trip may be far fewer than an equivalent up trip for the same number of people.

As may be seen, determining probable stops requires a knowledge of critical traffic in the building under consideration. Timing and the nature of the critical traffic for each type of building is discussed in later sections of this book. One form of traffic may be critical in many types of buildings or many types of traffic may occur in one building during various times of the day, so that probable stops will be related to the various types of elevator traffic.

In the following chapters we discuss the major traffic periods and the necessary time required for elevators to serve passengers during those periods. With this information and with a proposed or existing building, and an estimate of the type of tenants and their population, a reasonable estimate of the elevator system performance for a critical traffic period can be calculated. If elevators can handle the critical traffic, and if they are properly operated, they will also be adequate for the other traffic periods.

Incoming Traffic

INCOMING TRAFFIC

Incoming traffic provides one of the heavier traffic periods in office buildings and occurs to a greater or lesser degree in any building. The number of elevators required to serve a given number of people during this period is calculated and their operation discussed. In succeeding chapters two-way and outgoing traffic are similarly discussed.

Incoming traffic calculations are discussed in detail as this traffic type is important in any building. In addition, the method of calculation that is established in our study of incoming traffic can be extended to include any type of traffic. Providing sufficient handling capacity is the major consideration during incoming traffic periods.

Incoming or up peak traffic exists when everyone arriving at a lobby floor is seeking transportation to upper floors (Figure 4.1). Minimum down traffic is expected during this period and usually amounts to about 10% of the up traffic. A single elevator trip during this period consists of the following elements:

1. Loading time at the lobby.
2. Door closing time and running time to the next stop.
3. Door opening time and time to transfer part of the passenger load at that floor.
4. Door closing time and running time to the next stop.
5. Door opening time, transfer time, etc., until the highest stop is reached.
6. Door closing time and running time to a down stop.
7. One or two down stops and running time to the lobby.
8. Door opening time at lobby unload, and repeat loading time.

As may be seen, the elevator trip is made up of various time elements related to such factors as the number of people entering and leaving, the speed of the elevator, and the time to open and close the doors. Of prime importance is the

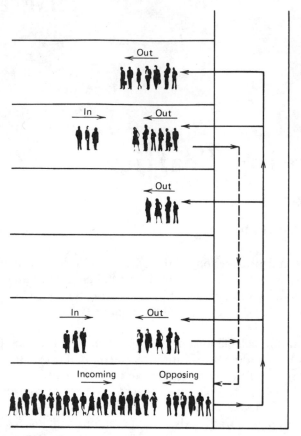

Figure 4.1.   Elevator people handling—typical incoming traffic flow.

number of stops the elevator will make on its up trip. Two methods are suggested for estimating the number of stops. The method selected will depend on the information available about the distribution of the building's population, working hours, and their activities within the building. The more that is known about these factors, the more practical and accurate the probable stop value will be. The less that is known, the more theoretical the probable stop value will be.

In a building in which a great deal is known about the per floor population it is possible to assign a stop value to each floor. Table 4.1 is based on an 11-story building. The population per floor is as indicated and everyone is expected to report for work at the same time. We assume that each elevator in this building has a nominal capacity of 10 people. For this table we have made the reasonable assumption that the expected number of stops (probable stops) is proportional to the number of people an elevator can carry. If we assume the top floor population as 5 instead of 100 as shown, an elevator would stop there only once out of every 20 trips instead of every trip. If people on the tenth floor, for

Table. 4.1.

Floor Number	Population per Floor	Expected Stop
11	100	1
10	100	1
9	100	1
8	100	1
7	50	$\frac{1}{2}$
6	50	$\frac{1}{2}$
5	25	$\frac{1}{4}$
4	25	$\frac{1}{4}$
3	10	$\frac{1}{10}$
2	5	$\frac{1}{20}$
	Expected stops	5.65

example, started work, say 15 min later, the expected stopping at the tenth floor would be greatly reduced during the time the other floors required service.

Seldom is as much information available as shown in Table 4.1. If it were, the task of elevatoring could be reduced, as is possible with a known building in which the elevators are manually operated and there is a desire to automate them. A study of the existing conditions can determine the average number of stops made each trip and, if sufficient elevators are available, the degree of improvement possible with automatic operation.

In estimating probable stops for a building under design we must give due regard to the real-life situation. We will not know that all starting times are the same, we will only know that so many square feet of area are available for the per floor population and we will not know the difference in attraction for each floor.

We can be reasonably certain, however, that in any elevator trip the elevator will stop at upper floors in proportion to the number of people in a car and the number of floors that elevator serves. This forms the basis of a second method of estimating probable stops.

If we assume that population on each floor is equal and that persons are entering a building in a random fashion so that all must be in place by a given time, there is a distinct possibility that one or more people will get off the elevator at the same floor within the same trip. A statistical calculation of the probable number of passengers leaving the elevator at a given floor at the same time provides the following formula:

$$\text{Probable stops} = S - S \left(\frac{S-1}{S}\right)^p \tag{4.1}$$

which becomes the formula for probable stopping of an elevator with a given passenger load wherein

S = the number of possible stops above the lobby

p = the number of passsengers carried on each trip

Chart 4.1.   Probable Stop Table

Passengers per Trip

Upper Floors Served	2	4	6	8	10	12	14	16	18	20	22	24	26	28	30
30	2	4	5.7	7.6	9.5	10.5	11.7	12.8	13.8	14.8	16.0	17.2	18.0	19.0	19.5
28	2	3.9	5.5	7.2	9.0	10.1	11.6	12.5	13.5	14.6	15.6	16.6	17.6	18.1	18.4
26	2	3.8	5.5	7.0	8.5	9.8	11.2	12.2	13.1	14.1	15.1	16.0	16.8	17.4	17.7
24	2	3.8	5.4	6.9	8.3	9.6	10.8	11.9	12.8	13.8	14.6	15.4	16.1	16.7	17.3
22	2	3.7	5.4	6.8	8.2	9.4	10.5	11.6	12.5	13.3	14.1	14.8	15.4	16.0	17.0
20	2	3.7	5.3	6.7	8.0	9.2	10.3	11.2	12.1	12.8	13.5	14.2	14.7	15.3	16.0
18	2	3.7	5.2	6.6	7.8	8.9	9.9	10.8	11.6	12.3	12.9	13.4	13.9	14.4	15.0
16	2	3.6	5.1	6.5	7.6	8.6	9.5	10.3	11.0	11.6	12.1	12.6	13.0	13.4	13.9
14	2	3.6	5.0	6.3	7.3	8.3	9.0	9.7	10.3	10.8	11.3	11.6	12.0	12.2	12.5
12	2	3.5	4.9	6.0	7.0	7.8	8.5	9.0	9.5	9.9	10.2	10.5	10.8	11.0	11.3
10	2	3.4	4.7	5.8	6.5	7.2	7.7	8.2	8.5	8.8	9.0	9.2	9.4	9.5	9.5
8	2	3.3	4.4	5.3	5.9	6.4	6.8	7.0	7.3	7.5	7.6	7.7	7.8	7.8	8
6	2	3.1	4.0	4.6	5.0	5.3	5.5	5.7	5.8	5.8	5.9	5.9	6	6	6
4	2	2.7	3.3	3.6	3.8	3.9	3.9	4	4	4	4	4	4	4	4
2	1.5	2	2	2	2	2	2	2	2	2	2	2	2	2	2

Formula 4.1 is the one used in planning elevators for buildings when little is known about the distribution of future population and each floor is assumed equal in population.

Applying the formula to a 12-passenger, 12-upper-floor elevator trip, a probable stop value of approximately 7.8 stops is obtained. Chart 4.1 shows approximate probable stop values for various numbers of passenger per trip and various numbers of upper floor stops.

Transfer Time

Table 4.2 shows the time required to transfer people from the lobby to a waiting elevator based on various numbers of people loading. There is usually a fixed time the lobby doors are held open, often referred to as dispatch time. In some operating systems this time is established after the first car call is registered and in other systems it is established after the car arrives and its doors are opened. The times represented are nominal and will be used in our calculations.

At an upper floor, if the elevator stops for only a car call, a certain predetermined dwell-time is established between the time the doors are fully opened and when they start to close. This time can be extended or shortened by the interceptions of a light ray when a passenger enters or leaves the car or when the

Table. 4.2.   Transfer Times

Lobby: Minimum 8 sec plus 0.8 sec per passenger over 8 passengers in, in and out, or out only.

Number of Passengers	8	10	12	14	16	18	20
Lobby Time (sec)	8	10	11	13	14	16	18

Car Calls	Dwell time	2 sec per stop
	Transfer time	Use 2 sec for first two passengers (dwell-time) + 1 sec per passenger over 2
		Example: 4 passengers, 1 stop time = 2 + 2 = 4 sec
Landing calls	Dwell time	4 sec per stop
	Transfer time	Use dwell time + 1 sec per passenger over 1 passenger entering
		Example: 4 passengers, 1 stop time = 4 + 3 = 7 sec

Note: Based on nominal 48-in. (1200-mm) wide center-opening entrances. For other widths and types of entrances see inefficiency associated with the various entrance widths and types, Table 4.3.

doors are reopened by operation of the door safety edge. The car dwell-time is usually about 2 sec and can be adjusted to be longer. A landing call dwell-time is usually about 4 sec and should be adjusted to reflect the time a person will require to walk from a likely waiting position near a landing button to the individual elevator. For our calculations we are assuming a minimum of 2 sec per car call and 4 sec per landing call.

The times given are based on a 48-in. (1200-mm) wide center-opening door. In Table 4.3, which lists door times, a list of increased (or decreased) transfer time and door time inefficiencies are given which should be applied to the total standing time (door time plus transfer time) an elevator spends at a particular stop.

Requirements for handicapped persons have been adopted as law or are reflected in building codes in many areas. Many of these requirements establish the amount of time that must expire between the time an elevator arrives at a floor or a landing lantern lights and the time the elevator leaves. These times are designed to allow a slow moving person to travel from a point near the landing call button to the waiting elevator and should be reflected in calculated transfer time if applicable.

After the elevator loads at the lobby floor it takes time for the doors to close and the car to start moving to the next landing. This time varies with the size

Table. 4.3.   Door Operating Time

Door Type	Width, in. (mm)	Open (sec)	Close (sec)	Total[a] (sec)	Transfer Ineffi- ciency[b] (%)
Single-slide	36 (900)	2.5	3.6	6.6	10
Two-speed	36 (900)	2.1	3.3	5.9	10
Center-opening	36 (900)	1.5	2.1	4.1[c]	8
Single-slide	42 (1100)	2.7	3.8	7.0	7
Two-speed	42 (1100)	2.4	3.7	6.6	7
Center-opening	42 (1100)	1.7	2.4	4.6[c]	5
Two-speed	48 (1200)	2.7	4.5	7.7	2
Center-opening	48 (1200)	1.9	2.9	5.3[c]	0
Two-speed	54 (1400)	3.3	5.0	8.8	2
Center-opening	54 (1400)	2.3	3.2	6.0[c]	0
Two-speed	60 (1600)	3.9	5.5	9.9	2
Center-opening	60 (1600)	2.5	3.5	6.5[c]	0
Two-speed, center-opening	60 (1600)	2.5	3.0	6.0[c]	0

[a] Includes 0.5-sec car start.

[b] Transfer inefficiency: Increase normal standing time inefficiency by this percentage to reflect delay in passengers passing through doors.

[c] When preopening can be used, these values can be reduced by 1 sec.

and type of hoistway doors as shown in Table 4.3. Repeated for each stop, these times become an important factor in round-trip calculations.

Once the door is closed it takes a bit more time to ensure that the door is locked and for the elevator motor to "build up" to run to the next floor. For this requirement, 0.5 sec is added to the door operating time in the table. More time should be allowed if it is known that door mechanisms are slower or if build-up time is longer than the value used.

Door closing time is established by safety codes based on kinetic energy and the values shown are within the 7 ft.-poundal (0.29 joule) kinetic energy code limitations for average weight, hollow metal doors.

Ignore for the moment the time it takes an elevator to travel from floor to floor; the next element in an elevator trip is the door opening at the next stop. Many elevators have doors that start opening as the elevator levels to the floor (preopening)* and open wide enough for passengers to start moving in or out by the time the car stops.

This procedure saves a small amount of time per stop. If adjustment is not maintained it can present a tripping hazard and the present trend is toward requiring the elevator to be stopped before the doors start to open. This is especially true with solid-state motor drives wherein various safety circuits ensure that the elevator is stopped before doors are allowed to open. Preopening is usually limited to center-opening doors and reduces door time about 1 sec per stop. Properly adjusted preopening limits the start of door opening so that the car is fully stopped before the doors are open wide enough for passenger transfer to start.

Table 4.3 shows the total door opening and door closing time for various widths and types of elevator entrances. The total time includes 0.5 sec for car start. A table of various inefficiencies based on different types of entrances is given. This value is in addition to the expected inefficiency of an elevatoring study based on a particular type of building which will be discussed in general in this chapter and specifically in later chapters.

In addition to the loading and unloading and door opening and closing, running time is the other element in the round-trip time of an elevator. Part of the running time is spent in accelerating the elevator up to speed and decelerating it to a stop.

Persons can feel changes of acceleration and deceleration but are not too conscious of a constant acceleration or deceleration; therefore elevator equipment must be able to overcome inertia and accelerate and decelerate the elevator smoothly. The machinery must be capable also of moving the car a distance of 11 to 12 ft (3 m), a floor height, in minimum time. Heavy duty equipment such as a gearless machine can move an elevator from floor to floor in about 4 to 5 sec. Lighter equipment, such as a geared or hydraulic machine, will require about 6 sec for the same distance, depending on the ultimate speed the machine can develop and its accelerating capability.

---

* Preopening is often erroneously called "premature."

Floor-to-floor speed seldom exceeds about 350 to 400 fpm (1.8 to 2 mps) regardless of the ultimate speed of the elevator. The heavy duty gearless equipment is necessary for minimum floor-to-floor time since average acceleration of 5 fps² (1.5 mps²) can be obtained versus 3.5 to 3.4 fps² (1 to 1.2 mps²) for geared equipment.

A sample set of time-distance curves for various speeds of elevators is shown in Chapter Seven, page 127. Note that the curve includes both acceleration and deceleration time so that the total time to travel the distance shown is indicated by a point on a curve.

Since time-distance curves vary with all elevator speeds, a means to approximate running time will be given which can be used easily for any speed and elevatoring situation.

Running Time

Typical floor-to-floor traveling time for elevators of various speeds and with average acceleration rates of 3.5 fps² (1.07 mps²) for 400 fpm (2.0 mps) and under, and of 4.5 fps² (1.37 mps²) for 500 fpm (2.5 mps) and above are given in Chart 4.2.

The above time is measured from when the car starts from a floor until the elevator stops level at the next adjacent floor. It includes acceleration, running at the maximum speed that can be attained in the given distance, and deceleration plus an allowance as shown for leveling. The formula for the development of the above times is given in detail on page 126 (Chapter Seven).

To calculate the time for a distance greater than the above, the following formula is used.

$$\frac{(\text{total travel } - \text{ distance } a \text{ from Chart 4.2}) \times 60}{\text{ultimate elevator speed in fpm}}$$

$$+ \text{ time from Chart 4.2 for distance } a = \text{total time}$$

For example: What is the time required to travel 100 ft (30.5 m) at 300 fpm (1.5 mps)?

1. From Chart 4.2, 30 ft at 300 fpm = 9.4 sec.
2. $\dfrac{(100 - 30) \times 60}{300} + 9.4 = 14 + 9.4 = 23.4$ sec.
3. Total time to travel 100 ft at 300 fpm is 23.4 sec.

If the equipment under consideration for a particular installation requires longer for acceleration for floor-to-floor travel, an extra time value must be considered. If equipment levels at slow speed such as two-speed ac elevators or if the leveling speed operation is too long, floor-to-floor operating time may be as long as 8, 10, or more seconds.

Chart 4.2. Running Times, Car Start to Car Stop, Seconds

Floor Heights: Feet	9	10	11	12	13	14	15	20	30	Each Additional 10 ft	Notes
Floor Heights: Meters	2.7	3.0	3.35	3.65	4.0	4.3	4.6	6.1	9.1	Each Additional 3 m	
Elevator speed											
100 fpm (0.5 mps)	7.6	8.2	8.8	9.4	10.0	10.6	11.2	14.2	20.2	6.0	a
150 fpm (0.75 mps)	6.7	7.1	7.5	7.9	8.3	8.7	9.1	11.1	15.1	4.0	
200 fpm (1 mps)	5.8	6.1	6.4	6.7	7.0	7.3	7.6	9.1	12.1	3.0	b
300 fpm (1.5 mps)	5.2	5.4	5.6	5.8	6.0	6.2	6.4	7.4	9.4	2.0	
400 fpm (2 mps)	4.8	5.0	5.1	5.2	5.4	5.6	5.7	6.5	7.0	1.5	
500 fpm (2.5 mps)	—	—	4.3	4.4	4.5	4.6	4.7	5.2	6.4	1.2	
700 fpm (3.5 mps)	—	—	4.3	4.4	4.5	4.6	4.7	5.2	6.1	0.86	
1000 fpm (5 mps)	—	—	4.3	4.4	4.5	4.6	4.7	5.2	5.8	0.6	

a Speeds of 100 fpm and 150 fpm include 0.75 sec for leveling.
b Speeds of 200 fpm and above include 0.5 sec for leveling.

Incoming Traffic Calculations

Suppose we want to know how many people a 16-passenger elevator at 500 fpm (2.5 mps) with 48-in. (1200-mm) center-opening doors, in an 11-story building with 12-ft (3.65-m) floor heights, can serve during a 5-min incoming traffic peak period.

1. Chart 4.1 shows that 16 passengers will make approximately 8.6 stops on the 10 upper floors in this building.
2. Time to load 16 passengers (Table 4.2)   14   sec

3. Time to open and close 48-in. (1200-mm) center-opening
    doors and start car (Table 4.3)   5.3 sec
4. Time to run from floor to floor in the up direction:
    10 floors $\times$ 12 ft = 120 ft
    120 ft $\div$ 8.6 probable stops = 14 ft
    from Chart 4.2, 14 ft at 500 fpm = 4.6 sec
5. Time to run down from top floor to the lobby

    $$\frac{(120 - 20) \times 60}{500} + 5.2 \text{ (time to run 20 ft)} = 17.2 \text{ sec}$$

6. Transfer time at upper floors
    8.6 probable stops $\times$ 2 sec per stop = 17.2 sec
    versus 16 passengers a 1 sec each, use 17 sec
7. Adding all the time factors together:
    a. Standing time
        Lobby transfer time   14   sec
        Upper floor transfer time   17   sec
        Door operation upper floors, 8.6 times
        plus lobby door operation, 1 time.
        Therefore 9.6 $\times$ 5.3 sec   51   sec
        Total time spent at floors   82   sec
    b. Normal standing time inefficiency 10%   8.2
        Adjusted standing time   90.2 sec
    c. Running time
        run from floor to floor up
        4.6 $\times$ 8.6 =   39.6 sec
        run down   17.2 sec
    Total round-trip time   147.0 sec

Elevator 5-min capacity: $\dfrac{16 \text{ passengers per trip} \times 300 \text{ sec}}{\text{round-trip time } 147 \text{ sec}} = 33$ people.

In other words the single elevator in our example can serve 16 passengers in 147 sec or a total of 33 passengers in 5 min.

Item 7(b), a 10% factor for inefficiency, is added to compensate for the rounding off of probable stops, door time, transfer time, and starting and stopping time as well as the unpredictability of people.

If a two-speed door, 48 in. (1200 mm) wide was used, the door time would be 7.7 sec per stop and the inefficiency would be the normal 10% plus 2% for the two-speed door or a total of 12% (see Table 4.3).

The elevator calculation format just developed will be used in Example 4.1 to show its application as follows:

Example 4.1. Total Incoming Traffic Calculation

Given: 15-story building, lobby 1 to 2, 20 ft; typical floor 12 ft

Required: elevatoring to accommodate 110 people during 5-min peak morning in-rush (assume 11% of 1000-person population)

Procedure:

Assume: four 500-fpm elevators, serving floors 1 to 15, 17 passengers per trip = 10 probable stops

Time to run up, per stop

$$\frac{20 \text{ ft (lobby)} + 13 \times 12 \text{ (to top floor)}}{10} = \text{rise per stop}$$

$$\frac{176}{10} = 17.6 \text{ ft} = 5 \text{ sec (from Chart 4.2)}$$

Time to run down,

$$\frac{(176 - 17.6) \times 60 \text{ sec}}{500 \text{ fpm}} + 5 = 24 \text{ sec}$$

Elevator performance calculations:

Standing time

Lobby time		=	15 sec
Transfer time, up stops 2 sec × 10 stops		=	20
Door time, up stops (10 + 1) stops × 5.3		=	53
Total standing time			88 sec
Inefficiency, 10%			8.8
		Total	96.8 sec

Running time

Run up (10 × 5)		=	50
Run down		=	24
Total round-trip time			171 sec

$$\text{Five-minute handling capacity (HC)} = \frac{17 \times 300}{171} = 29 \text{ people}$$

$$\frac{110}{29} = 4 \text{ elevators required for handling capacity}$$

Interval: 171/4 = 43 sec. Too long. Should be 30 to 35 sec

Observe: four cars can carry load but time between cars (interval) too long, need fifth car to improve interval

Recalculate:

Assume: five 500-fpm elevators, 11 passengers per trip = 7.8 probable stops

Time to run up, per stop

$$\frac{176}{7.8} = 22.6\text{-ft rise per stop}$$

$$22.6 \text{ ft} = 5.4 \text{ sec}$$

Time to run down,   = 24 sec
Elevator performance calculations:
Standing time
   Lobby time   =   11   sec
   Transfer time, up stops (2 sec × 7.8 stops)   =   16
   Door time, up stops (7.8 + 1) stops × 5.3   =   47
   Total standing time   74   sec
Inefficiency, 10%   7

                                                      Total   81   sec

Running time
   Run up (7.8 × 5.4)   =   42
   Run down   =   24
                                                            147   sec

$$HC = \frac{11 \times 300}{147} = 22 \text{ people}$$

$$\frac{110}{22} = 5 \text{ elevators required for handling capacity}$$

Interval: 147/5 = 29 sec

Full-Trip Probability and High Call Reversal

In the discussion of probable stops earlier in this chapter it was indicated that stopping is a function of the attraction each floor has to the passengers in an elevator. The probable stop chart (Chart 4.1) is based on an equal attraction per floor. Attraction can also be considered proportional to the population on each floor and experience has shown that an elevator, especially one operating during an incoming period, does not travel to the top floor each trip. The average trip will be a percentage shorter than a full trip to the top and advantage may be taken of this in elevatoring calculations.

As an arbitrary rule, if the population of any office building floor is less than 10% of the total population on the floors served by a group of elevators, the average trip can be reduced proportionally. For example, if the group of elevators is expected to serve 1500 people on 15 upper floors, the per floor population is 100. A 10% reduction is 150 people, hence the elevator trip and the basis of probable stops can be reduced from 15 floors to 14 floors. The total population served by the group of elevators will remain constant at 1500 people and the required percentage handling capacity sought will also remain the same.

If the top floor to be served contained some special attraction such as a cafeteria or restaurant, the advantage of this high call reversal should not be taken.

In residential buildings such as apartments and hotels, similar advantages of high call reversal can be taken. The percentage will be greater than in office buildings and would be between 15 and 25% of the population served. Additional discussion will take place in the chapter on residential buildings.

If Example 4.1 is recalculated on the basis of short trips, the following results would be obtained (see Examples 4.1A and 4.1B).

Example 4.1A

Probable stops, 13 floors, 17 passengers = 9.6
Time to run up, per stop

$$\frac{20 \text{ ft (lobby)} + 12 \times 12}{9.6} = \frac{164}{9.6} = 17\text{-ft rise per stop}$$

17 ft (from Chart 4.2)   = 5 sec

Time to run down,

$$\frac{(164 - 17) \times 60}{500} + 5 = 22.6 \text{ sec}$$

Elevator performance calculations:
Standing time

Lobby time, 17 passengers up	=	15 sec
Transfer time, up stops 9.6 × 2	=	19.2
Door time, up stops (9.6 + 1) × 5.3	=	52.2
Total standing time		86.4 sec
Inefficiency, 10%	=	8.6
	Total	95.0 sec

Running time

Run up 9.6 × 5	=	48.0
Run down	=	22.6
Total round-trip time	=	165.6 sec

Four 17-passenger @ 500 fpm elevators serving floors 1 to 15

	Calculated	Required
5-min capacity per elevator:	30 people	27.5 people
4 elevators; interval:	42 sec	30 to 35 sec

The four elevators do not provide an adequate interval even though handling capacity is sufficient.

Example 4.1B

Probable stops, 13 floors, 11 passengers = 7.6
Time to run up, per stop

$$\frac{20 \text{ ft (lobby)} + 12 \times 12}{7.6} = \frac{164}{7.6} = 21.6\text{-ft rise per stop}$$

21.6 ft from Chart 4.2   = 5.2 sec

Time to run down,

$$\frac{(164 - 21.6) \times 60}{500} = 5.2 = 22.3 \text{ sec}$$

Elevator performance calculations:
Standing time

Lobby time 11 passengers up	= 11 sec
Transfer time, up stops 7.6 × 2	= 15.2
Door time, up stops (7.6 + 1) × 5.3	= 45.6
Total standing time	71.8 sec
Inefficiency, 10%	= 7.2
Total	79.0 sec

Running time

Run up 7.6 × 5.2	= 39.5
Run down	= 22.3
Total round-trip time	= 130.8 sec

Five 11-passenger @ 500 fpm elevators serving floors 1 to 15

	Calculated	Required
5-min capacity per elevator:	25 people	22 people
5 elevators; interval:	26 sec	30 to 35 sec

When all other economic factors are considered, the foregoing may encourage the building of one more floor if five elevators are provided.

Choosing the Proper Elevator Capacity

The choice of the proper elevator capacity is a major decision in the design of the elevators in a particular building. Conflict between the space the architect wants to allow and the requirements of a proper sized elevator platform and the necessary space for the hoistway to contain the mechanical equipment to lift and guide the elevator will always be present.

The essential feature of a proper elevator is the interior square foot (or square meter) area of the car. This is based on the number of people required to

Table. 4.4   Area Versus Capacity—Suggested Standards

1. Inch/Pound Units

Capacity (lbs)	Platform (in.)		Car Inside (in.)		Area (ft²)	A17.1 Area Allowed (ft²)	Observed Loading (people)[a]	Maximum Loading (people)[b]
	Wide	Deep	Wide	Deep				
2000	72	60	68	51	24.0	24.2	8	16
2500	86	60	82	51	29.0	29.1	10	19
3000	86	66	82	57	32.5	33.7	12	22
3500	86	75	82	66	37.5	38.0	16	26
3500 (alt.)	96	66	92	57	36.4	38.0	16	26
4000	86	82	82	73	41.6	42.2	19	28
4000 (alt.)	96	75	92	66	42.2	42.2	19	28
4500	96	82	92	73	46.6	46.2	21	33
5000	96	86	92	77	48.2	50.0	23	35
6000	96	99	92	90	57.5	57.7	27	41

2. Metric Units

Capacity, kg (lb)	Platform (mm)		Car Inside (mm)		Area (m²)	Code Area Allowed (m²)	Observed Loading (people)[a]	Maximum Loading (people)[b]
	Wide	Deep	Wide	Deep				
1200 (2640)	2200	1550	2100	1300	2.7	2.8	10	19 (16)
1400 (3080)	2200	1700	2100	1450	3.05	3.24	12	22 (18)
1600 (3520)	2200	1900	2100	1650	3.5	3.56	16	26 (21)
1600 (alt.)	2450	1700	2350	1450	3.4	3.56	16	26 (21)
1800 (3960)	2200	2050	2100	1800	3.8	3.88	18 to 19	29 (24)
1800 (alt.)	2450	1900	2350	1650	3.9	3.88	18 to 19	28 (24)
2000 (4400)	2450	2050	2350	1800	4.2	4.2	20	32 (27)
2250 (4950)	2450	2200	2350	1950	4.6	4.6	22	34 (30)
2700 (5940)	2450	2500	2350	2150	5.1	5.32	25	39 (36)

[a] Based on 2.3 ft² (0.22 m²) per person for 4000 lb and above.
[b] Based on 1.5 ft² (0.14 m²) per person.
Numbers in parentheses in the metric maximum loading column are established in European Code EN81.

73

be carried per trip as established by elevatoring calculations and a minimum allowance of square feet (square meters) per person. The relation between area and load is established by elevator codes and is usually based on a maximum density of people in a given area. Actual loading is less than the weight allowed since people will not crowd that close together. Table 4.4, which is an expansion of Table 2.1, gives the relationship between weight and area versus a people density of 2.3 ft² (0.22 m²) per person for normal (observed) loading and 1.5 ft² (0.14 m²) for maximum loading. The relation given is linear for 4000 lb and above, i.e., the same density is used for the smallest to the largest elevator. In practice, the actual loading of a smaller elevator, 3500 lb (1600 kg) and below, is somewhat lower and is indicated as observed.

When elevator capacities are chosen for a particular building, consideration must always be given to the possibility of one elevator being out of service. For example, if calculations indicate that four 10-passenger elevators are required, which, by the table, would be four 2500-lb elevators, consideration of one car out of service is made by a simple calculation as follows:

$$4 \times 2500 = 10{,}000 \text{ lb}$$

If one elevator is out of service, $3 \times 3500 = 10{,}500$ lb would be required; therefore, 3500-lb elevators should be used instead of 2500-lb elevators. This is especially true where elevator service is vital such as in a hospital. The requirement can be modified if alternate transportation to the floors such as a service elevator, escalators, or overlapping elevators are available, so judgment must be employed. We will call this consideration the (N-1) rule.

The normal capacity shown in Table 4.4 is based on expected office-building-type loading. For apartments, hospitals, or hotels, where two-way traffic is generally the rule, normal loading is about one-half the value given. This allows for people with baggage, accompanying carts, packages, or other items which may be carried in addition to the passengers. The reduced loading also provides additional space for people to satisfy the one elevator out of service consideration. This will be specificially discussed in various examples.

INCOMING TRAFFIC PERIOD INTERVAL

Calculations of elevator requirements for incoming traffic yield an average loading interval or average time between elevators leaving the lobby floor. This is the average time a person who just misses an elevator has to wait before the next one leaves. As can be seen from the calculations, the interval includes the loading time of the elevator.

For efficient loading, a carload of people should arrive and enter the elevator during the time allowed for loading. Visualize the dynamic situation: people are arriving at random and waiting for an elevator. When an elevator arrives, a car will fill and additional people arrive so that, on the average, a carload of people

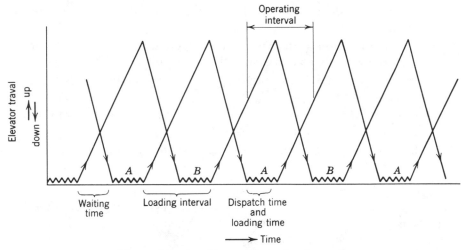

Figure 4.2. Loading and operating interval.

are waiting and their waiting time approximates the interval between car departures.

Incoming traffic interval differs somewhat from operating interval as described earlier in the discussion on a two-stop installation. With incoming traffic, loading interval represents the average wait for service at the lobby; operating interval is related to the wait for service at an upper floor. Loading interval includes lobby waiting time and both loading and operating interval are approximately equal. Operating interval, theoretically, is twice waiting time at an upper floor but in practice is about 60 to 70% of the interval. This distinction is shown graphically in Figure 4.2.

Figure 4.2 shows a two-car operation. With four cars, all other time factors remaining equal, loading time and loading interval are almost equal. The next car arrives at the lobby as the last car leaves and a person's wait for service is spent in the car. With eight cars and all other time factors remaining equal, two cars must be loaded simultaneously to avoid wasting carrying capacity.

INCOMING TRAFFIC OPERATION OF ELEVATORS

To provide a sufficient quantity and quality of elevator service during incoming traffic periods, elevator operation must conform to that traffic. For proper operation elevators must be controlled as follows:

1. Each car must depart from the lobby as soon as it is filled to an optimum percentage of capacity.
2. Loading should start as soon as the doors are opened and the first call is registered in the car, assuming the first person who enters operates a car button.

3. The elevator should travel no higher than required by calls registered in the car or at the highest landing.

4. The elevator should return to the lobby promptly to the extent that some landing calls may be bypassed if there is no elevator at the lobby.

5. Passengers should be encouraged to load as promptly as possible.

6. If the interval between elevators is less than the average loading time or if random distribution causes more than one car to be at the lobby at one time, more than one car should be loaded at one time.

7. Cars should not be held for loading beyond the time allowed for dispatching.

These requirements must be translated into automatic operations by the elevator operating system. Modern microcomputer operating systems do this with the necessary speed and accuracy. Automatic operations should include the features discussed in the following paragraphs.

In office buildings especially and in other buildings as well, the incoming traffic peak usually begins after a long period of quiet. If the operating system has shut down the elevators, that is, parked them with doors closed, they must be activated as traffic occurs in the building. Because traffic is now incoming, someone entering the building will operate a call button at the lobby floor to summon a car or activate a car parked there with the doors closed.

When the elevator responds, its doors open and one or more passengers operate call buttons in the car for their floors. If the car does not fill, these passengers should not be held at the lobby more than 10 to 15 sec. If the car fills quickly, as indicated by a load-weighing device on the elevator, the car should depart immediately and another elevator become activated since quick loading of a car signifies continuing incoming traffic.

An extremely important aspect of incoming traffic operation of elevators is the use of a switch or device which can measure the loading of an elevator car. Present practice is to measure the deflection of the elevator platform and trip a switch or sensor when a certain load in pounds (or kilograms) is attained. At this setting, the elevator is practically filled and should leave since additional people would disproportionally delay elevator service.

The load sensor should also initiate actions to return additional elevators to the lobby floor since a filled elevator is an indication that possibly more people are waiting. Continued load dispatch should also maintain a programmed up-peak operation of the elevators since such continuation is a good sign that this is major traffic especially when compared to a minimum number of down landing calls.

Development work is required to create load weighing or sensing devices that can determine if the elevator is filled on the basis of volume. Present designs are simply based on the assumption that the average person weighs about 150 to 175 lb (68 to 80 kg), and do not consider baggage carried or mail carts.

Once the car has started it should serve its car calls and return to the lobby immediately after serving the highest car or down landing call. A continuing flow of traffic into the lobby will be indicated by filled cars or a multiplicity of car calls in cars that are at the lobby, and by no appreciable landing call activity.

Cars should be controlled so that they depart from the lobby after a period of time equal to the expected loading time or depart as soon as they are filled to an adjustable percentage of load.

A lighted sign in the lobby should indicate to passengers the next car to depart. Even if it has not yet arrived at the lobby, its sign should be lighted so people gather at the entrance, ready to load promptly when the car does arrive. If more than one car is expected to load at one time, all cars that return to the lobby should have their doors open to encourage simultaneous loading but only one car should have a lighted sign.

Under some circumstances, especially when car loading is relatively light; i.e., from four to eight people per car, it may be desirable to close the doors of an elevator that is not selected for dispatching. This will concentrate the loading in one elevator and improve efficiency. If this is done, and a person enters the nonselected before the doors close, the doors should remain open if a car call is operated or reopen if a car call is operated after the doors close. The car should then become the next to be selected and dispatch timing should start when selection takes place. If the car becomes filled as measured by the load sensor whether selected or not, it should depart.

If there are no cars at the lobby and previous cars have left the lobby filled, a heavy incoming rush is indicated. During this period the down traveling elevators should automatically bypass down landing calls until at least one car has returned to the lobby for incoming passengers. Elevators can bypass landing calls whenever lobby traffic requires this priority or bypassing can be restricted to certain periods of the day. At lunchtime, when both down traffic and incoming traffic become heavy, elevator service should be shared equally and the lobby should not have priority.

Restricting operations and giving priority to incoming traffic during certain periods of the day can be accomplished by a time clock. Such clocks can be set to recognize days of the week and restrict operations to certain times on weekdays only. The clock can have the elevators in position for a known recurring incoming rush and also sustain this operation during momentary lulls.

Once rapid loading and a multiplicity of car calls at the lobby ceases for a time or upper-floor landing calls become numerous, the incoming rush has subsided and the mode of elevator operation should change. Depending on the intensity of this subsequent two-way or interfloor traffic, elevator operation should be changed, as is discussed in the following chapter.

HANDLING INCOMING TRAFFIC PROBLEMS

If service demand in a building has changed or if it had an inadequate number of elevators from the start, certain measures may improve the handling capacity of its elevators during the incoming traffic period.

The first step in such an improvement is to determine if any factors are interfering with the handling of incoming traffic. Such factors may include an upper-floor eating facility to which building occupants travel by elevator to have breakfast or coffee after checking in at their floors. Basements, which are secondary entrances or contain eating facilities, can also detract from proper incoming traffic handling. If management solutions, i.e., restricting cafeteria hours until after rush hour or eliminating secondary entrances do not help, other means such as staggering of working hours may have to be employed.

Another measure is the restriction of landing call service during certain periods. Only one or two elevators of the group would be allowed to pick up landing calls during a clock-controlled period, thus concentrating most of the elevator service at the lobby entrance floor.

Up-peak Zoning

A major drastic step only recommended as a last resort is up-peak zoning. This is accomplished by reducing the number of stops each elevator makes. If enough elevators are in a group they may be zoned. Some of the elevators may be designated and operated to serve the lower floors served and the others the upper floors. Separate means of operation must be provided for each subgroup and stopping for landing calls should be restricted. Special signs should be provided at the lobbies and the car buttons of the floors not served must be deactivated. Zoning should be restricted to specified periods and for incoming passengers at the lobby only with the elevators arranged for ready access from the passenger waiting space. Figure 4.3*b* shows a preferable elevator arrangement for zoning.

Zoning increases handling capacity at the expense of interval. For example, one six-car group would be replaced by two three-car groups. From the calculation in Example 4.2, the handling capacity of a typical group of elevators is

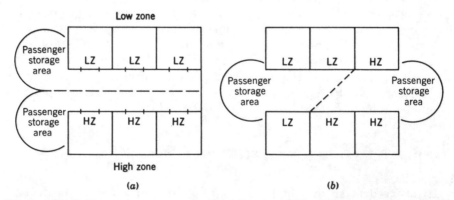

Figure 4.3. Lobby zoning arrangement for incoming traffic: (*a*) poor arrangement; (*b*) acceptable arrangement.

increased by 41% at the expense of a greatly increased (42%) loading interval. Zoning should never be used as the basis of establishing the number of elevators in a building.

Example 4.2.   Effect of Incoming Traffic Zoning

Original: six 3500-lb elevators @ 500 fpm, serving floors 1 to 12, rise 132 ft (12-ft floors)
Probable stops 8.6
Time to run up, per stop

$$\frac{132}{8.6} = 15.3\text{-ft rise per stop}$$

(from Chart 4.2) 15.3 ft $=$ 4.7 sec

Time to run down,

$$\frac{(132 - 15.3) \times 60}{500} + 4.7 = 18.7 \text{ sec}$$

Elevator performance calculations:
Standing time

Lobby time	=	14 sec
Transfer time, up stops 8.6 × 2	=	17.2
Door time, up stops (8.6 + 1) × 5.3	=	50.9
Total standing time		82.1 sec
Inefficiency, 10%	=	8.2
	Total	90.3 sec

Running time

Run up 8.6 × 4.7	=	40.4 sec
Run down	=	18.7
Total round-trip time		159.4 sec
Interval: 159/6	=	26 sec
HC: 6 cars = 6 × (16 × 300)/159	=	182 people

With zoning:
Low group: Three 3500-lb elevators @ 500 fpm, serving floors 1 to 7
Probable stops 5.7
Time to run up, per stop

$$\frac{6 \text{ floors} \times 12}{5.7} = \frac{72}{5.7} = 12.6\text{-ft rise per stop}$$

(from Chart 4.2) 12.6 ft $=$ 4.5 sec

Time to run down,

$$\frac{(72 - 12.6) \times 60}{500} + 4.5 = 11.6 \text{ sec}$$

Elevator performance calculations:
Standing time

Lobby time	=	14 sec
Transfer time, up stops 5.7 × 2 = 11.4 use	=	14.0
Door time, up stops (5.7 + 1) × 5.3	=	35.5
Total standing time		63.5 sec

Inefficiency, 10%   =   6.4

                  Total   69.9 sec

Running time

Run up (5.7 × 4.5)	=	25.7
Run down	=	11.6
Total round-trip time		106.2 sec

Interval: 106/3   =   35   sec

HC: 3 cars = 3 × (16 × 300)/106   = 135   people
High group: Three 3500-lb elevators @ 500 fpm, serving floors 1, 8 to 12
Probable stops 4.8
Time to run up, per stop

$$\frac{4 \text{ floors} \times 12}{(4.8 - 1)*} = \frac{48}{3.8} = 12.6\text{-ft rise per stop}$$

(from Chart 4.2) 12.6 ft = 4.5 sec

Up express run,

$$\frac{(84 - 12.6) \times 60}{500} + 4.5 = 13 \text{ sec}$$

Time to run down,

$$\frac{(132 - 12.6) \times 60}{500} + 4.5 = 18.8 \text{ sec}$$

Elevator performance calculations:
Standing time

Lobby time	=	14 sec
Transfer time, up stops 4.8 × 2 = 9.6, use	=	14
Door time, up stops (4.8 + 1) × 5.3	=	30.7
Total standing time		58.7 sec

Inefficiency, 10%   =   5.9

                  Total   64.6 sec

---

* Note: With the express run, i.e., traveling nonstop from the lobby to the eighth floor, the travel time to the first probable stop in the up direction is accounted for by the time to travel the express portion of the trip. To compensate, the number of up probable stops is reduced by one as shown.

Running time
Express run up	=	13.0 sec
Run up (local 3.8 × 4.5	=	17.1
Run down	=	18.8
Total round trip time		113.5 sec
Interval: 114/3	=	38 sec

HC: 3 cars = 3 × (16 × 300)/114   = 126   people
Group capacity: = 261   people
Gain by zoning: 79 people or 41%
Interval increase: 42% longer

Incoming traffic zoning has other limitations. Once a group of elevators is arranged for zoning, people must be conditioned as to what to expect. Once zoning is initiated it must be maintained for a period of time, since people do not pay close attention to signs and become confused if zoned and unzoned operations alternate spasmodically. One subgroup, even though it may momentarily have idle elevator capacity cannot temporarily aid the other subgroup. An alert lobby attendant can only temporarily control initiation or discontinuance of zoning where it is required.

An additional negative aspect of up-peak zoning is that interfloor traffic, i.e., traffic between high- and low-zone floors requires two elevator trips whereas only one would be required without zoning. A person wishing to go to a low-zone floor from a high-zone floor, or vice-versa, must travel to the lobby. If the high-zone cars are allowed to stop at the low-zone floors, or vice-versa, when car calls are registered, the purpose of zoning would be defeated and only restricted landing call service would result. A transfer floor, say at the top stop of the low zone and the lowest stop of the high zone would have doubtful value since an extra stop would probably be created on both groups of elevators, therefore detracting from the expected gain.

High- and Low-rise Elevators

It is a consequence of zoning that the concept of high- and low-rise elevators came into being but with both the criteria for interval and handling capacity being met. Instead of having all the elevators in a building serve all the floors, the elevators were divided into permanent groups, one group serving the lower floors of the building, a second group the next higher floors, and so on. The grouping of elevators is a characteristic of almost every major building of any height built today.

Each group of elevators has its own operating system and is calculated in a manner similar to the up-peak calculations. The essential difference is that sufficient elevators must be provided in each group to maintain a good operating interval. To serve the higher floors in the building, the time required to transverse the express run of the hoistway is reduced by higher-speed elevators. In

40-and 50-story buildings the highest rise groups of elevators (those serving the uppermost floors) may have speeds of 1000 fpm (5.0 mps) or 1200 fpm (6.0 mps).

Utmost efficiency in providing elevators may be gained by establishing various-sized groups of elevators in a given building. The performance of each group should be equalized by adjusting the number of elevators and floors served so that the interval and handling capacity is essentially equal among all groups.

Multiple-Entrance Floors

A building with multiple-entrance or lobby floors requires extra elevator capacity. If ample capacity is not available, limiting incoming traffic to one entrance floor can improve elevator service.

Reviewing the calculations for incoming traffic handling capacity, we note that only one time allowance was included for the lobby or loading floor. If passengers are expected to board elevators at more than one floor, this time allowance must be increased in proportion to the number of passengers expected to enter at each floor.

For example, if a building has two entrance floors and half the incoming passengers are expected to arrive at each, two lobby stops are necessary. The time to make a round trip will automatically be increased by at least 15 sec in order to operate to the lower level. To ensure full use of elevator capacity, each elevator must stop at both loading levels. The loading time spent at each loading level will be equal and will represent the time needed to load about one-half the capacity of the elevator. The trip back up will require additional time so that a minimum of 15 sec extra is required to serve the second terminal.

If elevators are filled at the lower level and bypass people on the upper level, the latter will soon discover that they must go to the lower level for reasonable service. Escalators or shuttle elevators should then be provided. Preferably, the escalators or shuttle elevators should carry all passengers to the upper level, thus eliminating an extra elevator stop, reducing the distance traveled, and increasing the efficiency of the entire system. If stairs or escalators are used between building entrance levels, a shuttle elevator must be used to provide the necessary service for handicapped persons.

Garage Floors

Similar shuttle elevator arrangements are helpful for multilevel garages above or below the main entrance level. To provide good service for passengers to and from the garage levels, all elevators must stop at those levels. Each stop adds 15 to 20 sec to the normal incoming traffic round-trip time, but a shuttle elevator system would avoid this loss in elevator efficiency. Passengers could then enjoy the convenience of parking in the building and the benefit of improved elevator service to their floors. Building security is enhanced since nontenant parkers do not need to use the tenant elevators.

Parking floors above the lobby level but below the office floors or other usable space creates more serious problems. If parking is used by tenants in the building each parking level represents a potential loading level in calculating elevator round-trip time and handling capacity. A time factor proportional to the number of stops expected at parking levels must be added. If parking is open to both tenants and transients, still another time factor must be included proportional to the number of people expected to leave the building from the parking levels.

As an example, if floors 2 through 6 are parking floors and about 50% of the spaces are to be used by transients, and remaining parking patrons represent 50% of the tenant population, the following additional stops may be expected. If each elevator is expected to carry 12 people on a trip, 6 people who work in the building are expected to enter the elevator at the ground floor and 6 parkers who work in the building are expected to enter on floors 2 through 6. The latter 6 people will create about three additional stops during the up elevator trip. In addition, as the elevator returns to the lobby floor it must stop at least two or three more times to pick up the 6 transient parkers leaving the building. These people must be unloaded before the elevator can begin to serve incoming passengers again. In all, the five garage levels add about five additional stops (about 50 sec) to the round trip of an elevator, plus extra unloading time at the lobby.

The best solution is shuttle elevators between the lobby and parking levels with main elevators running directly from the lobby to the office floors. This improves tenant security as well as elevator service. People cannot enter the building or leave without passing through the main lobby. The parking area can be used in the evening whereby access to the upper floors of the building is controlled and use of its main elevators is minimized.

If the parking areas must be served the controls for all elevators should be provided to minimize the impact of the extra stops. The number of stops any elevator makes at the garage levels should be limited to only a proportional measure of service. Limiting the number of elevators that will answer down calls at garage levels is a means of limiting elevator service in that area. How extensive such limitations are depends on the particular building.

DOWN TRAFFIC DURING UP-PEAK

Due to increased horizontal transportation congestion and as a convenience to employees, many organizations are employing "flextime," or staggering starting and quitting times. Flextime is a concept whereby employees can establish their own starting and quitting times, within limits, provided that they put in the prescribed hours per day and all personnel are at work between fixed times. This may naturally happen in multitenanted buildings where employees soon learn to change their starting and quitting times to avoid the rush on the elevators.

This practice imposes a critical aspect on the operation of the elevators during the incoming traffic peak, since, while people are coming in, others are about their early chores and traveling either down or between floors on the elevators. No longer can the landing calls be temporarily ignored to solve a traffic problem nor can use of up-peak zoning be considered to reduce lobby congestion. The building must accommodate the opposing traffic and make provisions to serve it.

When traffic calculations are made to determine the number of elevators for a particular building, a factor for down stops during up-peak traffic periods should be built in. Observations have shown that this amounts to about 10% of the peak up traffic. For example, if the elevator carries 16 people on each up trip, an average of 1.6 persons will probably ride on its return trip to the lobby. This traffic will create two to three additional elevator stops and a subsequent increase in round-trip time. It often has an effect on the total number of elevators needed for the building.

Redoing Example 4.1 with the 10% down traffic consideration will change the results of the calculation as follows.

Inspecting the results of Example 4.1, it can be noted that the five elevators each with an 11-passenger capacity were equal to the required handling capacity. By adding down stops, the handling capacity will be reduced; hence, the example will be redone as Example 4.3, elevators each with a 14-passenger capacity.

Example 4.3

Given: five elevators at 500 fpm, 14 passengers up per trip, 1.4 passengers down per trip, 9 probable stops up, assume 1.4 probable stops down

Time to run up, per stop

$$\frac{176}{9.} = 19.5\text{-ft rise per stop}$$

(from Chart 4.2) 19.5 ft $= 5.2$ sec

Probable stops down 1.4
Time to run down,

$$\frac{176}{(1 + 1.4)^*} = \frac{176}{2.4} \qquad = 73.3 \text{ ft}$$

$$\frac{(73.3 - 19.5) \times 60}{500} + 5.2 = 11.7 \text{ sec}$$

---

* Note: The uppermost "up" stop is also the highest "down" stop. To compensate, one is added to the down probable stops to obtain a correct down time-distance calculation.

Elevator performance calculations:
Standing time
Lobby time 13 + (1.4 × 0.8)	=	14.1 sec
Transfer time, up stops (9 × 2) = 18	=	18.0
Door time, up stops (9 + 1) × 5.3	=	53.0
Transfer time, down stops 1.4 stops × (4 + 1)	=	7.0
Door time, down stops 1.4 × 5.3	=	7.4
Total standing time		99.5 sec
Inefficiency, 10%		10.0
	Total	109.5 sec

Running time
Run up 9 × 5.2	=	46.8
Run down 2.4 × 11.7	=	28.1
Total round-trip time		184.4 sec

$$HC = \frac{(14 + 1.4) \times 300}{184} = 25 \text{ people per elevator}$$

Elevators required: (110 + 11)/25 = 4.84, use 5.
Interval: 184/5 = 36.8 sec versus a 35-sec criterion

The five elevators will meet the handling capacity requirements with a slight excess. The interval is slightly deficient, but is close enough to the 35-sec maximum criterion when considered together with the ample handling capacity. The final elevator design should include elevators of at least 19-passenger capacity so that sufficient handling capacity would be available if one car is out of service. The final recommendation then becomes five 4000-lb (1800-kg) elevators at 500 fpm (2.5 mps).

GENERAL CONCLUSIONS

Incoming traffic problems are usually those of handling capacity and often result in increasing time delays. The foregoing gave examples of means to reduce time delay and increase handling of incoming passengers.

But maximizing incoming traffic handling by zoning or other means that reduce the number of elevators may create a critical traffic situation at some other time of the day. Calculations of incoming traffic must be compared with other types of traffic in reaching final elevatoring decisions. Down traffic during peak incoming traffic periods is especially prevalent and must be a major elevatoring consideration.

The foregoing chapter has introduced elevator time-study calculations. Comparing time factors in a calculation with those observed in actual installations aids in a personal evaluation of elevatoring problems. Means can be discovered to speed up a trip: better door opening operation, faster floor-to-floor time, means to get doors closed and a car started sooner, or eliminating unnecessary stops.

Each opportunity should be utilized to investigate various phases of the elevator trip as well as the loading and unloading of people from the elevator car. The result will be improved pedestrian transportation.

Analysis was begun with incoming traffic because it is not complex and its time factors can be calculated with relative ease. The following chapters will discuss two-way, interfloor and outgoing traffic, rounding out the treatment of elevatoring calculation methods presented in this chapter.

CHAPTER FIVE

Two-Way Traffic

IMPORTANCE OF TWO-WAY TRAFFIC CONSIDERATIONS

Once a building is occupied its residential or working population will want to enter or leave the building at various times during the day. Some people may be entering at the same time as others are leaving, as at lunchtime (two-way lobby traffic), or people will be going from one floor to another (two-way interfloor traffic), or there will be a combination of these traffic patterns.

In many buildings the peak elevator traffic period is two-way traffic. In a hotel guests are often checking out or going to activities while other groups are checking in or returning from meetings.

In buildings other than hotels the peak traffic may often be between floors. In some hospitals, for example, traffic can be heaviest when patients are being transferred for treatment, doctors are making visits, volunteers are tending to patients' needs, and other personnel are moving from floor to floor. In other hospitals, peak traffic may occur when one shift is leaving and another is taking its place together with visitor traffic to and from the lobby.

Traffic situations like these require elevators to make stops in both the up and down directions during the same round trip.

Two-way lobby traffic may be relatively simple; people enter the elevator at the lobby floor and leave at various stops during the up trip. On its down trip the elevator picks up passengers at various floors and lets them out at the lobby floor. Compared with purely incoming traffic, the round trip will now be considerably longer but the elevator will serve many more people, the up passengers plus the down passengers on each round trip.

The two-way interfloor traffic trip is more complex. During an up trip, for example, an elevator makes one stop to pick up a passenger and another stop to let the passenger off. Each landing stop for a typical floor waiting passenger usually requires two stops for the elevator. The round-trip time is therefore long in relation to the number of passengers served.

During extremely heavy interfloor traffic, as in schools during class change periods and in department stores during rush buying seasons, it is conceivable that each elevator will make every stop up and every stop down!

Calculations for two-way lobby and two-way interfloor traffic follow the course outlined for incoming traffic. During these periods elevators should be operated to equalize waiting time for all elevator users both on the upper floors as well as at the lobby.

Fortunately, during periods of two-way traffic people will generally tolerate longer waiting times than during the up-peak incoming traffic periods. This will be discussed in detail later in the chapter under the subject of interval and waiting time.

TWO-WAY AND INTERFLOOR TRAFFIC REQUIREMENTS

To calculate how many elevators are required to serve a two-way or interfloor traffic situation, we begin by estimating passenger demand. Intensity of demand is a function of the building population, whereas its complexity depends on the distribution of that population among the floors of the building.

Determining population should be an easy matter. The building has been designed to accommodate so many office personnel, so many hospital beds and related staff, so many students, or so many sleeping rooms. These factors determine the population.

Next, we need to know how many of these people will require elevator service during the critical period.

As with incoming elevator traffic, we are concerned with a critical 5-min period. A 5-min period is convenient to use; it has been evaluated in all types of buildings and, if longer periods of traffic demand persist, 5 min can indicate average intensity over the longer period. In schools the entire peak may occur during a 5-min period since this is all the time the students may be allowed to change classes. Should 10 min be allowed, elevator traffic will peak in the middle 5 min of the longer period.

For each building type discussed later in this book, a qualifying percentage of population is given for the critical traffic period. The percentage, representing expected elevator passenger demand during a critical 5-min period, may vary from a low of about 6% of the resident population for apartments to a high of about 40% for classroom buildings.

Characteristics of two-way or interfloor traffic depend not only on the building types but also on the relative attraction of each floor in a building. In a hotel, for example, the location of meeting rooms or ballroom floors and their relation to the lobby or guest rooms have a direct influence on the number of stops an elevator makes. In a school, hospital, or office building the cafeteria location may greatly influence the stopping on each elevator trip. This chapter gives guidance of a general nature; more specific considerations of various building types appear in later chapters.

Interfloor traffic can be a problem in an office building. If a single tenant or organization occupies many floors, traffic between various divisions of such an organization may adversely affect elevator service.

Under certain conditions, a separate interdepartmental special service elevator may prevent overburdening the main building elevators. In a large organization mail distribution, for example, may require almost continuous use of an elevator or cart lift. Mail handling conveyors may also be used to reduce elevator use. A later chapter will give details of various systems.

Changing habits are making two-way traffic more critical than the incoming peak in office buildings in many areas. When people drive to the office their arrival is influenced by the accessibility of parking rather than by the time they are due at work. Arrivals may be spread over a longer period, which reduces up-peak elevator traffic. During the lunch period half the building population may be leaving while the other half is returning. The percentage of building population the elevators must serve during a peak 5 min of the luncheon period may be far greater then during a 5-min incoming traffic period.

Factors like these must be considered in calculating elevator requirements for two-way and interfloor traffic. The consequence of insufficient elevators for this period is impaired service for passengers. They must wait too long for an elevator and, once aboard, may face an unduly long trip marked by excessive stops for entering and leaving passengers.

Interval and Waiting Time

Capacity is seldom the problem; waiting time and riding time are the more critical aspects so that elevators must necessarily pass the floors in a building with sufficient frequency in order to provide prompt service to all passengers.

An earlier European device, the paternoster, has a series of continuously moving up and down platforms in a dual hoistway (Figure 5.1). Passengers must leap onto a moving platform and off at their floor of destination. If they pass their floor they can stay on and make the full cycle up and over or down and under, or get off at the next floor and take the platform in the opposite direction. Although the paternoster is a relatively slow-speed device [about 60 fpm (0.3 mps)], its use demands agility and its installation is no longer allowed, although some are still in operation in Europe.

With conventional elevators passengers can be transported only if the cars are intercepted, slowed, and stopped. A passenger's wait for service becomes a function of the elevator operating interval. To provide service with acceptable average waits, the operating interval must be established based on the maximum wait that passengers will tolerate.

In hurried, commercial atmospheres this waiting tolerance seldom exceeds about 40 to 50 sec. In a more relaxed residential atmosphere, a wait of about 60 to 75 sec is tolerated before tempers rise and people complain. Interval is related to average and maximum waiting time in the following manner. If we establish a desired maximum waiting time of 40 sec, the average wait will be one-half or 20 sec. Waiting time is about 60% of interval as discussed in Chapter 2, therefore, the interval should be $20 \div 0.60 = 33$ sec. The operating or two-way

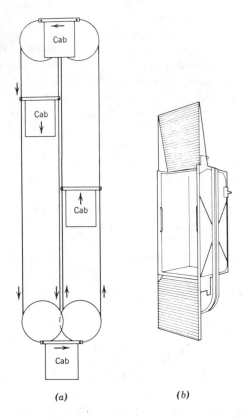

(a)   (b)

Figure 5.1. "Paternoster elevator:" (*a*) schematic; (*b*) paternoster car; (*c*) actual installation (Courtesy of J. E. Hall Ltd., United Kingdom and *Elevator World Magazine*).

interval in a diversified office building should therefore be about 35 to 40 sec and about 65 to 70 sec in a residential building. It also must be recognized that some waits will exceed the desired maximum and this aspect will be discussed later in this chapter.

Calculations for Two-Way Traffic

Calculating the round-trip time of elevators serving two-way lobby and two-way interfloor traffic can follow the format used for incoming traffic. The time for each stop is established and multiplied by the number of expected stops, to which is added lobby time and running time. The total is an approximate round-trip time.

To establish the number of expected stops, a passenger load per trip in each direction must be determined. The total of up passengers plus down passengers times 300 sec (for 5 min) divided by the round-trip time per trip gives an elevator's 5-min capacity:

$$\text{5-min handling capacity} = \frac{\text{(up plus down traveling passengers)} \times 300 \text{ sec}}{\text{round-trip time in seconds}}$$

Operating interval or average time between elevators passing a given upper floor in the building in either direction is found by dividing the round-trip time of a single elevator by the number of elevators in the group.

$$\text{Interval} = \frac{\text{average round-trip time per elevator}}{\text{number of elevators in group}}$$

Probable Stops. The number of stops an elevator makes in any trip is a function of the number of passengers it carries, the number of floors the elevator serves, the relative attraction of each floor, and the relationship of each floor to the others in the building. Establishing probable stops for a building requires a great deal of evaluation and judgment.

Prevailing occupancies and observations of people in known buildings can help in this evaluation and give an indication of the number of stops elevators will make during two-way trips. People, for example, usually go to lunch at noon or to a coffee break in midmorning. In specialized buildings, such as hospitals or hotels, activities take place that can be measured, evaluated, and projected to the next hospital or hotel to be built.

In any building certain floors will be expected stops for each elevator trip. The lobby is one such floor; the cafeteria floor would be another.

Entrances to the building from parking areas on several levels will cause elevator stopping relative to the attraction of each such floor. During two-way traffic, the probability of passengers going to or from the top floor of a building is proportional to the relative population attraction on that floor.

As a beginning, we have at least two certain stops in each two-way or interfloor elevator trip—the lobby and some upper floor in the building. Stopping at other floors is a function of the number of people using the car during each trip. For two-way lobby traffic, because we are concerned with people boarding the car at the entrance floor or floors, the number of up stops is approximately the same as for incoming traffic. Provisions for highest call reversal should be taken based on the relative attraction of the upper floors.

On the down trip, during periods when there is approximately equal traffic up and down, the number of probable stops is usually less than the number made during the up trip. This is especially true in office buildings when people usually go to lunch with companions or due to the general accumulation of people at various upper floors waiting to go down. The percentage of stops in the down direction compared to up stops may be arbitrarily taken as 75% based on actual values from 70 to 80%.

The probable stop values for two-way traffic then become:

> Up two-way probable stops = up probable stops from Chart 4.1
> Down two-way probable stops = 0.75 (up probable stops)

When many floors are to be served, the likelihood of each passenger causing a stop is increased. If, for example, an elevator is expected to carry 10 passengers in a trip and can make, say, 30 stops, it is almost certain to make 10 stops. Similarly, if an elevator is expected to carry 10 passengers in each direction and can make only 5 stops, it is almost certain to make every stop up and down. These later considerations will prove especially valuable in approximating two-way trips in apartment buildings or any situation in which minimum car loading is expected.

Elevator capacity must be ample for two-way lobby and two-way interfloor traffic. We can assume that the elevator car is filled equally in both directions although this is seldom the case in actual situations. Essentially, the platform must be large enough to serve the required number of people in each direction in each trip. As a rule, the car should not be filled to more than about one-half to two-thirds of its observed capacity as shown in Table 4.4 in two-way traffic calculations. With full loads the trip time usually becomes much too long, the transfer time at each stop is unduly lengthened, and no reserve capacity remains for momentary traffic surges or imbalance between up and down traffic.

Chart 4.1 developed for up-peak probable stops can also be used for two-way probable stops. Based on given number of floors in a building and a certain number of passengers per trip either up or down, the probable stop value is chosen. Interpolation is used for odd numbers of passengers or odd numbers of floors. It is assumed that each floor has an equal attraction and that the elevators do not invariably travel to the top floor on each trip.

Earlier versions of two-way probable stop tables were predicted on the elevators making full trips from the bottom to the top terminal and based on the formula

$$\text{Probable stops} = S - (S - 1)\left[\frac{S - 1}{S}\right]^p$$

where S is equal to the number of possible stops above the lobby and p is equal to the number of passengers traveling in a single direction. The formula in Chapter 4 changed the first $S - 1$ term to S which had the effect of reducing the number of stops and eliminating the constant top-floor stop.

For interfloor traffic, determining probable stops under extreme conditions is quite simple. Because each passenger requires two elevator stops, one to board and one to disembark, the number of stops will be twice the number of passengers in each direction less one for the highest stop. For example, if two people are expected to travel up and two people down, an elevator will make about seven stops. We say "about" because one or more of those stops may be coincidental, that is, a person will get on where a passenger made a car stop since both the car and landing call for the same floor are in the direction the car is traveling. Stated otherwise, some landing calls are expected to be coincidental with car calls. A coincidence factor can be established but because it must apply only to a particular building we will be conservative and assume none.

As may be seen, interfloor traffic trips can be time consuming and if a large number of people are traveling up and down, the elevators will make almost every stop in each direction. This is particularly true in educational situations during class changes. In offices and hospitals the general average is about every other stop up and down with the elevator filled to about 50% of its nominal capacity.

Transfer Time. Approximately the same time is required for passengers transfering at the lobby or at upper floors as was discussed for incoming traffic.

At the lobby, the transfer time will consist of both passengers unloading from the down trip and loading for the up trip. The average of 0.8 sec per passenger is valid. At an upper floor, transfer out of the elevator at a stop for which a car call is registered will include the same minimum door-hold open time (dwell time) of 2 sec. If the stop is for a landing call in the up direction, as would possibly occur during periods of interfloor traffic, the dwell time for a landing call of 4 sec must be used. Likewise, dwell time in the down direction must be based on whether a car call or landing call is registered, as well as the number of people expected to transfer at that floor.

During interfloor traffic trips there will be occasions when people are transfering out of the elevator and others are entering. In that event, the transfer time will be the longer landing call transfer time plus the additional time required for the number of passengers in and out. Care must be taken to consider local code requirements for handicapped people which may require transfer time to be considered from the time the elevator stops and the doors are fully opened plus a certain minimum time, depending upon the location of the elevator in relation to the landing call button location. This minimum time is for walking time of 2 fps which must be added after the car has stopped and car doors are opened. In some code jurisdictions, this time is measured from the time the landing lantern illuminates and a gong sounds. This advance lantern time may be as long as 4 sec before the elevator doors are fully opened, depending upon company design, and will serve to reduce the total time spent at the floor.

As discussed for the incoming traffic situation, time compensation adjustment should also be allowed for narrow or wide doors different from the 48-in. (1200-mm) standard center-opening entrances. Additional time should be allowed for other than standard-shaped platforms. If elevator lobbies are exceedingly wide, still additional transfer time will be required.

This is accomplished by increasing the percentage of inefficiency over the 10% allowed in our calculations for office buildings. As an approximate guide, the door time table, Table 4.3, shows certain plus inefficiencies for different widths and types of doors. For narrow and deep platforms the inefficiency should be increased from 5 to 10% depending upon the increase in platform depth. In addition, if the elevator lobby is excessively wide (or narrow with entry restrictions) another 5% inefficiency should be added.

SAMPLE CALCULATIONS

Examples 5.1 and 5.2 show the complete calculations for two-way and interfloor traffic. Both moderate car loadings and full car loadings are shown.

For two-way traffic, handling capacity is simply defined, as all people have obvious destinations. Handling capacity is measured in terms of people traveling to and from the lobby.

In the interfloor traffic examples, handling capacity is the net number of people transferred to and from the lobby and does not account for additional people carried from floor to floor.

It is essential that the standard of elevator demand be established.

If the demand is defined as people carried from floor to floor, elevators effectively carry many more people during interfloor traffic than is shown by only a lobby count. This is best explained by a sketch:

	Up	*Down*
Fl. 4	20 ←	→ 20
	20 ←	← 20
Fl. 3	20 →	→ 20
	20 →	→ 20
Fl. 2	20 ←	← 20
	20 pass.	20 pass.
Fl. 1	in →	out →

In and out at first floor per trip,
40 people.
In and out at all floors per trip,
120 people.

Note that the total interfloor trip only carried 40 people to and from the lobby but 120 people from floor to floor. The essential point is to define and measure demand in terms of the problem.

For simplicity, demands for interfloor traffic will be expressed in terms of lobby traffic in and out of the building. If between-floor traffic must be analyzed, percentages of passengers traveling one, two, three, or more floors as well as the percentage of passengers traveling to and from the lobby must be ascertained. Generally, if a high interfloor traffic demand is expected, escalators are the best way to serve that demand.

Example 5.1. Two-way Lobby Traffic

A. Given: two-way traffic peak of 120 people in 5 min, 12-story building, 10-ft floor heights

Assume: four 500-fpm elevators—8 passengers up, 8 passengers down

Probable stops up 5.9

Probable stops down $0.75 \times 5.9 = 4.4$

Time to run up, per stop $\dfrac{11 \times 10}{5.9} = 18.6$-ft rise per stop

Time to run up 18.6 ft = 5 sec

Time to run down, $\dfrac{11 \times 10}{(4.4 + 1)} = \dfrac{110}{5.4} = 20.4$ ft = 5.2 sec

Elevator performance calculations:

		Up Transit Time
Standing Time		
Lobby time 8 in + 8 out	= 14 sec	8 sec
Transfer time, up stops (5.9 × 2)	= 11.8	11.8
Door time, up stops (5.9 + 1) × 5.3	= 33.4	33.4
Transfer time, down stops (4.4 × 4)	= 17.6	—
Door time, down stops (4.4 × 5.3)	= 23.3	—
Total standing time	100.1 sec	53.2
Inefficiency, 10%	= 10.0	5.3
Total	110.1 sec	58.5 sec
Running time		
Run up 5.9 × 5	= 29.5	29.5
Run down 5.4 × 5.2	= 28.1	—
Total round-trip time	167.7 sec	88.0 sec

HC: 4 cars $= 4 \times \dfrac{(8 + 8) \times 300}{168}$

$\qquad = 116$ people in 5 min vs. 120 desired

Interval: $168/4 = 42$ sec vs. $35 - 40$ sec desired

B. Given: two-way traffic peak of 250 people in 5 min, 12-story building, 10-ft floor heights

Assume: 500-fpm elevators, 16 passengers up, 16 passengers down

Probable stops up 8.6

Probable stops down $0.75 \times 8.6 = 6.5$

Time to run up, per stop $\dfrac{11 \times 10}{8.6}$ = 12.8-ft rise per stop

Time to run up 12.8 ft = 4.5 sec

Time to run down, $\dfrac{11 \times 10}{(6.5 + 1)} = \dfrac{110}{7.5}$ = 14.7 ft = 4.7 sec

			Up Transit
Elevator performance calculations:			Time
Standing time			
Lobby time 16 in + 16 out $(8 + .08) \times 24$ =		27.2	16.0
Transfer time, up stops (8.6×2)	=	17.2	17.2
Door time, up stops $(8.6 + 1) \times 5.3$	=	50.9	50.9
Transfer time, down stops (4×6.5) =			
26 + 9.5*	=	35.5	—
Door time, down stops (6.5×5.3)	=	34.5	—
Total standing time		165.3 sec	84.1
Inefficiency, 10%	=	16.5	8.4
	Total	181.8 sec	92.5 sec
Running time			
Run up 8.6×4.5	=	38.7	38.7
Run down 7.5×4.7	=	35.3	—
Total round-trip time		255.8 sec	131.2 sec

$\text{HC} = \dfrac{(16 + 16) \times 300}{256}$ = 38 per elevator

Elevators required = 250/38 = 6.6, use 7

Interval: 256/7 = 36.6 sec vs. 35 to 40 sec desired

In Examples 5.1A and 5.1B we introduce the concept of up transit time by separating all the time factors involved for the theoretical last passenger to exit on an up trip of the elevator. Up transit time is the third factor in judging good elevator service as described in Chapter 3. In Example 5.1A, this time is calculated to be 88 sec which is quite acceptable. In Example 5.1B, the up transit time is 131 sec which is over 2 min and approaching an arbitrary limit of 2.5 to 3 min (150 to 180 sec) which is about the longest a passenger will tolerate riding an elevator in a single trip without a degree of irritation, depending upon the individual's patience.

If a situation depicted in Example 5.1B were to be faced, judgment may suggest that rather than a single group of seven elevators, two groups of four elevators, low-rise and high-rise, be considered. The other factors such as incoming traffic, population, building design, and the economics of the eight elevators versus seven, as well as the architectural treatment of the core area would be part of the judgment factors affecting the final recommendations.

---

*Rule: 4 sec for first passenger per stop plus 1 sec per passenger over 1. 16 passengers, 6.5 stop = (4×6.5) = 26 for first 6.5 passengers plus 9.5 sec for remaining 9.5 passengers.

Example 5.2.   Two-way Interfloor Traffic

A. Given: two-way traffic requirements 90 people in 5 min measured to and from the lobby, 8-story hospital, 12-ft floor heights, 48 in two-speed doors
Assume: four 400-fpm elevators, 6 passengers up, 6 down
Assume: elevators make every other stop
Probable stops up: assume every other floor, 7 upper floors, use 4 probable stops
Probable stops down: 4 every other floor = 3 probable stops

Time to run up, per stop $\dfrac{(7 \times 12)}{4} = \dfrac{84}{4} = 21$-ft rise per stop,

(from Chart 4.2) 21 ft at 400 fpm = 6.7 sec

Time to run down, $\dfrac{84}{(3 + 1)} = \dfrac{84}{4} = 21$ ft = 6.7 sec

Elevator performance calculations:			Up Transit
Standing time			Time
Lobby time (6 in + 6 out)	=	11	6
Transfer time, up stops (4 × 2)	=	8	8
Door time, up stops (4 × 1) × 7.7	=	38.5	38.5
Transfer time, down stops (3 × 4) = 12 +3	=	15.0	—
(The additional 3 sec is for the top down stop)			
Door time, down stops (3 × 7.7)	=	23.1	—
Total standing time		95.6 sec	52.5 sec
Inefficiency, 10% normal + 2% doors + 5% hospital = 17%	=	16.3	8.9
	Total	111.9 sec	61.4 sec
Running time			
Run up 4 × 6.7	=	26.8	26.8
Run down 4 × 6.7	=	26.8	—
Total round-trip time		165.5 sec	88.2 sec

HC: 4 cars = $4 \times \dfrac{(6 + 6) \times 300}{166}$

= 88 people vs. 90 required

Interval: 166/4 = 41.5 sec vs. 40 sec desired

B. Given: interfloor traffic requirement 250 persons in 5 min (lobby count), 6-story school, 12-ft floor heights, 54-in. center-opening doors
Assume: 300-fpm elevators, 23 passengers up, 23 down
Probable stops: each floor up and down, 5 up and 4 down

Time to run up, per stop $\dfrac{(5 \times 12)}{5} = \dfrac{60}{5} = 12$-ft rise per stop

(from Chart 4.2) 12 ft at 300 fpm = 5.8 sec

Time to run down, $\dfrac{(5 \times 12)}{(4 + 1)} = \dfrac{60}{5} = 12$ ft = 5.8 sec

Elevator performance calculations: Up Transit
Standing time   Time
    Lobby time 23 in + 23 out (8 + 38 × 0.8) = 38.4 sec   19.2 sec
    Transfer time, up stops (5 × 2) = 10 + 13 = 23   23
    Door time, up stops (5 + 1) × 6 = 36   36
Transfer time, down stops (4 × 4) = 16 + 19 = 35   —
    Door time, down stops 4 × 6 = 24   —
    Total standing time   156.4 sec   78.2 sec
Inefficiency, 10% normal + 10% school =
20%   = 31.3   15.6
                  Total 187.7 sec   93.8 sec
Running time
    Run up 5 × 4.8 = 29   29
    Run down 5 × 5.8 = 29   —
    Total round-trip time   245.7 sec   122.8 sec

$$HC = \frac{(23 + 23) \times 300}{246} = 56 \text{ people per elevator}$$

Elevators required = 250/56 = 4.5, use 5
Interval: 246/5 = 49.2 sec vs 50 sec acceptable for school
Five elevators required for acceptable handling capacity and interval, 5000-lb capacity minimum.

ELEVATOR OPERATION DURING TWO-WAY AND INTERFLOOR TRAFFIC

When the number of elevators for two-way interfloor and lobby two-way traffic situations are calculated, a primary objective is to minimize waiting time by providing a short interval of service. Interval depends largely on the effectiveness of elevator operation and is mainly the result of the sophistication of the group supervisory system provided.

Minimizing the time an average passenger would spend on the elevator is of secondary concern. Riding time is a function of elevator layout and arrangement as well as the number of floors served and the size of each car.

With modern automatic elevators all decisions to start or "dispatch" an elevator are made by electronic devices. The simplest will start an elevator whenever a call is registered with little regard for closeness to the car ahead or behind or "elevator spacing." As a result, cars often operate in close proximity and the group of elevators offers service not much better than one large elevator.

An on-call operation of this type can only be effective when traffic is light and elevators are sufficient so that each car can operate almost independently of the others. Our calculations have indicated that if only a few people are seeking service, even the most elementary operating system will suffice.

Where many people are seeking service at a given time, elevators require more intricate and educated automatic operation to maintain a proper spacing in time and an ability to respond to a myriad of traffic conditions to achieve an optimum mode of operation that is constantly changing. Each prospective passenger will then have an equal opportunity to board the elevators and the individual's average wait will not exceed that of any other passenger.

To begin our discussion of elevator operation let us first classify elevator traffic into three categories as follows:

1. Light traffic, when the number of passengers seeking service at a given time is no more than two or three times the number of elevators available to give service.
2. Moderate traffic, when the number of passengers seeking service at one time will fill the elevators to less than one-half their nominal capacity.
3. Heavy traffic, when the number of people seeking service will fill elevators beyond 50% of their nominal capacity.

By assuring the most efficient operation during moderate and heavy traffic situations the number of elevators required is minimized.

Light Traffic

During light traffic elevators may often park and wait for the next landing call. The cars should be distributed among the floors of the building so that they are within minimum response time of a next landing call. It is also essential that the cars move promptly to serve an existing call and to park at a strategic location in anticipation of subsequent calls.

An arrangement for light traffic operation is shown in Figure 5.2*a* and *b*. Waiting cars are parked in zones (1 and 2) that have been established on the basis of population of the various floors or the expected activity caused by a single tenant on a group of floors. Each car should move to the next call in its zone with minimum delay and be free to travel either up or down from that call as passengers desire.

One car is stationed at the lobby, where the probability of a call is high. If that car leaves another car should promptly take its place and the former lobby car should park, after completing its car call, in the zone of the car it displaced or to occupy a vacant zone.

Because elevators can give individual service, each car should only operate to the highest or lowest call in the zone it covers, remaining in that zone to answer the next call. If a car is taken from one zone to another, and the car in the latter zone is also moved to another zone, the former car should assume responsibility for the new zone. If for any reason a zone is temporarily unoccupied, an elevator in an adjacent zone should be free to answer calls in the unoccupied zone.

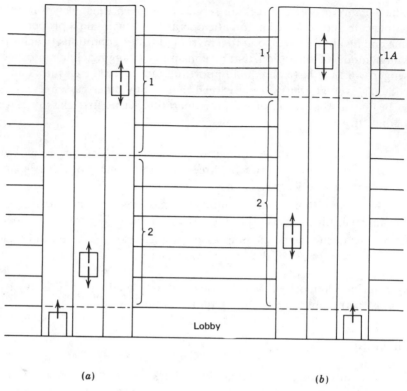

(a)   (b)

Figure 5.2. Distribution of population for effective zoning: (*a*) population zone 1 = population zone 2; (*b*) population zone 1 = population zone 2 or zone 1*A* is a single tenant.

Moderate Traffic

When traffic increases and elevators must be deployed to answer frequent landing calls at various floors, there is little opportunity to park a car in a zone and allow it to remain idle. Various strategies must be employed to reduce the overall waiting time between the registration of a landing call and the prompt response of an elevator to that call.

The light traffic parking strategy tends to space elevators throughout the building in the best locations to answer the prejudged next landing call for service. When one or two calls per zone exist, an elevator can serve them and quickly be restored to the strategic parking position. When there are a number of calls, the elevators are moving, answering calls, and a tendency increases for calls to become "long wait." In anticipation of this condition, the elevator operating system needs to take certain actions to prevent excessive long waits.

One such action is to activate any parked elevator to help out in a zone when either a certain number of calls are waiting or the number of landing calls and their accumulated waiting time reaches a predetermined maximum. Another action is to prevent any excessive travel to the main floor unless a car or landing call is registered for that floor.

There is a tendency for elevator operating systems to park at least one car at the main entrance floor. This practice should be temporarily abandoned until such time as the level of traffic is reduced to a minimum to allow such parking.

Figure 5.3 shows the dynamics of a moderate elevator traffic situation. In general, elevator traffic is not continuous but comes in spurts. During lulls the elevators should be parked in the various zones throughout the building to be in a better position to serve the next traffic requirement. A typical parking arrangement is shown in Figure 5.3*a*. This is in opposition to some elevator operating systems which allow last call parking, which is often at the lobby floor since most elevator trips terminate there. Figure 5.3*b* shows a wave of traffic in two parts. The symbols designated "1" are the initial calls and those designated "2" the subsequent wave. As can be appreciated, the pattern of calls and elevator movements are constantly changing and we can only view a "snapshot."

The various actions that take place are as follows: Car A being parked at the floor with a down landing call immediately answers that call. Car B responds to the car call registered in it for the third floor. When car B enters the zone where car D is parked, car D is immediately released to answer a call in car A zone or to replace car B at the main lobby. We show it answering the down landing call at the fifth floor. Car C moves to answer the up landing call behind car A and, in doing so, is in a position to answer the subsequent down landing call on floor 10.

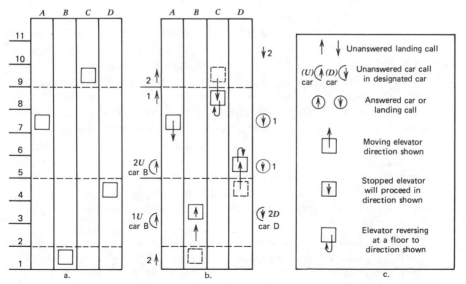

Figure 5.3. Elevator deployment during zoning: (*a*) elevators parked during a traffic lull; (*b*) elevator movements: 1, first calls; 2, second calls; (*c*) key to symbols used.

Any variety of traffic situations can occur and as long as the elevators can answer all the calls within moderate time of an arbitrary 30 sec or less, there is little need for sophisticated operating systems. It is when calls start intensifying and waiting times increase that the need for sophisticated operations is apparent. This can occur in the best elevatored building when, for example, one car is out of service for repair or is handling an alteration or moving job. In other buildings such as hospitals and schools, the elevators are frequently overwhelmed and special operating considerations need to be made for this heavy traffic.

Modern approaches to elevator group supervisory systems include microprocessors which can gather all the available information such as: the location, direction and number of landing calls; the number of car calls in each elevator; the loading of each elevator as indicated by the load sensors; the direction each elevator is traveling or whether it is stopped at a floor to discharge or receive passengers, or free to be sent to another floor. This information is processed and various strategic decisions are made which should be designed to minimize the waiting time of landing calls and to cause elevators to travel to areas where they are needed to serve the current traffic. If the supervisory system fails to do this the elevators can be hopelessly overwhelmed and heavy traffic will exist where only moderate traffic existed before. As can be appreciated, elevator traffic is dynamic—it is always changing—and if it is not served when it occurs passengers will back up at every floor, where they will register both up and down calls in frustration and make the situation worse. A few seconds can change the entire aspect of a system from good to bad; hence everything possible must be done to serve such traffic and the response of the elevators to it.

A summary of the operations required for moderate two-way and interfloor traffic should include the following items.

The supervisory system must have the ability to measure the traffic a group of elevators is expected to encounter and dispatch each elevator accordingly in the proper direction. Such traffic factors measured would include car calls in each car, the loading of each car, the number of landing calls that each car may encounter, the spacing between one car and the preceding car, the relation between up and down traffic, the location of all the cars in the system, and the time to service the existing car calls to the destination floor or floors.

In addition, the group supervisory system must decide whether the elevators should be quickly changed from concentration on down or up traffic to traffic in the opposite direction, whether to travel an elevator through to the main landing or short-trip that car, where to send cars to be fed into prevalent traffic demand, and when to concentrate elevators in zones where long-wait calls may exist.

All of the foregoing constitutes a formidable task that microprocessors can accomplish by gathering, storing, and processing all the necessary information. Once the calls are counted they must be weighted. Elevators are not traveling at constant speed but rather running, stopping, or starting; people's habits are not predictable, and one call can change the entire aspect of an elevator system in less than a second. We have dealt with averages in our computation. In real life discrete trips and people must be accomodated.

Heavy Traffic

Once traffic becomes exceedingly heavy in both directions, we find the cars filling on the up trip (either at various floors or at the lobby similar to up peak operation) and filling in the down direction. The cars will probably make every stop up and they should go no higher than the highest up landing call or car call, and be immediately reversed. On the down trip the cars may make every stop down and perhaps bypass the lower of the down landing calls because of capacity load in the car and load bypass arrangement. The time the elevator spends at the lobby should be the minimum required for passenger transfer.

Landing calls that are bypassed should be given a measure of preference. An upper limit of waiting time can be established, and any car in the proper direction, with capacity and without intervening car calls, can be bypassed to the call that has waited overly long.

Bypassing must be used with discretion. Any bypassed call becomes a potential long-wait call—with indiscriminate bypassing, many calls could wait overly long and a hopeless situation could develop. To avoid this it is necessary to do everything possible to deploy elevators at the first indication of anticipated traffic rather than wait for a traffic demand situation to develop.

Overwhelming situations can occur in any building. It is most noticeable in schools during class change periods and in hotels during the time when convention meetings break up. Everyone in the building wants to go some place at the same time, and if enough elevators were provided there would be little room for usable area on the building's floors. The potential problem should be recognized when the pedestrian planning is done for the building and means incorporated to alleviate the possible problem.

One of the most appropriate means is to use supplementary continuous-flow systems such as escalators or large stairways. Only one floor should be designated to load the heavy incoming crowd and all people directed to that floor.

Additional efficiency can be gained by providing means to call more than one elevator to a designated landing floor if it is other than the main dispatching floor of the group of elevators. On other than the main dispatching floor, if one elevator stops for a landing call, all other elevators are electrically prevented from stopping unless someone registers a car call for that floor. If it is known that at certain times a floor other than the main dispatching floor is going to have excessive crowds, electrical provisions can be made to have more than one elevator stop or to temporarily change the system so that a secondary dispatching floor is created at that busy floor.

Another solution to the overwhelming traffic problem is to restrict service to incidental floors such as basements or penthouses. These floors could be manually or automatically cut out and lighted signs could inform passengers to use

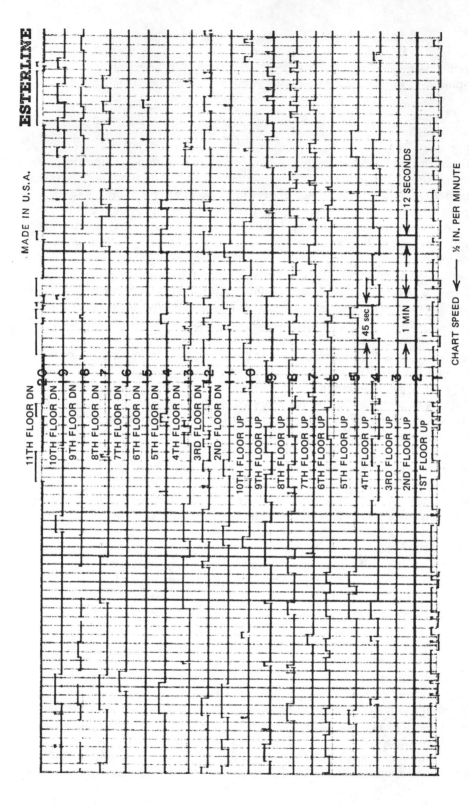

Figure 5.4. Sample of landing call waiting time recording.

stairs. In a building that is initially underelevatored or where a change in the building use has created handling capacity problems, the feasibility of adding additional elevators should be considered. Permanently removing time-consuming and poorly productive elevator stops such as basements or garage floors should be accomplished by installing an additional single shuttle elevator to assume that function.

Waiting Time for Landing Calls

The effectiveness of an elevator group supervisory system's operation during periods of lobby two-way and two-way interfloor traffic is measured by the waiting time for landing calls. To do this, a recording meter is usually employed which consists of pens actuated to mark a continually running calibrated paper chart to indicate when the call is actuated and when response takes place. The paper chart moves at a given speed, usually about 1/2 in. (13 mm) per minute, so that each 1/10 in. (2.5 mm) represents 12 sec (see Figure 5.4). One pen is provided for each landing call in each direction on each floor.

By tabulating the number and duration of landing calls a chart of the distribution of landing call waiting times can be developed (see Figure 5.5). With well-designed elevator systems most calls are expected to be answered within

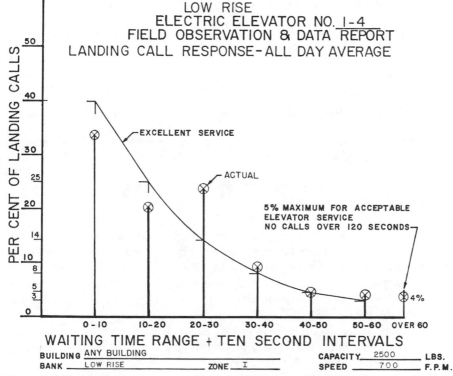

Figure 5.5   Actual observed waiting time distribution (Courtesy of Jaros, Baum and Bolles).

the design interval for the average two-way traffic. An excess percentage of calls over that time is cause for concern and further investigation. Such investigation may determine faults with the supervisory system or discover elevators being excessively held out of service.

The relation between operating interval and the normal expected excess waiting time for landing calls is given by the following example.

As discussed earlier in this chapter, average waiting time is about 60% of the operating interval. Assume that it is desired to know the percentage of landing calls that will wait an excessive amount of time when the elevator system is designed for a certain operating interval. The distribution of waiting times will generally follow a curve developed on a basis of the natural number e which is derived from an infinite series and is equal to 2.71828 (to 5 decimal places).

For a given operating interval the percentage of landing calls expected to wait over a given limit can be calculated by the formula

$$P = e^{-t/T}$$

where P is a number less than 1 which can be converted to a number representing the percentage of calls waiting over a given time t.

t = the arbitrary time limit
T = the average waiting time assumed to be 60% × the operating interval

For example, if the interval is 40 sec, and $T = 24$ sec, what percentage of landing calls will wait over 60 sec (t)?

$P = e^{-t/T}$
$P = e^{-60/24} = e^{-2.5}$
$P = 0.0821 \times 100 = 8.21\%$ of the landing calls wait longer than 60 sec

Figure 5.6 shows various values of P calculations for intervals of 20, 30, and 40 sec.

GENERAL CONCLUSIONS

Two-way lobby and two-way interfloor traffic can be the most serious of the traffic situations encountered in any building. Not only must sufficient elevators be provided to serve the expected traffic but also they must be operated in such a way that their most efficient use is assured. Failure to do so will create a temporary heavy traffic condition where, perhaps, only an extended moderate traffic should exist.

All the inefficiencies of poor elevator shape, size, layout, and so on, as described for incoming traffic, will be amplified during the heavier two-way traffic situations. Poor operating intervals will cause immeasurable employee time loss by riding and waiting passengers. When we consider the number of

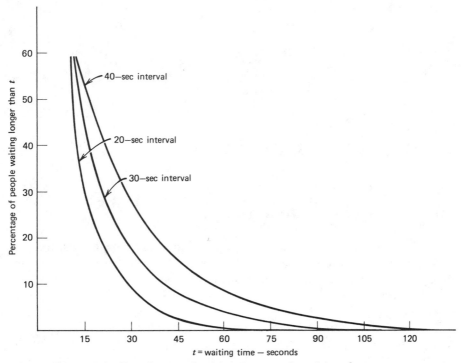

Figure 5.6. Distribution of waiting times—various intervals.

passengers riding and waiting at a given time, we see that 10 to 20 sec extra time spent waiting and on each trip will add up to a considerable number of man-hours lost by inefficient elevator service each year. This cost can only be avoided by proper initial design of the elevator system along with subsequent concern of the installed operating system.

As with all the examples and calculations given in this book, these are offered as a guide to the philosophy of elevatoring.

CHAPTER SIX

Outgoing Traffic

IMPORTANCE OF OUTGOING TRAFFIC

When the quitting bell sounds or, in some buildings, when lunch time arrives, a building's occupants are ready to leave and will demand elevator service. During these outgoing traffic peaks passengers, by sheer numbers alone, may measurably overwhelm an elevator system unless provisions are made for this contingency.

The most elementary of these provisions is sufficient elevator capacity to serve the outgoing traffic demand expected. In office buildings, especially, the outgoing 5-min peak traffic may exceed any other traffic peak by 40 or 50%. Proper design and operation of an elevator system can provide this capacity without more elevators than are normally required to serve other traffic periods.

Proper design requires a reliable means to prevent an elevator filled to capacity from responding to landing calls. For this load weighing and bypassing operation, automatic elevators can be equipped with any of several sensing devices, the present most sophisticated of which will measure only the live load on the platform. Devices of this nature should be capable of adjustment; average capacity load in one building may never be attained in another but the problem of efficient outgoing traffic operation will remain. Future development should include capacity loading based on volume rather than weight.

Since demand is usually scattered among the various floors and an elevator is loaded promptly when it arrives at each floor, advantage may be taken of this in calculating outgoing traffic performance. Because people are on their own time, so to speak, and anxious to get out of a building, they willingly load elevators to near, and often beyond normal capacity. Loads can exceed the average nominal loading experienced during incoming traffic periods, so much so that elevator safety codes require that the platform area in relation to elevator capacity in pounds or kilograms be designed to accommodate 125 lb (60 kg) per square foot (0.093 m^2). In addition, the elevator brake must be able to slow down and stop a down traveling car filled to 125% of rated elevator capacity. Figure 6.1 shows heavy loading of an elevator during a down peak period. The elevator shown

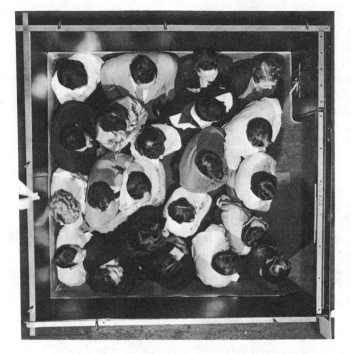

Figure 6.1.   **Heavy elevator car loading.**

[3000 lb (1400 kg)] in the photo has a nominal capacity of 12 to 14 people, but 22 people are overcrowded into the car at approximately 1.5 ft^2 (0.14 m^2) per person.

OUTGOING TRAFFIC REQUIREMENTS

Normally an elevator system should be capable of evacuating the population of a building within 25 to 40 min. This period may be lengthened or shortened depending on particular tenant needs. Although the primary need is quantity of service (handling capacity), and quality of service (interval) is secondary, service should be available at every floor within 60 sec. A sophisticated elevator operating system will have a means to provide equal service to all floors once a multiplicity of down landing calls and an absence of up landing calls or cars loading at the main lobby is apparent. The sequence of operations might be as follows:

If all floors are receiving outgoing traffic demands, the elevators should be equally spaced in time in the down direction, unloaded in minimum time at the lobby, and immediately returned up into the building to serve additional outgoing traffic. Elevators should be allowed to serve up landing calls as long as these calls do not interfere with service to the outgoing rush.

As the cars fill at upper floors they tend to bypass landing calls in the lower section of the building. When this occurs all but one or two elevators in the group should then bypass up landing calls.

As the waiting time of calls in the lower section of the building increases beyond a predetermined value, some elevators should be assigned to serve only the lower portion of the building.

This can be accomplished by various means, one of which is to have the first up traveling car reverse in the next upper zone, etc. Another means is to use a timed-out zone approach where the waiting time is measured in each zone either as a total or totalized by the number of down landing calls times the number of seconds each call is waiting, and to assign a car to that timed-out zone. Up landing calls can be restricted to those cars that are assigned to the higher zones. These procedures will tend to equalize the elevator capacity available to all floors of the building.

If traffic continues to increase until landing calls tend to wait beyond a satisfactory period of time, a priority operation should be instituted. After elevators are unloaded at the lobby they are directed to any floor at which passengers (as indicated by a down landing call) have waited beyond a predetermined period of time. At that call the elevator reverses and, if it is not filled, it continues to pick up down landing calls until it fills or reaches the lobby. Under extreme conditions with this operation, elevators may become, in effect, two-stop elevators, completely loaded at one floor and completely unloaded at another. This is the most efficient operation that can be gained from any elevator.

Obviously, if extreme conditions for down traffic occur with any frequency in a building, service to any unusual floors, up landing calls, basements, and so on, should be temporarily suspended. Any traffic opposed to the outward rush will reduce the system's efficiency and should be considered in that light. Naturally, if up service is necessary and outgoing traffic demands all the available elevator-handling capacity, staggered quitting times, or other approaches to the problem, which will be discussed later in this chapter, are necessary.

Calculations for Outgoing Traffic

Intensity of outgoing traffic is seldom the primary consideration in elevatoring a building. Under certain circumstances, however, it can be exceedingly important. For example, if elevators serve an upper-floor meeting room or auditorium they require sufficient capacity for the orderly evacuation of that facility. In theaters, classroom buildings, stores, and so forth, evacuation ability may be of prime interest in elevatoring. For upper-floor "sky lobby" arrangements discussed in a later chapter, shuttle elevators should be configured based on the down-peak capacity of the local elevators they serve.

Determining the capacity of an elevator system during outgoing traffic periods can proceed in a manner similar to incoming traffic with the following essential differences.

The mode of operation must be established. Will the peak be severe enough to warrant two-stop operation (enough elevators must be provided)—loading at one floor and unloading at a lower floor, or are many stops in the down direction expected?

Elevators may carry greater loads than during the incoming peak; an increase by one-eighth to one-third of the nominal car loading in Table 4.4 can be considered.

Loading can be efficient. An average of 0.6 sec per person can be used rather than the 1 sec established for incoming traffic period landing calls, provided there is sufficient door width and a reasonably shaped platform. We presume passengers will be attentive to hall lanterns and be at the entrance when an elevator arrives at a floor.

The time that passengers take to exit at the lobby will be correspondingly shorter and will be based on a value of 0.6 sec per person (or about 75% of the incoming traffic transfer time) as shown in Table 6.1.

Table. 6.1.

Lobby:	Minimum 6 sec plus 0.6 sec per passenger over six passengers out only						
Number of Passengers	8	10	12	14	16	18	20
Lobby Time (sec)	7	8	9	10	11	13	14
Car Calls	(Same as incoming and two-way traffic) Dwell time 2 sec per stop. Transfer time. Use 2 sec for first two passengers (dwell time) plus 1 sec per passenger over 2. Example, 4 passengers, 1 stop : time = 2 + 2 = 4 sec						
Landing Calls	Dwell time 4 sec per stop. Transfer time. Dwell time of 4 sec will include transfer of up to six passengers, use 0.6 sec per passenger over six. Example, 8 passengers, 1 stop: time = 4 + 2 × 0.6 = 5.2 sec						

Times in Table 6.1 are based on 48-in (1200-mm) center-opening doors. Use the inefficiency shown in door time Table 4.3 plus normal inefficiency.

Other time factors that are used in calculations as well as calculations for outgoing handling capacity and interval will be the same as incoming traffic.

$$5\text{-min handling capacity} = \frac{\text{number of passengers down} \times 300\,\text{sec}}{\text{round-trip time}}$$

$$\text{Average interval} = \frac{\text{round-trip time}}{\text{number of elevators}}$$

The maximum interval for average outgoing traffic conditions should not exceed 40 to 50 sec. This is predicated on the intention of offering all passengers service within 60 sec during the peak outgoing traffic 5-min period.

The intensity of outgoing traffic will vary in all types of buildings. In office buildings a reasonable goal is to provide sufficient elevator service to evacuate the total population in a 25- to 40-min period. The elevators should be able to serve about 18% of the population during an outgoing peak 5-min period. The shorter transfer time, reduced number of stops, increased car loading, as well as an efficient outgoing or down-peak operation will generally make this feasible with the same number of elevators required to serve the incoming peak traffic. This figure may be modified for particular buildings to a lesser or greater percentage or to provide sufficient capacity to evacuate the building within a time criterion which may be established by a tenant or design requirement.

Probable Stops

The number of stops an elevator makes during the outgoing traffic period depends on how much it fills at each stop. If sufficient passengers are waiting at a floor to fill any car that stops there, it usually descends directly to the lobby. This situation can be determined from information about a particular building: the population of each floor, the nature of population (executive or clerical, for example), the quitting time or reason for leaving that floor. An outgoing traffic schedule can then be established and elevator service calculated accordingly.

When the exact nature of the occupancy cannot be known in advance, as in designing a building for general office use, probable stopping must be estimated. We assume that quitting times are sufficiently staggered so that the peak will not exceed the 5-min average. We also assume that each floor has an equal possibility of originating outgoing passengers. With these assumptions, probable down stopping will be approximately 75% of incoming probable up stopping. As outlined in the discussion of two-way traffic, people tend to leave in groups or with a partner, and experience has shown that average stopping is between 60 and 80% of up-peak or incoming traffic stopping.

Chart 4.1 gives the values for probable incoming up stopping for all conditions of loading. These values should be reduced to 75% as discussed for two-way traffic and the lobby stop accounted for.

Advantage of highest call reversal or travel to the highest down landing call can be taken on the same basis as discussed for incoming traffic depending upon the type of building and the percentage of total population on each floor.

In the following examples, 1.2 times the normal up-peak loading of the elevators will be used to approximate the maximum down-peak loading. As always, judgment must be employed when elevators are related to a particular building or the local practices considered which would warrant less than that factor.

SAMPLE CALCULATIONS

The calculations on the following pages exemplify a few situations to which outgoing traffic studies may apply. Example 6.1A may be considered a typical situation in a diversified office building. Note that if sufficient cars are provided to serve incoming and two-way lobby traffic demands, outgoing traffic is served with comparative ease. If the combination of increased loading and the minimum delay in passenger transfer operation of the elevators were to be applied to Example 6.1A as in Example 6.1C, outgoing capacity could be substantially increased.

In Example 6.1B an evacuation situation is shown. The elevators described can do a creditable job in getting students out of the building. This may not be the prime consideration in elevatoring a classroom building, however; class change periods with their tremendous interfloor as well as lobby traffic may be the factor that determines the number of elevators.

Example 6.1C gives one means of solving an extreme outgoing traffic problem and relates actual elevator operation to the traffic served. Stated otherwise, evacuation of a building can be expedited by minimizing the number of stops elevators make, provided that each floor in the building has equal access to vertical transportation. Dependence on performance such as is shown in Example 6.1C would require an elevator operating scheme that is responsive to outgoing traffic and designed to assign each elevator as it leaves the main landing into a zone of the building. Each zone should be so designated to represent equal traffic demand and the elevators arranged so that each zone would have an elevator assigned to it. Assignments would be changed each trip to provide equal service to all floors.

Example 6.1. Outgoing Traffic

A. Given: building population 1000 persons, 15 stories, five 500-fpm elevators, rise 176 ft, up-peak capacity 110 people (see Example 4.1).
Determine: down-peak capacity. Will the elevators accommodate 16.8% of the population during 5 min of down peak?

1. Example 4.1 was developed on the basis of 11 passengers per elevator. Calculating the elevator size needed to meet the $(N - 1)$ rule would be done by finding the four elevators equivalent to five 11-passenger elevators. From Table 4.4, 11-passenger elevator = 2500 lb, 5 $\times$ 2500 = 12,500 lb $\div$ 4 = 3100 lb = 3500 lb The recommendation should be five 3500-lb elevators.

2. Nominal capacity of a 3500-lb elevator 16 people times 1.2 gives a maximum down-peak capacity of 19 people.

3. Probable stops, 19 people, 13 upper floors from Chart 4.1
Probable stops = 10.1 (Chart 4.1) $\times$ 0.75 = 7.6

Time to run down, per stop $\dfrac{176}{7.6}$ = 23.2 ft rise per stop

(from Chart 4.2) 23.2 ft = 5.4 sec

Time to run up, $\dfrac{(176 - 23.2) \times 60}{500}$ + 5.4 = 23.6 sec

Elevator performance calculations:
Standing time

Lobby time 19 passengers	=	14 sec
Transfer time, down stops (7.6 × 4)	=	30
Door time, down stops (7.6 + 1) × 5.3	=	45.6
Total standing time		89.6 sec
Inefficiency, 10%	=	9.0
	Total	98.6 sec

Running time

Run up	=	23.6
Run down 7.6 × 5.4	=	41.0
Total round-trip time		163.2 sec

$$HC = \dfrac{(0 \text{ up} + 19 \text{ down}) \times 300}{163} = 35 \text{ people per elevator}$$

For 1000 people:
5-min outgoing capacity with 4 elevators 140 = 14%
5-min outgoing capacity with 5 elevators 175 = 17.5%
Five elevators required to meet 16.5% requirement
Interval: 163/5 = 33 sec vs 40 to 50 sec desired

B. Given: school, 6 stories, 500 students on floors 2 through 6, 12-ft floor heights

Determine: size and speed of elevators to evacuate school in 20 min

Procedure: outgoing 5-min average traffic = $\dfrac{500 \text{ Students}}{4 \text{ (5-min periods)}}$ = 125

Probable stops, maximum 5 (stop at each floor)

Time to run down, per stop $\dfrac{12 \text{ ft} \times 5}{5}$ = 12-ft rise per stop

Assume 300 fpm, from Chart 4.2, 12 ft = 5.8 sec

Time to run up, $\dfrac{(60 - 12) \times 60}{300}$ + 5.8 = 15.4 sec

Assume 5000-lb elevators 23 × 1.2 = 28 passengers per down trip, 54 in. center-opening doors

Elevator performance calculations:
Standing time

Lobby time 28 passengers 6 + (22 × 0.6)	=	19 sec
Transfer time, down stops 5 × 4 = 20, 0.6 × 4 = 2.4	=	22.4 sec
Door time, down stops (5 + 1) × 6	=	36.0
Total standing time		67.4 sec
Inefficiency, 10%	=	6.7
	Total	74.1 sec

Running time

Run up		=	15.4
Run down 5 × 5.8		=	29
Total round-trip time			118.5 sec

$$HC = \frac{28 \times 300}{119} = 71 \text{ people per elevator per 5 min}$$

Elevators required $\dfrac{125}{71} = 1.76$, use 2

Interval: $119/2 = 59.5$ sec interval

C. Given: four 4000-lb elevators @ 500 fpm in a 10-story office building, 12-ft floor heights

determine: greatest number of people that can be moved out of building from all floors in a 5-min period

Procedure: at least one trip to each floor must be made during each 5 min

Divide floors into zones to maximize service:

Zone 1—floors 2,3,4   Zone 2—floors 5,6

Zone 3—floors 7,8   Zone 4—floors 9,10

1. Calculate zone 4.

Probable stops 2, 1 included in down run to lobby

Time to run down, per stop 12 ft = 4.4 sec

Time to run down, floor 9 to lobby $\dfrac{(7 \times 12) \times 60}{500} + 4.4 = 17.5$ sec

Time to run up, $\dfrac{(8 \times 12) \times 60}{500} + 4.4 = 15.9$ sec

4000-lb elevator = 19 passengers × 1.2 = 23 down passengers, assume 48-in. center-opening doors

Elevator performance calculations:

Standing time

Lobby time 23 passengers = 6 + (0.6 × 17)	=	16.2 sec
Transfer time, down stops 2 × 4 = 8 + (15 × 0.6)	=	17.0
Door time, down stops (1 + 2) × 5.3	=	15.9
Total standing time		49.1 sec
Inefficiency, 10%	=	4.9
	Total	54.0 sec

Running time

Run up express	=	15.9
Run down per stop	=	4.4
Run down express	=	14.5
Total round-trip time		88.8 sec

$$HC = \frac{23 \times 300}{89} = 76 \text{ people in 5 min}$$

2. Calculate zone 3.

Probable stops 2, 1 included in down run to lobby

Time to run down, per stop 12 = 4.4 sec

Time to run down, floor 7 to lobby $\dfrac{(5 \times 12) \times 60}{500} + 4.4 = 11.6$ sec

Time to run up $\dfrac{(7 \times 12) \times 60}{500} + 4.4 = 14.5$ sec

Elevator performance calculations:

Standing time

Lobby time 23 passengers 6 + (0.6 × 17)	=	16.2 sec
Transfer time, down stops 2 × 4 = 8 + (15 × 0.6)	=	17.0
Door time, down stops (1 + 2) × 5.3	=	15.9
Total standing time		49.1 sec
Inefficiency, 10%	=	4.9
	Total	54.0 sec

Running time

Run up express	=	14.5
Run down per stop		4.4
Run down express	=	11.6
Total round-trip time		84.5 sec

$$\text{HC} = \frac{23 \times 300}{85} = 81 \text{ people in 5 min}$$

3. Calculate zone 2.

Probable stops 2, 1 included in down run to lobby

Time to run down, per stop 12 ft = 4.4 sec

Time to run down, floor 5 to lobby $\dfrac{(3 \times 12) \times 60}{500} + 4.4 = 8.7$ sec

Time to run up $\dfrac{(4 \times 12) \times 60}{500} + 4.4 = 10.2$ sec

Elevator performance calculations:

Standing time

Lobby time 23 passengers 6 + (0.6 × 17)	=	16.2 sec
Transfer time, down stops 2 × 4 = 8 + (15 × 0.6)	=	17.0
Door time, down stops (1 + 2) × 5.3	=	15.9
Total standing time		49.1 sec
Inefficiency, 10%	=	4.9
	Total	54.0 sec

Running time

Run up express	=	10.2
Run down per stop	=	4.4
Run down express	=	8.7
Total round-trip time		77.3 sec

$$\text{HC} = \frac{23 \times 300}{77} = 90 \text{ people in 5 min}$$

4. Calculate zone 1.

Probable stops 3

Time to run down, per stop 12 ft = 4.4 sec

Time to run up to floor 4 $\dfrac{(2 \times 12) \times 60}{500} + 4.4 = 7.3$ sec

Elevator performance calculations:

Standing time

Lobby time 23 passengers 6 + (0.6 × 17)	=	16.2 sec
Transfer time, down stops 3 × 4 = 12 + (5 × 0.6)	=	15.0
Door time, down stops (3 + 1) × 5.3	=	21.2
Total standing time		52.4 sec
Inefficiency, 10%	=	5.3
Total		57.6 sec

Running time

Run up	=	7.3
Run down 3 × 4.4	=	13.2
Total round-trip time		78.1 sec

$$HC = \frac{23 \times 300}{78} = 88 \text{ people in 5 min}$$

Total HC, four elevators = (76 + 81 + 90 + 88) = 335 people per 5 min

$$\text{Average interval} = \frac{(89 + 85 + 77 + 78)}{4} = 80 \text{ sec}$$

Therefore four 4000-lb elevators @ 500 fpm serving floors 1 through 10 can serve approximately 335 in 5 min, outgoing. Average interval will be approximately 80 sec.

Recalculate Example 6.2C with 4 cars serving all floors. Probable stops, 19 up passengers × 1.2 = 23 down passengers, 9 upper floors = 8.4 × 0.75 = 6.3

Time to run down, per stop $\dfrac{(8 \times 12)}{6.3}$ = 15.2 ft = rise per stop from Chart

4.2 = 4.7 sec

Time to run up $\dfrac{(96 - 15.2) \times 60}{500} + 4.7 = 14.4$ sec

Elevator performance calculations:

Standing time

Lobby time 23 passengers = 6 + (0.6 × 17)	=	16.2 sec
Transfer time, down stops 6.3 × 4	=	25.2
Door time, down stops (6.3 + 1) × 5.3	=	38.7
Total standing time		80.1 sec
Inefficiency, 10%	=	8.0
Total		88.1 sec

Running time

Run up express	=	14.4
Run down 6.3 × 4.7	=	29.6
Total round-trip time		132.1 sec

$$HC: 4 \text{ cars} = 4 \times \frac{23 \times 300}{132} = 208 \text{ people in 5 min}$$

Interval: 132/4 = 33 sec

As can be seen from Example 6.1C, the complete zoning of the elevators provides a substantial increase in handling capacity over having all elevators available to serve every floor. Of course, the increase is not without a penalty. The interval, hence average waiting time, will be extremely long with many people waiting on each floor lobby. Interestingly, the one car per floor or two-stop shuttle operation would be the expected way an elevator operating system would respond if calls started to wait overly long at each floor.

SPECIAL OPERATIONS

Nature of the Traffic

Heavy outgoing traffic situations can occur in any building and from various floors. An excellent example is a hotel with a ballroom located two or three floors above the street entrance. Another example is a cafeteria on an upper floor of an office building, where on nice days many people will lunch and then seek the outdoors, creating a heavy down traffic from that floor. In such situations special considerations must be given to the operating program of elevators serving floors which generate heavy traffic.

The nature of the traffic must first be determined: are the passengers all expected to head for a building exit floor or to distribute themselves among that and other floors? Is the traffic from the floor in question all down (or up) with little or no traffic in the opposite direction? In the latter case the loading floor should be considered as an upper "lobby" with all elevators arranged to return to that floor. In addition, the operating system should be designed to allow the loading of more than one elevator at that floor simultaneously as opposed to normal operating systems which only allow one elevator at a time to stop at an upper floor for a single landing call.

Extra time for inefficiency must be included in situations where passenger discipline and familiarity with elevators is not great. In hotels, hospitals, department stores, and other public places 10 to 20% inefficiency in transfer is to be expected in addition to normal inefficiency. This should be reflected in time considerations and in providing ample platform area as well as sufficient door width.

Escalators should be considered when outgoing traffic is expected to be as great as 200 or more people in a 5-min period. Escalators have capacities of from 200 to 500 people in 5 min depending upon width and speed of the steps. This will be further discussed in the chapter on escalators and moving walks.

Escalators can be used effectively to funnel outgoing or incoming pedestrian traffic to an area from which they can distribute themselves to other traffic generators such as check rooms, lounges, and refreshment stands. When paired escalators are provided to serve two-way traffic, both escalators can be operated in the direction of heavy traffic and other traffic diverted to stairs or elevators. The layout of escalators must be planned for this contingency. As with elevators, sufficient area must be available for persons to enter and leave.

It is futile and dangerous to transport a stream of persons to an area faster than they can leave it. An escalator is an unforgiving passenger conveyor—one cannot turn back! Measures of building design and operation such as revolving doors or exit doors should prevent such contingencies, but should they arise the simplest correction is to shut down temporarily or reduce the capacity of the transportation system and have people stay where they are.

Emergency Evacuation

Under certain circumstances it may be necessary to evacuate a building which can be done with the elevators if a plan is developed and emergency personnel are trained to operate elevators in their most efficient mode. Such circumstances could include a bomb scare, a remote section of the building on fire, or an explosion.

In the event of a fire in the building, the elevators should not initially be used by the occupants and should be automatically returned to a main landing for use by emergency personnel and supervised evacuation. Occupants should be trained by means of fire drills to use stairways.

Modern elevator systems have features for such emergency recall of the elevators as will be discussed in a later chapter. Once the elevators have been returned to a main landing and while elevators not returned are being accounted for, certain elevators will be commandeered for necessary emergency work. The remainder can be used by emergency personnel to evacuate various floors, one at a time, to effectively create a two-stop elevator operation, loading at one floor and exiting at another. The emergency personnel must give proper attention to handicapped personnel on the various floors; an essential prerequisite is for building management to know at what floors such persons are on and for responsible people on that floor to be aware of their presence or absence each day.

Automatic operations to accomplish an emergency evacuation do not appear to be feasible since it is an almost impossible task to program all the contingencies such as elevators out of service, the location and nature of emergencies, the time when they may occur, and the danger that people requiring special aid may be ignored. Trained police, fire, and building personnel can provide the most important aspect of judgment and supervision.

GENERAL CONCLUSIONS

Consideration should always be given to the outgoing traffic in a building. Expected elevator performance should be calculated and checked against the expected outgoing traffic demand and necessary adjustment made in elevator capacity and speed.

Outgoing traffic requirements are often effectively met by incorporating suitable features in the elevator operating system. One of the most useful of these procedures is efficient load bypass operation of the elevators as well as means to optimize elevator unloading time at the lobby and to speed their return up into the building to continue the outgoing operation.

Additional operating features include dispatching means so that elevators provide equal service to all floors in the building or to areas of equal demand. To guard against long waits, priority and zoning service should be provided if required by traffic intensity. In extreme down traffic situations such as in schools, "zoning" service as demonstrated in Example 6.1C can be employed and automatically effected by clock or operational control. As an alternative, an elevator system with provisions for future addition of these features should be considered. Such an approach is practical for any future expansion within a particular building.

Escalators should be considered for outgoing traffic of substantial magnitude from a limited number of floors. Two-way escalator service allows both escalators to be operated down if their capacity is needed. Clear signs to direct passengers to transportation as well as ample means to leave the exiting area are necessary. Fire drills and other emergency exercises should be practiced to ensure orderly evacuation in the event of an emergency.

CHAPTER SEVEN

Elevator Operation and Control

OPERATION AND CONTROL SYSTEMS

Once the number, size, speed, and location of the elevators are determined for a particular building, an adequate operation and motion control system must be provided so that those elevators operate to serve expected traffic. This was alluded to in the previous chapters in discussing various types of traffic. Tied in with serving the predominant traffic period is the ability of the elevators to operate in the best manner to serve that traffic and to perform with a minimum average waiting time for passengers. These terms must be translated into specifications so that the supplier of the elevator equipment knows what is expected.

Operations is the designation as an inclusive term for all the electrical decisions designed into an elevator system to control the sequence of movements an elevator or elevator group will make in response to calls for service. For ease of identification, operating systems have been given broad titles that are classified in the A17.1 Elevator Code and discussed later in this chapter. Individual manufacturers have added their own features and attractions to these operating systems, some providing more than others. The differences between various manufacturers are difficult to define in concise terms, since each has a different philosophy of elevator operation which will be discussed later in this chapter. Various suppliers are more than willing to discuss their approaches to solving elevator problems, either by demonstration on previous installations or by verbal and written assurances.

Motion control is the designation for the equipment that determines individual performance characteristics of an elevator; how quickly it can travel from floor to floor, the means and speed of door opening and closing, built-in time factors for passenger transfer, its ability to level swiftly and accurately, how the elevator displays hall lantern signals, are all functions of the control applied to an elevator. This control can and will be modified to a slight extent by the operating system but in general it is sufficiently distinct to be discussed separately.

Motion Control

An important part of the control system is how power is applied to the elevator to control its starting, acceleration, running, deceleration, leveling, and stopping, which will be referred to as motion control.

As stated in the first chapter, there are three main types of motion control, single or two-speed alternating current, generator field control utilizing a motor generator, and various types utilizing solid-state electronic devices. Further distinction can be made between motion controls using a direct feedback, sometimes referred to as "closed loop" and "open loop" wherein the motion of the elevator is programmed to follow and established pattern. This will be discussed in detail later in this chapter.

Alternating current resistance control is used to start the pump motor of a hydraulic elevator and is used in a traction elevator to control the starting of the hoist motor in the desired direction. A step or steps of resistance may be inserted between the motor and the line to reduce the in-rush current as the motor is started. The resistance is shorted out when the motor attains normal speed (see Figure 7.1a). In some controls, a three-phase motor may be connected in a "Y" configuration and switched to a "Δ" (delta) configuration when up to speed, which will also reduce the starting in-rush current. Alternating current control is used for relatively low-speed elevators of up to 100 fpm (0.5 mps). Stopping is accomplished by removing power from the motor and applying the brake, resulting in floor stopping accuracy which may vary plus or minus 2 in. (50 mm), depending upon the loading in the elevator.

Refinements of alternating current motor control include the use of two-speed motors with ratios of from 2:1 up to 6:1 between the low-speed and high-speed windings. The elevator runs with the high-speed winding and is switched to the low-speed winding when it is near the floor stop. Final stopping is done with the brake and stopping accuracy may be improved to plus or minus 1 in. (25 mm). Other approaches use the low-speed winding for releveling, i.e., the elevator is switched to the low-speed winding for a floor stop and if it overshoots floor level, the motor is reversed and the elevator is leveled back to the floor at low speed. Floor stopping accuracy can then be guaranteed within the limits of the dead zone, i.e., the distance above and below floor level that the elevator must be out of level before releveling takes place. Normal dead zones are usually ½ to ¾ in. (13 to 19 mm) above and below floor level.

The application of elevators with alternating current motion control is generally limited to apartment houses of up to about six floors, slow-speed freight elevators, and dumbwaiters. It was quite popular from about 1920 to 1960 as the accepted motion control in low-rise apartment houses and is an accepted

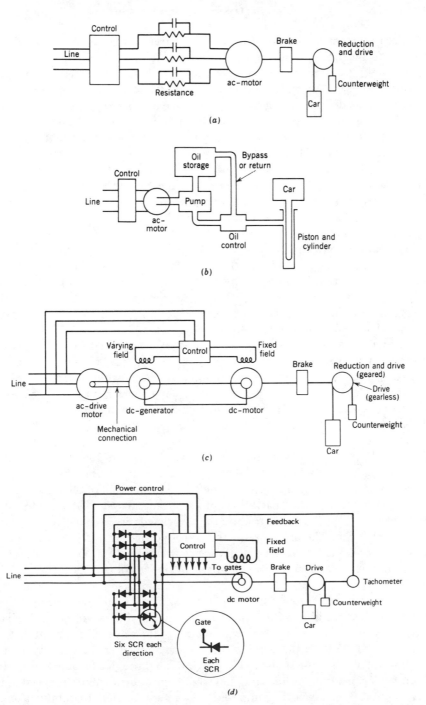

Figure 7.1. Elevator motion control systems: (*a*) alternating current; (*b*) hydraulic; (*c*) generator field control; (*d*) SCR (silicon controlled rectifiers) motion control.

standard in most low-rise European applications. It must be provided with leveling to meet most handicapped requirements.

Alternating current elevators have been generally superseded by hydraulic elevators for lower-rise applications. The hydraulic elevator covers the lower speed range as do alternating current traction elevators, is universally designed with leveling, and is, in general, more economical to install than alternating current traction elevators. The alternating current elevator, since it is a counterweighted elevator, requires far less horsepower, hence energy consumption, than the hydraulic which requires horsepower sufficient to lift both the dead weight and the carrying load of the elevator. The hydraulic elevator usually has a lower initial installation cost since it does not require a car safety device, counterweight, and counterweight rails, and it imposes very little load on the building structure.

Hydraulic elevators use resistance or Y/Δ for the single-speed motor used to drive the pump supplying oil to the piston and pumping in the up direction only. Stopping is accomplished by throttling and then cutting off oil flow (Figure 7.1b). Leveling is accomplished by restricting oil flow or by bypassing a measured quantity of oil. Down control is accomplished by opening a down valve and allowing oil to return to the storage tank.

Hydraulic elevators are restricted in top operating speed to about 200 fpm (1.0 mps). For speeds of 200 fpm (1 mps) or greater, the generator field motion control of electric elevators is superior and it has become the accepted standard of elevator performance. This may change in the near future as solid-state alternating current motor drives are introduced and perfected.

Generator field motion control consists of providing a varying voltage to a direct current elevator drive motor. The characteristics of the direct current drive motor are that it has the torque to move the elevator load smoothly up to speed and can absorb the inertia of the moving load by regenerating to stop the elevator with smooth deceleration. Stopping is independent of the brake with all the energy being absorbed back through the electrical system. The system consists of the dc drive (hoist) motor and the associated direct-connected or geared drive mechanism. The armature of the drive motor is directly connected to a source of controlled dc voltage such as the armature of a dc generator (Ward-Leonard system, see Figure 7.1c) or a bank of silicon controlled rectifiers (SCR, see Figure 7.1d).

With solid-state motion control, the most common currently being the silicon controlled rectifier (SCR), the source of the varying dc voltage is the bank of SCRs wherein the gate controls the amount of the wave of ac current that can be conducted. This, in turn, limits the voltage applied to the dc drive motor. The control of the firing signal to the gate is done through a feedback loop. The tachometer generator develops a signal which is proportional to the speed of the drive motor. This signal is compared to a reference signal contained in the control circuitry and further modified by signals which indicate that the elevator should either be speeding up, slowing down, or running at constant speed.

The gates are then energized to fire at the precise instants that allow the controlled amount of current to flow to the drive motor.

When the elevator is traveling up or down with sufficient car or counterweight load to cause overhauling, the dc hoist motor acts as a generator. This voltage is fed back through the SCR bank through reverse gating and back into the power feeders.

The bank of 12 SCRs shown in Figure 7.1d are arranged so that when the ac power flows toward the machine in the positive wave only three SCRs conduct; in the negative wave, three others conduct and when current is flowing away from the machine, one set conducts at each instant to match the incoming power wave characteristics to create ac power from the regenerated dc power.

The SCR system shown in Figure 7.1d is a closed loop, i.e., the tachometer feeds back a signal to the control to vary the input to the dc hoist motor armature depending upon the speed of the elevator compared to a desired pattern. The dc generator field control (Ward–Leonard) shown in Figure 7.1c is an open loop, i.e., additional control circuitry consisting of a floor selector or switches in the hoistway responds to the position of the elevator and, by contact closure, drops out or pulls in relays to control the direction and amount of dc current being fed to the driving machine.

As the voltage is increased on the drive motor armature, the elevator or load is accelerated up to speed. As the voltage is decreased speed is reduced until the elevator comes to a complete stop, and the brake is then applied to hold the car at the floor. Releveling requirements due to changing load at the floor are accomplished by lifting the brake and applying small voltages to the drive motor.

Operating speeds of generator field or SCR-drive elevator-motion control systems are available in any speed up to the present high of 2000 fpm (10.0 mps). For gearless machines, performance is from 4 to 5 sec for a one-floor run of 12 ft (3.6 m) and leveling accuracy under all conditions of loading is from plus or minus ¼ to ½ in. (6 to 13 mm) with minimum time penalty incurred for leveling. For lower-performance geared machines, floor-to-floor time is from 5 to 6 sec.

Gearless machines are greatly capable of acceleration rates of 4 feet per second per second (written as 4 fps^2) (1.2 mps^2) with generator field control and can be made to accelerate faster with SCR motion control. Geared machines are generally capable of acceleration rates of 3 fps^2 (0.9 mps^2) and also can be made to accelerate faster. The limiting factor is not the accelerating rate, but the rate of change of acceleration (sometimes referred to as jerk) which is felt by the riding passenger. This is a matter of personal tolerance but, in general, an upper limit of 8 feet per second per second each second (written 8 fps^3) (2.4 mps^3) is usually the maximum.

As an example of the calculation of the expected running time of an elevator for a given distance, a given top speed, and with an acceleration and deceleration rate of 4 fps^2 (1.2 mps^2), the following derivation is given.

Example 7.1

Given: 300 ft (91 m) express run from the ground floor to floor 24, 700 fpm (3.5 mps) elevator speed. What is the time from start to stop? (See Figure 7.2)

Assumptions: Between t_0 and t_1, acceleration = constant at 4 fps². Between t_1 and t_2, the product of the velocity and the acceleration Va = a constant. The time, t_1 is when approximately 60% of full speed is attained.

During deceleration the same conditions apply. 0.5 sec is added to allow for leveling into the floor at the end of the slowdown.

$$a = 4 \text{ fps}^2 \qquad V_1 = \frac{0.60 \times 700}{60} = 7 \text{ fps}$$

$$V_2 = \frac{700}{60} = 11.67 \text{ fps}$$

$$t_1 = \frac{V_1}{a} = \frac{7}{4} = 1.75 \text{ sec} \qquad S_1 = \frac{V_1^2}{2a} = \frac{7^2}{8} = 6.125 \text{ ft}$$

$$t_2 = \frac{V_2^2 - V_1^2}{2V_1a} + t_1 = \frac{11.67^2 - 7^2}{2 \times 7 \times 4} + 1.75 = \frac{87}{56} + 1.75 = 3.3 \text{ sec}$$

$$S_2 = \frac{1}{3a}\left(\frac{V_2^3}{V_1} - V_1^2\right) = \frac{1}{3 \times 4}\left(\frac{11.67^3}{7} - 7^2\right) = \frac{1}{12}\left(\frac{1589}{7} - 49\right)$$

$$= \frac{1}{12}(227 - 49) = 14.8 \text{ ft}$$

$$\begin{aligned} S_1 + S_2 &= 6.125 + 14.8 &&= 20.925 \text{ ft} \\ 2(S_1 + S_2) &&&= 41.85 \text{ ft} \\ S_3 &= 300 - 41.85 &&= 258 \text{ ft} \\ t_3 - t_2 &= \frac{258}{11.67} &&= 22.1 \text{ sec} \end{aligned}$$

$$\text{Total time} = 22.1 + 2(3.3) + 0.5 \qquad\qquad = 29.2 \text{ sec}$$

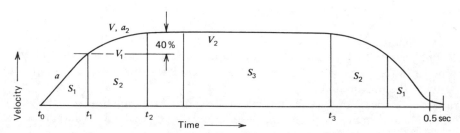

Figure 7.2. Elevator travel—velocity versus time.

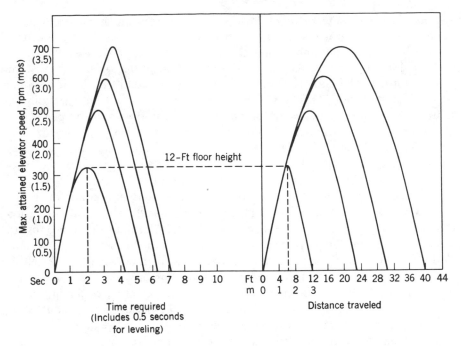

Figure 7.3. Elevator speed versus time and distance.

Speed regulation between no load and full load with a generator field or SCR system should be plus or minus 5%. Any tendency to overspeed or underspeed should be governed and corrective measures applied. The A17.1 Elevator Code requires an elevator to be electrically stopped if its running speed exceeds its rated speed by varying percentages of 10% or more depending upon the rated speed of the elevator. A graphic description of the time elevators require at various speeds to get up to full speed from a stop and immediately slow down to a stop is shown in Figure 7.3. These times include 0.5 sec for leveling.

Door Control

Operation from floor to floor or getting the elevator up to speed and stopping is the major function of the motion control system. A secondary, similar motion control system is the control of door operation. Like the elevator, the doors must be accelerated, decelerated, and stopped during both their open and close cycles. The A17.1 Elevator Code limits the force of the closing doors to 7 ft poundal (in the European code, EN81, this force is stated as 0.29 joules) and the thrust of the closing doors to 30 lb (13.6 kg), i.e., 30 lb are required to prevent the doors from closing. These forces are for doors with door protective devices such as: a safety edge which is a mechanical device that, when displaced, activates a switch to reopen the doors; photoelectric devices which will reverse the doors when closing; or other devices, such as ultrasonic or electronic proximity devices, which will sense a presence in the path of the closing door and reverse the door to minimize impact.

If no door edge protective device is used or if the protective device fails, the closing force of the door is required by the A17.1 code to be 2.5 ft poundal (0.115 joules) or less. This is accomplished by reducing closing speed and is featured by a nudging device which operates if the doors are held open excessively long (15 to 20 sec) in addition to a reasonable time for passenger transfer of from 2 to 5 sec. Nudging can be found in two forms, one that continues to try to close the doors with a slight forward movement and pressure, and the other that stops the doors until the obstruction is removed.

The operating system should establish the time the doors remain open at a floor which is called "dwell time." These times are usually about 2 to 3 sec for a stop made in response to a car call, and from 3 to 5 sec for a stop in response to a landing call depending upon the relation between the location of the elevator and the location of the landing button fixture. A further, separate adjustable time should be established for loading or unloading at a main terminal floor. The time for loading at a main terminal should be varied depending upon traffic. For example, during heavy traffic periods the elevator doors should start to close about 5 to 15 sec after a car call is registered depending upon the expected loading of the elevator, or sooner if the load sensing device detects a predetermined load.

Additional refinements of the adjustment of dwell-time include the shortening of the time the doors remain open after a passenger transfers into an elevator after being detected by a photoelectric device in the entrance way or the extending of dwell-time for a car call if the elevator is filled to a predetermined percentage of capacity load to allow additional time for passengers to leave. Advanced systems have an electronic proximity device or an ultrasonic detector on the elevator doors which allows the doors to be closed within the minimum transfer time.

The time the doors remain open, in an installation designed for heavy traffic, should be adjusted to the load in the elevator. Transfer to and from a lightly loaded car can be faster than from a filled car. The adjustment of the dwell-time should permit this distinction. Similarly, transfer time for loading or unloading at the lobby will be longer and should be recognized. Any of the foregoing transfer times should be independently adjustable to suit the requirements of the building in which the elevators are installed as well as the characteristics of the traffic they are expected to serve. Separate car-stop and landing-stop dwell-times should be specified for any building to match expected passenger transfer time with the operation of the elevator.

In addition to the time required to open and close the doors plus the dwell-time, there is usually a slight delay between the time the doors are fully closed and the elevator starts. This time is referred to as "lock time" or "build-up time" and is usually about 0.5 to 0.8 sec and consists of the time necessary to ensure that the door is closed and locked and to start the mass of the elevator.

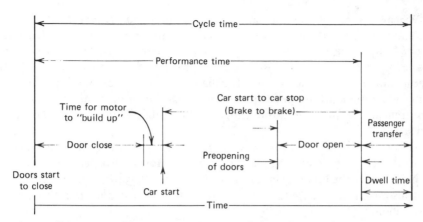

Figure 7.4. Elevator time elements—floor-to-floor performance.

Doors can be started to open as an elevator levels to the floor provided the various aspects of control can ensure that the doors will not be open wide enough for passenger transfer prior to the time the elevator stops. This slight preopening can improve elevator performance time and can be employed with motion control systems designed to ensure that such preopening can be done within a minimum distance of floor level.

The combination of all the times discussed to this point adds up to a performance time for an elevator on a one-floor run or a trip (Figure 7.4). One of the means of assuring a quality elevator installation is to specify and to measure the performance time of elevators as installed. The performance time or the time from the start of door closing at one floor until the doors are fully opened at another floor is a measure of quality and can easily be determined. The measurements should be consistent under all conditions of loading, from full load to no load in the car and in either the up or down direction.

Leveling

An extremely important aspect of an elevator's performance is the accuracy with which it levels to the floor and maintains floor level as passengers enter and leave. Leveling operation should be direct, i.e., the elevator should come to a stop at a floor within the prescribed leveling tolerance of plus or minus ¼ in. (7 mm) for a quality installation without overshooting and leveling back or stopping short of floor level and leveling into the floor (often called spotting). Once the elevator is at floor level and the passenger load enters or leaves, the floor level should be maintained with minimum perceptible movement which should always be toward floor level, never away. Suitable provisions should be made in the motion control system to maintain an elevator stopped at floor level under any adverse conditions which may include building sway, to be discussed in a later chapter.

When lanterns are provided at the landings they should light in advance of a car's arrival at the floor and in the direction the car is expected to leave that floor. Adequate advance lighting of a lantern will give prospective passengers a chance to approach the elevator and board it promptly, thus saving valuable seconds in transfer time. Recent legislation suggests the combination of advanced landing lantern operation and door-hold open time to meet designated requirements to provide ample time for the mobility impaired to board the elevator.

There are many other things an individual elevator should do. They may be best summed up by requiring any elevator to "inform" prospective passengers. If they observe the lantern signals, board the right elevator, and operate the car buttons promptly, the elevator should take them swiftly and smoothly to their destination.

OPERATION SYSTEMS

The ability of elevators to move in response to demands for service is solved by the operation system. With early elevators, hydraulic or steam-driven, the operating device was a continuous rope that ran the length of the hoistway and actuated a valve in the basement. To go up, the rope was pulled down to admit steam or water to the elevator driving means. To go down, the rope was pulled up and the driving machine was reversed. The pull in the opposite direction of travel had a distinct advantage; by sending the rope through a restricted hole in the car platform a stop ball on the rope could be placed at the extremes of elevator travel. The continued upward or downward movement of the car caused the rope to be pulled by the car and would stop the elevator (Figure 7.5).

Hand-rope or "Shipper Rope" operation, as it was called, had many advantages and continued in vogue for many years, and early electric elevators employed a hand rope to open and close contactors to start, stop, and reverse the elevator. Some elevators still have it although it has been generally outlawed by local safety codes. In the early days when hoistways were open or enclosed by waist-high gates, one could reach in, pull the rope, and summon the elevator, an "automatic" albeit very unsafe operation.

Car-Switch Operation

With the introduction of electric elevators, the natural step was to put a switch or lever in the car which when operated would electrically start the car in the up or down direction and if centered would stop the elevator. The electric impulses to the motor room from the moving elevator were transmitted by a traveling cable. This is an electrical cable that hangs from some point near the middle of the hoistway and is attached to the underside of the car. With electric elevators and increased concentration on the safety of passengers, such things as hoistway door interlocks and car gates were introduced and subsequently re-

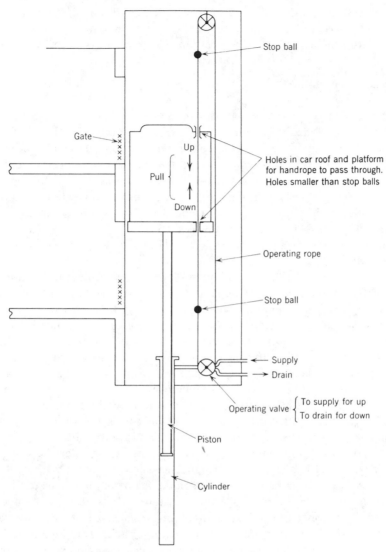

Figure 7.5.   **Hydraulic elevator—handrope operation.**

quired by safety codes. The interlock prevents the elevator from being operated unless the door is closed and locked. Electrical signaling devices such as annunciators and call bells were also introduced as hoistways became more fully enclosed and an efficient means to signal the elevator operator was required.

Car-switch operation, as it was known, was the chief operation system available in office buildings until the early 1920s. Also developed were elaborate signaling systems to inform the operator where he was to stop as well as buzzer systems so that the elevator starter at the main floor could direct the elevators.

To aid the starter in the main lobby in this task, supervisory or "starters" panels with lights were added to indicate where landing calls were waiting, as were either dials or lights to show the position of the elevators in the hoistways.

Automatic Operation

Parallel with the development and refinement of car-switch operation for office buildings came the introduction of the simple automatic electrical operating system for single elevators. This was primarily intended to make the elevator acceptable to the small residential apartment house where the traffic did not warrant a full-time elevator attendant. Safety was established by the use of interlocks on the hoistway doors and contacts of the car gate to ensure that both were closed before the elevator could be run. An automatic operation system known as Single Automatic was introduced and became the accepted apartment house operating system in the United States for the period from about 1890 to 1920.

Single Automatic Push-Button operation or Single Automatic operation as it is presently known consists of single buttons at each landing and a button for each floor in a car operating panel. The car can be called from any floor provided the door and gate are closed and no one is operating the car. Once the car is intercepted, entering it, closing the hoistway door and car gate and operating a car button ensures the riding passenger exclusive use for that trip. It is a light-service operation; the elevator can serve only one call at a time and the next passenger must wait until it is free before he or she can use it. To indicate availability, an "in use" light is generally placed in the landing call fixture. When the light is out the passenger can call the car. To make single automatic operation work properly, sequence switching and timing must be provided so that the passenger who enters will have a chance to register the call before another passenger can call the car to another floor.

Single Automatic operation remains a valuable operation in light-traffic buildings where exclusive use of an elevator is desired. It is ideal for garages and factories where only one vehicle can fit on an elevator at one time. Single Automatic operation is the preferred operation for apartment houses in Europe where elevators are small and people prefer to wait to gain exclusive use.

Signal Operation

As elevator speeds were increased past 500 fpm (2.5 mps) the skill of the elevator operator on a car-switch elevator had to increase so that the operating handle could be dropped precisely and the next stop made with fair accuracy. Various schemes to overcome this difficult precision were tried.

A light would flash just before the next required stop and the operator had to respond with split-second accuracy. Leveling devices were applied to eliminate jockeying at the floor.

These approaches were satisfactory until taller buildings required speeds in excess of 700 fpm (3.5 mps). No operator could recognize when to stop when floors were passing at about a second apart and even then the decision to stop had to be made at least 20 to 30 ft from the desired floor.

The need for an improved operating system for fast elevators became apparent and signal control was introduced. Signal control is a push-button operating system in which the elevator operator registers the call for the floors at which the passengers wish to stop and operates a button or lever to start the car. Acceleration, response to floor stops, deceleration, and leveling are all done automatically. Similarly, response to landing calls is in the same manner; the operator does not know what landing calls the elevator will respond to until the car starts to slow down.

Signal control, or signal operation as it is presently known, removed the speed limitation from elevators. Speeds of 1000, 1200, and 1400 fpm (5, 6, and 7 mps) were attained and all the notable buildings of the late 1920s and the 1930s had signal-control elevators. These include the Empire State and Chrysler Buildings and most of Rockefeller Center as well as hundreds of others. Most buildings of that era have been modernized and Group Automatic Operation has now been provided.

Technology also came to the aid of the elevator starter. In addition to indicator lights to show waiting landing calls and the position or motion of each elevator, electrical scheduling systems were introduced. These systems attempted to operate cars on schedules maintained by the operator's response to signal lights in each elevator. The light indicated when to start and was timed from the departure of the previous car to maintain a time spacing of elevators on both the up and down trips.

The objective of spacing elevators throughout a building to render the best possible service to waiting passengers was only imperfectly realized. Elevator operators saw little point in rushing when traffic seemed light and could not resist the opportunity to stop at a floor and enjoy a smoke or conversation. Electrical dispatching gave way to human dispatching and many such systems were abandoned.

Collective Operation

While signal operation was being introduced for the taller office buildings, the demand was increasing for automatic operation for apartments, hospitals, small office buildings, and any building that required elevators but had only limited traffic. The need for better service than the Single Automatic elevator could provide led to the introduction of the collective elevator and improved means of automatic elevator door operation.

Collective operation, as the name implies, is a means to collect or remember and answer all the calls in one direction, reverse the elevator, and then "collect" and answer all the calls in the opposite direction.

The common collective operation is Selective Collective Operation. With full selective collective both up and down landing call buttons are provided. In addition to car calls, the elevator will only stop for up landing calls in the up direction and down calls in the down direction, all calls being remembered until answered. A variation of selective collective is Down Collective when only down landing buttons are provided at each upper floor. This is acceptable in most apartment houses where people generally want service up to their apartments from the lobby (up car call) and down to the lobby from their floors (down landing call). Another variation is Single-Button Collective where the landing call button will stop either an up or down traveling elevator (sometimes referred to as interceptive collective).

Automatic operation of elevators is enhanced by the power operation of the doors and ample protection to keep the closing doors from striking the passengers. Earlier automatic elevators had manual doors and gates, usually equipped with spring or weighted closers. The passenger had to open both once an elevator answered a landing call. With power operation, the doors opens automatically, remains open for a time interval, and automatically close.

If the leading edge of the door touches or is about to touch a person or object, a sensing device on the door edge should actuate the reopening of the door and recycle the closing operation. A photoelectric device projecting across the entrance can be used in the sensing device to ensure that the door will remain completely open until the entrance is clear. With provisions for the handicapped many elevator codes are requiring photoelectric devices.

Selective collective is the accepted operation for single elevators in any type of building, and for service elevators in major office buildings. Its utility has been enhanced by added features such as: home landing so that the elevator is always returned to a given floor on completion of any trip; load sensing devices, which prevent the car from stopping for additional landing calls if the weight in the car exceeds a predetermined percentage of capacity; independent service operation, wherein the car can be used to handle special loads and not respond to landing calls; and attendant operation so operation is manually controlled for security and freight-handling requirements.

Two-car selective collective operation is commonly known as Duplex Collective. This operation can be either full selective or down collective and consists of the following operations.

One of the two cars is designated as the home-landing car. It can be either car and may change after each trip. The other car is called "free" and may park at its last call or some designated landing (middle free car parking), or in a designated upper area of the building. The home-landing car is designated to respond to any landing call at the lobby or above it and below the "free" car. The free car is designated to respond to any up landing call above it or to any down landing call.

If the home-landing car is taken to an upper floor by a car call, the free car becomes the home-landing car. If the free car is set in motion by a car or landing call, and a call occurs behind the free car (either an up or down landing call below an up traveling free car, or an up or down landing call above a down traveling free car), the home-landing car can start to "help out" the free car. In addition, the home-landing car should start if any landing call, behind or not, remains unanswered beyond a predetermined time or if the free car does not move within a predetermined time. This latter feature was known as the milkman circuit and was developed because the milkman would block the elevator with his carrying rack while he delivered to the various apartments. Since tenants could be registering calls ahead of the free car, for example, if the milkman's car was at the second floor, the home-landing car might never start, hence the timing logic had to be developed.

Duplex Collective will adequately serve the normal expected traffic in an apartment house if the elevators have sufficient capacity to serve the needs of the building. Duplex Collective operation has no means to separate elevators if they are both traveling in the same direction and the cars will "bunch."

If heavier traffic persists for any period of time because of, say, an under-elevatored situation, Duplex Collective operation is undesirable as a system for heavier trafficked buildings.

As we stated, with duplex collective a call behind the free car starts the home-landing car. With a busy situation, one or many calls behind may appear immediately. The former home-landing car is traveling right behind the free car and in the same direction. With many calls the elevators are "leap-frogging," that is, both will arrive at the lobby about the same time, will leave again, and start the cycle over—a condition known as "bunching." Instead of "two-car coordinated operation", the operation is "two cars bunched", or not much better than a single large elevator.

A number of steps must be taken to improve two-car operation as well as any group operation for heavier traffic. Some of these steps have been described in the discussions of the various types of traffic in previous chapters. The application of various traffic-handling features to elevators results in another class of operation known as Group Automatic Operation.

Group Automatic Operation

In viewing the history of elevator operations, the almost parallel development of signal operation with attendant for the larger, busier building, and the collective both single and duplex, nonattended operation for the residential and lighter-traffic buildings can be observed. In about 1949, a number of developments occurred which started the rapid elimination of elevator attendants in any type of building.

At that time the completely automatic operatorless elevator was introduced. This innovation consisted of a number of steps such as the development of an operating system that caused the elevators to respond to changing traffic demands, the refinement of the door protective devices, and the establishment of timing and scheduling systems logic circuits that maintained elevators on operating schedules, in response to traffic requirements and without the necessity of human interference.

The demand for good automatic operation for larger groups of elevators was apparent for some time. Various attempts to meet this demand were made using three- and four-car selective collective operations which were generally unsatisfactory for the changing heavier-traffic situations such as incoming, two-way, and outgoing traffic. With a collective system and with increased traffic, serious bunching can and does occur. However, the use of automatic collective operations plus the fact that people had been operating elevators by themselves in various types of buildings for years allayed any fear that they would be unwilling to do so in a large office building.

All the developments came together in 1949 when the Otis Elevator Company introduced, through the efforts of their Chief Development Engineer Mr. William Bruns, a fully automatic elevator system for an office building in Texas. In addition, there was a city-wide strike of elevator operators in Chicago, which provided the marketing awareness. The total system provided the necessary scheduling system which made the decisions to start elevators on schedules, introduced a proximity system of door protection, and developed the various load sensing and protective devices to replace the other functions of an elevator attendant.

The closing of the car and landing door was the major task of elevator attendants. It was their duty to ensure that no one was coming and to close the doors to start the elevator. Reliable electric power door operators were available, and to provide good automatic operation, it became essential to provide safe protection against the doors striking a person and a means to reverse such doors swiftly if a person or object was encountered during closing. This task was aided by a reduction in the speed, hence the kinetic energy of the closing doors which resulted in only a slight time penalty over the speed at which they were being closed by an attendant. Otis introduced an electronic proximity device as a door protection device, whereas other elevator companies improved the movable door safety edge and photoelectric devices by making the edge retractable and altering dwell-time after the light beam was interrupted by passenger transfer.

The proximity type of door detection was developed based on the natural phenomenon that all people and objects have a degree of electrical capacitance to ground. An electronic field generated by a device at the edge of the elevator car door reacts to this capacitance to create an electrical signal which reverses the closing doors. The electronic field can be adjusted in sensitivity to include the edge of the hoistway door in addition to the car door so protection against both the hoistway and car doors could be provided. Unfortunately, if the sensitivity is increased too greatly, false reversal can occur, particularly during periods of increased humidity. The further development of ultrasonic detectors may provide more reliable and less sensitive door protection.

Safety edges are simply a movable strip on the leading edge of the car door which is deflected when it touches a person or object. This deflection actuates a switch to cause electrical reversal of the door. The photoelectric device projects a beam of light across the entrance which, when intercepted, also provides a signal to reverse the closing doors. The photoelectric device can also provide an electrical signal to decrease the amount of time the doors remain open (dwell-time). This signal plus additional operational logic to determine if the passenger is moving out of the elevator (stop in response to a car call) or boarding the elevator (stop in response to a landing call) can be helpful in minimizing the time spent at a floor and improve elevator service. Further refinement of logic is to relate the time the doors remain open to the number of people in the elevator through load sensing. If the car is filled, additional time can be allowed over the dwell time allowed for a lightly loaded elevator.

The elevator attendant had one additional task in relation to door operation. If the elevator was delayed the attendant's job was to enforce the closing of the doors. With automatic elevators this is accomplished by electrical logic which can initiate door closing after a certain elapsed time. If someone is holding the door an excessive length of time of about 15 or 20 sec, the protective edge will only stop the closing door and not cause reversal or, alternatively, the doors start to close at a greatly reduced speed to attempt to force closing. Such operation is referred to as "nudging" and is done in conjunction with the sound of a harsh buzzer.

Another task of the operator was to count the number of people in an elevator to limit loading and to leave the loading floor when sufficient passengers were on board. If the car was filled at an upper floor, the attendant could also bypass additional landing calls. These functions are replaced by load sensing devices. To reduce the danger of overcrowding, loading restrictions were developed by limiting the area of the car in relation to its capacity in pounds or kilograms to limit the number of people in the car. Another function of the attendant was to register the passenger calls, which they can easily do themselves.

The final task of the operator was to pay attention to the elevator starter or to an electrical signal. This was replaced by initiating door closing and car starting by electrical signals and by relying on the door safety edges to protect the passengers if transfer was still in progress.

The elevator starter in a main lobby generally provided the signals and the schedule of elevator operation in relation to the various demands for service as noted from a landing call and car position indicator in the lobby. Specialized electrical computers were developed to weigh all the factors a starter must plus many more that human limitations did not allow. Based on these factors, the various electrical signals for starting, stopping, and reversing elevators are given and the system operates completely unattended.

Group Automatic Operating systems were developed by the various elevator manufacturers based on their differing philosophies of elevator operation and logic. During the period from about 1950 to 1982, the Otis Elevator Company introduced Six Program Autotronic, Four Program Autotronic, Basic Auto-tronic with Multiple Zoning, VIP 260, Elevonic 101 and, in 1982, Elevonic 401.

Westinghouse's names were Selectomatic Six Pattern, Selectomatic Four Pattern, and Selectomatic Mark IV and Mark V. Haughton, now Schindler Haughton, called its systems Auto Signamatic, 1090, 1092IC and Aconic. Dover's system is called Traflomatic and Montgomery's is called Miprom. Each represented some degree of change or development and most systems in the 1980s employ some form of microprocessor using programmable computers and solid state electronic circuitry.

These group automatic elevator operating systems were all designed to accommodate the various differences between incoming, two-way, and outgoing traffic in busy buildings. Success of these efforts to gain public acceptance of automatic elevators in any type of building is evidenced by the fact that every elevator installed today is completely automatic with attendant features only for special services.

Group Automatic Operation is merely the means to provide the most effective operation of a fixed number of elevators to serve the varying traffic requirements encountered as the building functions during its intended use. The results are most important and can be simply stated as providing vertical transportation with the minimum waiting and riding time which optimizes the traffic-handling capability of the system. Stated otherwise, a grouping of elevators in a building has a maximum people-handling "quantity." The group operations system provides this "quantity" at optimum "quality" of service.

Different elevator manufacturers use different philosophies in approaching this task and the following will be a critical look and comparison of some of the logic ("software") of the decisions used to effect elevator operation for various traffic situations as well as some background about the electronic ("hardware") systems.

Group Automatic Supervisory Systems

Perhaps the most dynamic aspect of the changes occurring in modern elevator operating systems has been the introduction and use of microprocessors with programmable features. The elevator operational controllers need no longer be massive and hard-wired with a myriad of relays to translate the built-in design logic required to respond to many calls for service as a result of changing conditions. The microprocessor affords the ability to establish a basic design and the programmable feature allows relatively easily made changes to adapt that basic logic to the particular requirements of a building.

Computer buzz words such as ROM, CPU, RAM, PROM, REPROM, LSI, burn in, algorithm, and others are abounding among elevator technicians and, through it all, the same basic requirements remain, i.e., making elevators provide their optimum handling capacity with the optimum quality of service. The microprocessor allows the collection of more information at an extremely high speed so more meaningful and timely directions can be given to the individual elevators to best deploy them for the immediate and constant changing tasks of serving people.

The inputs to the microprocessor can be divided into three classes. The primary inputs are those generated by people which are either the prospective or riding passengers. There are car calls, landing calls, and the loading of each elevator. The loading is represented by the passenger weight on the car platform. The secondary inputs are the fixed and readily recognizable aspects of individual elevators and the elevator system. These are: the location and current operation of each elevator whether it is stopped, with or without passengers (parked), with its doors open or closed; its direction of travel; the number of floors served; the elevator's capacity and speed and the time factors for door operation; dwell-time; and performance from one floor to the next floor or two or more floors away. The tertiary inputs are programmable and can consist of such inputs as the time of day, the day of the week, basic strategies such as the monitoring of the time individual calls wait, selected parking positions of the elevators, decisions as to whether a car will stop at or bypass a particular landing call, the predicted time to reach the highest call for reversal, the time to reach the main lobby, or other fixed decisions which represent the philosophy of the manufacturer as to how best to serve particular traffic situations.

These inputs must be transmitted either on an individual wire from the source to the microprocessor or by multiplexing, i.e., each group of inputs is converted to digital signals and sent along a single wire and again converted at the receiving source. This later is advantageous from the moving elevator car to the machine room wherein all signals have to be transmitted by traveling cable. Multiplexing can serve to reduce the number of wires in the traveling cable (hence cost and weight) substantially.

With microprocessors, solid-state electronic devices are employed and all electrical signals are usually at a very low voltage level of about 5 V. Such a low level requires critical attention to wiring details and can often be subject to spurious electrical noise or voltage variation. One of the important aspects in employing microprocessing is that all inputs must be verified. This is usually done by repeating the signal a number of times and comparing each repeat to ensure that they are the same. As can be imagined, such comparisons can increase in geometric proportion when a number of signals are following each other on the same wire microseconds apart. Failure to verify can lead to totally wrong logic decisions and may stop elevator service when it is needed most.

Since there is no obvious movement of relays with microprocessors, a means of checking proper operation must be provided for maintenance personnel. This may be done by indicator lights on the controller, by printing the sequence of events on a teletype printer, or by special test kits. The intricacies of solid-state circuitry require that the entire electronic system be sectionalized and each placed on a separate circuit board so that each can be readily replaced while the defective board is being repaired (see Figure 7.6).

The ANSI/ASME Safety Code for Elevators A17.1 defines group automatic operation as follows:

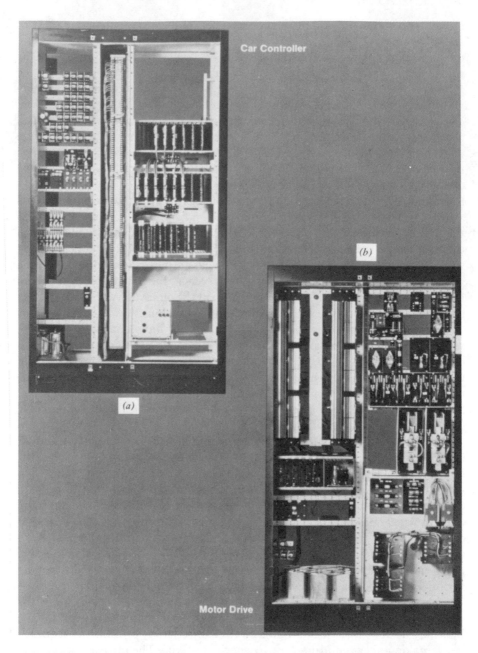

Figure 7.6. Elevator controllers with solid-state devices and microprocessors: (*a*) car controller—operational; (*b*) motor drive—motion controller (Courtesy Westinghouse Elevator Company).

. . . automatic operation of two or more nonattendant elevators equipped with power operated car and hoistway doors. The operation of the cars is coordinated by a supervisory control system including automatic dispatching means whereby selected cars at designated dispatching points automatically close their doors and proceed on their trips in a regulated manner. It includes one button in each car for each floor served and up and down buttons at each landing (single buttons at the terminal landings). The stops set up by the momentary actuation of the car buttons are made automatically in succession as a car reaches the corresponding landing irrespective of its direction of travel or the sequence in which the buttons are actuated. The stops set up by the momentary actuation of the landing buttons may be accomplished by any elevator in the group, and are made automatically by the first available car that approaches the landing in the corresponding direction.

The key word in the definition is "regulated" which implies that the elevator is directed to perform certain operations based upon data gathered by the group supervisory system at the instant the elevator is started (dispatched). Dramatic changes can take place from instant to instant as people are operating landing and car buttons and the elevators are changing positions. For this reason, the initial directions given an elevator as it starts may change before the elevator makes its first stop. A view of some traffic situations will attempt to show some of these changes.

Starting at a point in time when there is no traffic, which can be some time during the course of the working day and usually before starting time and after quitting time, elevators are parked. The Otis philosophy is to park them throughout the building in zones, each of which consists of a floor or number of floors either with equalized population or because of some importance such as an executive floor. The lobby is a separate zone and the number of zones is generally equal to the number of elevators in the group so each elevator can be stationed in a zone. The section of the floors making up each zone provides a basis for later operational logic decisions.

The Westinghouse approach is to park elevators at the place where they completed their last call and to move them from that point to where an elevator is needed for a later requirement. Westinghouse also divides the building into zones, but based on measuring demand rather than designated parking. With Westinghouse the floors the elevators serve are divided into one, two, or three or more floor zones for down service and usually two zones for up service above the lobby.

With both Otis and Westinghouse, one car is usually parked at the lobby. When a demand for service appears, which must be a landing call if the elevators are inactive, an available car proceeds to answer it. With Otis, an available car is one that is in the zone where the call was placed, and with Westinghouse, it can either be the closest elevator or one that is designated as next available.

If traffic remained such that no more than the number of elevators available were needed to serve the demand at one time, there would be little need for sophisticated operations. It is when traffic exceeds that amount or when the traffic occurs in a recognizable pattern such as incoming, outgoing, or two-way that the logic of the group supervisory system becomes most important.

A group automatic system should be capable of being programmed to suit the expected traffic activity in any type of building. For example, if the number, speed, and capacity of the elevators are chosen to serve a specific type of peak traffic, the group operations must initially include the necessary features to serve that traffic and have the capability of being adjusted or reprogrammed to accommodate future changes in traffic patterns or requirements.

TRAFFIC DEFINITIONS AND OPERATIONS

In the discussion of two-way traffic in Chapter 5 traffic was categorized into light, moderate, and heavy traffic and preliminary definitions given. Those definitions are modified as follows.

Intensity of Traffic

Light traffic occurs when the number of people riding or requiring elevator service at one time does not exceed the number of elevators in the group. Traffic is light, for example, if there are four cars in a group and no more than four calls, either car or landing, are expected to be registered at one time.

Moderate traffic occurs when the number of people riding or seeking elevator service at a given time is such that the available elevators in a group must be shared by more than one person and the average loading is not expected to fill any elevator beyond 50% of its capacity.

Heavy traffic occurs when the demands on the elevator system are such that the available capacity must be equalized among many passengers and priority of service may have to be given to passengers riding in one direction over those seeking to travel in the opposite direction.

Various traffic conditions require various operations as follows:

Incoming Traffic.   Light incoming traffic periods require at least one elevator parked at the loading lobby to receive passengers. Operation should be to the highest required call and immediately return to the lobby if no other elevator is at the lobby to receive passengers.

Moderate incoming traffic periods require that all the elevators should be at the loading lobby or returning to the lobby to receive passengers. The elevator travels to the highest required call and returns immediately to the lobby. More than one elevator should be able to be loaded at the same time, and any filled car should be dispatched.

Heavy incoming traffic periods require all the foregoing operations plus a system to give priority to lobby traffic during designated periods of working days. Service to upper-floor up and down landing calls can be temporarily denied if no car is available at the lobby for loading.

Two-Way Traffic.   Light two-way traffic periods require the elevators to be available throughout the building either at rest or in motion so that a car is either at or traveling toward the next expected call. One approach is to station elevators at various zones of the floors served and to operate them as individual units during periods of minimum demand.

Moderate two-way traffic periods require operation of all elevators in a predetermined regulated manner to minimize the waiting time for landing calls and to provide concentrated elevator service in the direction of heaviest traffic.

Heavy two-way traffic periods require that in addition to the predetermined operation to minimize landing call waiting time, priority operations for longer waits and for potentially long-wait calls should be instituted. Means should be provided to move elevators from areas of light demand into areas of heavy directional demand. For intense situations a system that bypasses some calls to deploy elevators to floors with potential long-wait landing time should be employed.

Interfloor traffic operations include all the actions required for two-way traffic with the elevators concentrating on floors of heaviest demand. Trips to terminal floors should be restricted unless definite demands are registered for travel to those floors.

Outgoing Traffic.   Light outgoing traffic periods should reduce time spent by any elevator at the unloading terminal. This is true for any outgoing situation. Elevators should return up into the building to serve down demand with travel no higher than necessary.

Moderate outgoing traffic periods require elevators to minimize time spent at the unloading landing, and bypass landing calls if the elevators are filled up to or beyond a predetermined loading.

Heavy outgoing traffic periods require all of the foregoing plus a system to restrict service to up landing calls if down landing calls are being bypassed during a predetermined time each day. In addition some means should be provided to divide the available elevator service to minimize the possibility of lower floors being bypassed by cars filling at upper floors. Priority should be given to outgoing traffic limited to a predesignated time at the end of a working day. A system to automatically restrict up landing call service when passengers only wishing to go down operate both up and down landing calls is often required.

Nighttime and Traffic Lulls. (Off-Peak)   With any group automatic operating system is should be decided how the elevators will be stationed during nighttime and periods of traffic lulls which can occur at any time during the day. As previously mentioned, this may be done by parking the elevators in zones or at other strategic locations.

Operational Strategy

The Otis Elevator Company uses the zone parking approach to provide the starting point for the operation of elevators during various traffic patterns. Both Otis and Westinghouse use load sensing devices in the cars and landing call button information to change the operation of the elevator group during certain distinct traffic patterns. For example, an absence of substantial landing call demand and filled elevators leaving the lobby is a definite indication of incoming traffic activity. Conversely, a heavy down landing call demand and filled cars traveling in the down direction indicate heavy outgoing traffic.

Continued down traffic with elevators bypassing down landing calls and down landing calls waiting beyond a predetermined time is a signal that Westinghouse uses to cause elevators to travel directly without intermediate stopping to down landing zones (predesignated zones consisting of two or three floors), with down landing call or calls exceeding a predetermined time. Otis employs a similar strategy of distributing the elevators among the various zones or concentrating service in the zone with the heaviest demand.

A microprocessor and the integration of many data inputs can be used for additional strategies of matching car and landing calls. By matching a car call for a particular floor with a landing call at the same floor in the direction the car is traveling, improved elevator efficiency will result. As can be appreciated, rapid changes and new innovations will be constantly introduced in the years to follow and the ability to reprogram elevator operating systems should afford building owners a substantial management opportunity to maintain their elevators at their optimum traffic-handling ability.

Providing for Contingencies

A group automatic operating system should consider all contingencies. If, for example, a car is delayed at a floor beyond a predetermined period of time, some means to disengage that car from group operation should be provided. This will ensure that the delayed elevator does not interfere with other elevators in the group or prevent other elevators from stopping at that floor in response to landing calls. A signal at some central location should be given of the delay or operating failure.

If for any reason the group operating system fails to function and elevators are not moving as required, auxiliary means should be provided. This auxiliary means should operate the elevators in a random fashion and provide a signal to inform a responsible person of failure.

A third possible serious failure could occur if for any reason the landing buttons failed to function. Passengers would be stranded on floors, unable to summon elevator service. A means should be provided to detect such failure and operate the elevators to make predetermined stops so that at least one car stops at each floor. A warning signal should occur in some central location so that necessary corrective action can be taken. If an emergency occurs such as a fire as indicated by a detecting system, such random stopping should be aborted and all elevators returned to a designated floor for use by emergency personnel.

Elevators are complex combinations of electrical and mechanical elements. There may be thousands of electrical signals interchanged during the course of a single trip. Elevators have been proven extremely reliable with minimum down time. The probability of a passenger being trapped is extremely remote and if it does occur, adequate means to inform someone are required by elevator safety codes. The common information system is an alarm bell button located in the car operating panel. Many buildings have either a two-way communication system in each elevator or a telephone connected to the building office. Others have alarm systems connected to central protection agency offices. In any building where it is possible for only one person to be in the building and riding an elevator, this latter consideration is essential.

Passengers in a stalled elevator are safest staying where they are. If they try to get out without help, the elevator may move and they may be injured. Elevator safety codes require immediate emergency lighting in the car. A low-voltage electric fan operated from the same emergency batteries can provide some air circulation. The alarm bell should also be connected to the emergency power source. In buildings with standby generators, the car lights and fan should be connected to the standby power source when the generator is started. Adequate switching should be provided since it is desirable to turn off the car lights when the elevator is parked by normal means. Operation of elevators during a power failure when standby power is available will be discussed in a later chapter.

SPECIAL OPERATING FEATURES

Many of the operations and features described in the preceding pages may be applied to any type of elevator operation and are not restricted solely to group automatic operation. When elevators are specified for a particular building the desired features should be described in detail and the elevator supplier required to supply them or offer an acceptable substitute.

In addition to normal operating features, it is desirable to take any elevator out of a group and use it for special service. This is necessary in an office building during moving or in a hospital to transfer a patient. At that time it is required to have a system of operation from the car buttons only, bypassing landing calls. A key switch should be located in the car operating panel, the actuation of which will cancel all existing car calls and cause the car to bypass landing calls. Direction can be established by operating the car button for the desired floor and the car is started by operating an appropriate car start button. A key switch, master keyed to the building key system, is preferred to avoid abuse by unauthorized persons, and making the key removable only in the off position ensures that the car will be restored to normal service when the special operation is complete. Such a system of operation is generally referred to as independent service, emergency service, or hospital service.

In certain buildings it is desired to summon an elevator to a particular floor to give priority service to an executive, emergency personnel, of for other reasons. A key switch can be located at that floor which, when operated, will call an available car to that floor. The question often to be answered is, Should the car be allowed to complete its existing calls or should the passengers who are on that car be made to travel to the priority call. For emergency personnel or in a hospital the answer is obvious, but what would happen if a vice-president called the car and the president happened to be riding? Nevertheless, once the question about existing car calls is answered, the key switch will call the car to the floor of call and from that floor the car may be operated as described for independent service. Quite obviously such special service must be limited or conflicts could arise. Elevator operation during emergencies will be discussed in a later chapter.

A third form of special service is to operate one or more of the elevators in a group from a separate or independent riser of landing buttons. When on this service, the elevator responds only to the separate button riser and the other cars respond only to the normal landing buttons. Independent riser service can be abused if its use is unsupervised. One solution is to have the separate riser operate an annunciator so that an attendant on the separated car can control its use. In addition, a lighted sign in the landing button faceplate can inform the users of the elevator's availability for this special operation.

How to serve basements and special floors can often be a problem. Extra time allowances must be made for special stops if they are to be served during any traffic situation. In addition, all the elevators in the group should serve the special floors or basements unless they are to be served only during independent service operations. Failure to do so impairs the accessibility to and from those floors.

With a single basement, a suggested criterion is that if less than 25% of the critical traffic on the elevators is expected to originate from the basement, the basement should be served only on call. Any car not selected for dispatching or a down traveling car should travel to the basement only if a car or landing call for basement service exists. If more than one basement is served the car should not travel any lower than the lowest up landing call in the basement area and return up.

If more than about 33% of the critical traffic is expected from the basement, consideration should be given to making the basement the lower dispatching landing and providing additional stopping considerations for the elevator so that more than one or any car not filled will stop at the main floor above the basement. If the basement is a cafeteria and only open at lunch time, the dispatching landing for the group of elevators should then be switched from the lobby to the basement by either a clock-controlled switch or automatic means.

Floors above the main group of floors the elevators serve should be treated similarly, in effect like upsidedown basements.

Many other situations could arise in applying the proper operations to a group of elevators, the foregoing being some of the most common. Odd situations should be recognized early in building design and sufficient additional operations engineered into the group automatic operating system to meet the expected problems.

ADJUSTING OPERATION TO TRAFFIC

Initiating the proper operations of the elevators at the proper time is a task that must be accomplished automatically. In earlier systems an attendant was required to change the mode of operation or program, as it was known. Too often it was set on one operation and not changed so that when it was time to go home, the passengers had to fight elevators that had been left on "up-peak" operation. Modern systems should be completely responsive to the traffic situation. Sophisticated systems that utilize programmable microprocessors to determine where elevators can be directed to best serve the prevailing traffic are available and offer immeasurable opportunity.

With the increasing availability of advanced computer circuitry and microprocessors that can process more information, greater utilization of elevator capacity should become practicable and elevators will be able to provide service of the highest quality. In preceding chapters the basis of calculating required elevator capacity and quality of service was shown. The elevators chosen for a given building and the operating scheme furnished should be capable of providing that required performance.

NEEDS FOR FUTURE DEVELOPMENT

As sophisticated as elevator operating systems are becoming, there is still vast opportunity for new development and improvement. No elevator company at present has the means to determine if an elevator is full in volume although devices that measure weight in the elevator car are common. A means to determine if the space within an elevator is fully occupied would be especially valuable in a hospital elevator where one cart or stretcher can fill up the elevator which makes it inefficient to stop for additional landing calls until that load is removed.

Elevator companies should recognize the difference between operation and motion control. It should be possible to obtain the most sophisticated group operation on a low-rise hydraulic elevator installation as on a high-rise downtown building. This is not possible with present technology and is an injustice to the owner of a suburban office building. It is simple to separate operation from motion control and amazing that it has not been practiced by elevator manufacturers.

Another area for research and development is a means to determine how many people are waiting for elevator service at a lobby or particular floor. One landing call can represent one person or ten people. Elaborate systems have been tried to provide buttons for each elevator stop at each floor and to provide elaborate illuminating signs over each elevator to indicate the floors at which they will stop. As can be surmised, the confusion of finding and standing in front of the right elevator where six or eight elevators stop at a common lobby can be overwhelming. A reliable electronic crowd counter could be a great asset to improve elevator service.

The era of the microprocessor and programmable control applied to elevators is just beginning. The key will be to ensure the traditional reliability that has made the elevator an accepted part of everyone's life with minimum concern about safety and practically no concern about availability of service.

CHAPTER EIGHT

Space and Physical Requirements

SPACE FOR ELEVATORS

Architects of successful buildings recognize that good internal and external pedestrian circulation is an essential feature of a well-designed building. Individual floor plans may readily be changed to suit individual requirements but the facilities for pedestrian circulation between floors in the building is fixed. Good vertical transportation is one of the first aspects of good building design.

Good internal transportation includes all aspects of the circulation within the building—the proper lobbies, the right-sized corridors, sufficient elevators of proper size, and adequate stairs. Elevator and escalator planning must consider both the space allocated for the equipment and the space provided for people to use that equipment. This becomes essential on the lobby floor where people not only must be able to easily use elevators or escalators but also must enter and leave the building with comparative ease. Lobby amenities such as stores and information areas must be properly located so as not to interfere with pedestrian circulation. Ample queuing space for escalators and waiting space for elevators are both important elements in building design as are adequate exits to accommodate the crowds that elevators may discharge.

In Chapter Two elevator lobby space with its essential contribution to efficient elevator utilization was discussed. Traffic handling by elevators emphasizes the importance of sufficient elevators of ample size and speed as well as providing the proper door opening width on those elevators. In this chapter building space to accommodate elevators is emphasized.

REQUIRED SPACE

To make an elevator platform travel up and down smoothly and safely certain structural requirements must be met. Each elevator must have guides or rails which keep it in a substantially vertical path and provide the columns from which an elevator may be supported if, for any reason, its continued operation becomes unsafe. This latter consideration is called safety application and consists of a device below the elevator platform which will stop and hold it on the rails if it overspeeds in the down direction for any reason. During safety application, the entire weight of the elevators is arrested and transferred to the building structure through the rails. Stopping an elevator that has a structural weight of 3 or 4 tons (3600 kg) plus a passenger load of 2000 to 4000 or more pounds (1800 kg) which is traveling at hundreds of feet (1 mps or more) per minute by clamping it on the building structure through the guide rails will require a sizable rail and rail supports. Space for rails, therefore, is a necessity (Figure 8.1*a*).

Rail Support

In addition to the rails adequate rail supports or brackets must be provided. The brackets are fastened to the building steel by the elevator contractor or, in a concrete building, by means of inserts set in the concrete by the concrete con-

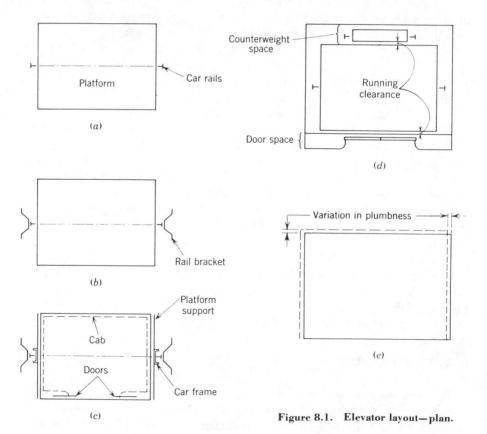

Figure 8.1. Elevator layout—plan.

tractor or by the elevator contractor (Figure 8.1*b*). The vertical space between brackets is critical since the rail has a design stiffness and can only be allowed to deflect about ⅛ in. (3 mm). Elevator rails are designed so that fastening at floor levels is generally sufficient when floor levels are up to about 14 ft (4.2 m) apart. If longer spans are expected additional supporting beams should be provided in the structure, heavier rails should be used, or additional horizontal space to reinforce the elevator guide rails and provide the necessary stiffness should be provided. Heavier rails with a larger cross section are stiffer and can have a greater bracket spacing depending on the total load of the elevator.

The maximum vertical distance between brackets, related to the size of elevator rails and the expected load on those rails, is established in the A17.1 Elevator Code. The latest edition of the code should be checked for the requirements. If the elevator is to be installed in a seismic risk zone, additional considerations are necessary and both A17.1 and local requirements need to be followed.

Platform and Enclosure

The elevator must be enclosed. Because the interior should create a comfortable surrounding and is seen by everyone entering and leaving a building it merits the best architectural design and finest craftsmanship. This requires space and creates weight. In addition the enclosure must be completely isolated, both for sound and vibration, from the structure that supports it. Such isolation, usually accomplished by elastomer pads, may require space. The supporting structure, called the car frame, must be constructed to withstand the force of a safety application and the lifting forces on the elevator as well as the weight of the mechanism that will operate the doors (door operator), switches, cams, and so on. All this weight requires a structure that may become quite heavy and requires space (Figure 8.1*c*).

This space, if measured from the side of the hoistway to the side of the car platform, amounts to 8 in. (200 mm) for elevators up to 500 fpm (2.5 mps) and to about 10 to 12 in. (250 mm) for higher-speed elevators, double-deck elevators, or elevators with capacities of 5000 lb (2300 kg) or more.

Access to the elevator from each floor is guarded by a hoistway door. This door and entrance assembly is required by local building codes to be of fire-labeled construction and must be certified for the type of wall used, whether masonry or dry-wall. In addition a mechanism must be provided to lock the door safely when the elevator is not at the floor and automatically open it when the car is leveling to the floor. Space for both the doors and its mechanism is required (Figure 8.1*d*).

The distance from the front of the hoistway at the floor line (since it is assumed that the door sills will be placed on top of this edge) is 1 in. (25 mm) for sill overhang and 1¼ in. (31 mm) running clearance or a total of 2¼ in. (56 mm). The doors on each floor will require front space in addition to the running clearance and sill overhang which will be discussed shortly.

In a traction-type elevator the structural load and part of the passenger load, from 40 to 50% of the rated load, is counterweighted, and the counterweights travel up or down in the opposite direction from the elevator. Counterweight guides are usually at the rear of the elevator, although some installations such as elevators in hospitals have counterweights at the side. Horizontal space for the necessary cast iron or steel counterweight is required (Figure 8.1*d*).

The total space for counterweights behind the elevator as measured from the back hoistway wall to the back edge of the car platform is approximately 15 in. (380 mm) for counterweights without safety devices and 17 in. (430 mm) for counterweights with safety devices as would be required if space that could be occupied was located below the elevator hoistway. This should be determined before space is allocated. A counterweight with a safety device will also require a heavier rail. If seismic requirements must be fulfilled, additional bracketing and support space of about 4 in. (100 mm) should be provided. If a hydraulic elevator is being considered, all the foregoing requirements except the space for the counterweight at the rear must be provided. In lieu of the rear counterweight space, a minimum allowance of 2 in. (50 mm) is provided between the rear of the platform and the back wall of the hoistway.

Construction Tolerance

An elevator must run up and down as plumb as possible. If it does not it may experience extra wear on the elevator guides. Buildings are often not plumb; tolerance may vary from a fraction of an inch (25 mm or less) to an inch or more. Elevator contractors ask for 1 in. (25 mm) plus or minus plumbness in an elevator shaft, which means that a difference of 2 in. (50 mm) may exist between the top and bottom in either the front-to-back or side-to-side direction or both. In higher-rise buildings, the building may be built with a slight lean to compensate for prevailing winds or may sway with a calculated deflection and period due to wind loading. This should be ascertained and necessary provisions made for these contingencies in the design of the elevator system (Figure 8.1*e*).

With all the foregoing specifications, the net usable platform area available in any elevator layout is 60 to 75% of the hoistway required for the elevator. Details of the reason for allocating this space are shown in Figure 8.2. Industry effort to reduce this nonusable space is continuous, including use of stronger steel, new approaches to door arrangement, and new fastenings. Since the safe operation of an elevator is predominant and most of the space required at present is for safety reasons, the space must be provided so that the optimum size car inside can be had.

To "rough out" the required elevator hoistway for an elevator of a given platform size or, conversely, to determine what platform size can be accommodated in a given hoistway the following steps can be used. All dimensions are approximate.

Rated Load (mass) (kg)	Maximum Available Car Area (see note) (m²)	Maximum Number of Passengers	Rated Load (mass) (kg)	Maximum Available Car area (see note) (m²)	Maximum Number of Passengers
100	0.40	1	975	2.35	13
180	0.50	2	1000	2.40	13
225	0.70	3	1050	2.50	14
300	0.90	4	1125	2.65	15
375	1.10	5	1200	2.80	16
400	1.17	5	1250	2.90	16
450	1.30	6	1275	2.95	17
525	1.45	7	1350	3.10	18
600	1.60	8	1425	3.25	19
630	1.66	8	1500	3.40	20
675	1.75	9	1600	3.56	21
750	1.90	10	1800	3.88	24
800	2.00	10	2100	4.36	28
825	2.05	11	2500	5.00	33
900	2.20	12			

Beyond 2500 kg, add 0.16 m² for each 100 kg extra.

(a)

Maximum Inside Net Platform Areas for the Various Rated Loads

Rated Load (lb)	Inside Net Platform Area (ft²)	Rated Load (lb)	Inside Net Platform Area (ft²)
500	7.0	4,500	46.2
600	8.3	5,000	50.0
700	9.6	6,000	57.7
1000	13.25	7,000	65.3
1200	15.6	8,000	72.9
1500	18.9	9,000	80.5
1800	22.1	10,000	88.0
2000	24.2	12,000	103.0
2500	29.1	15,000	125.1
3000	33.7	18,000	146.9
3500	38.0	20,000	161.2
4000	42.2	25,000	196.5
		30,000	231.0

To allow for variations in cab designs, an increase in the maximum inside net platform area not exceeding 5 %, shall be permitted for the various rated loads.

1 lb = 0.454 kg
1 ft² = 0.0929 m²

(b)

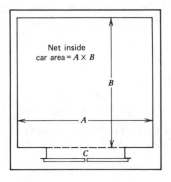
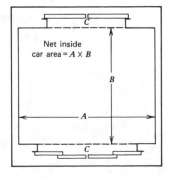

(c)

Figure 8.2. Elevator capacity vs. inside car area: (*a*) European EN81 Code; (*b*) North America A17.1 and Canadian B44 Codes; (*c*) measurement diagram [A × B = A17.1 and B44 areas, (*A* × *B*) + area *C* = EN81 area.

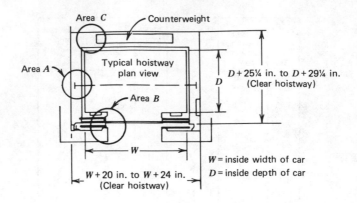

Area C — Counterweight

Area A

Typical hoistway
plan view

$D + 25\frac{1}{4}$ in. to $D + 29\frac{1}{4}$ in.
(Clear hoistway)

D

Area B

W

$W + 20$ in. to $W + 24$ in.
(Clear hoistway)

W = inside width of car
D = inside depth of car

Detail area A

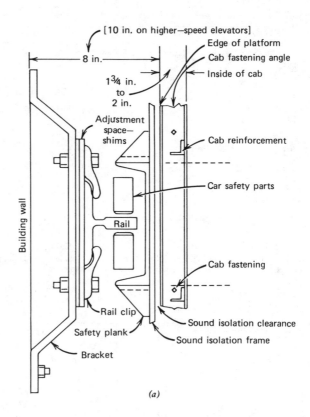

[10 in. on higher—speed elevators]
Edge of platform
Cab fastening angle
Inside of cab

8 in.

$1\frac{3}{4}$ in.
to
2 in.

Adjustment
space—
shims

Cab reinforcement

Car safety parts

Rail

Cab fastening

Building wall

Rail clip

Safety plank

Bracket

Sound isolation clearance
Sound isolation frame

(a)

Figure 8.3. Elevator construction: (a) Detail of Area A;

Detail Area *B*

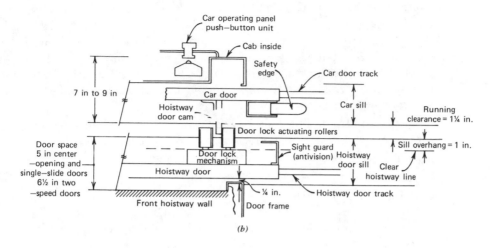

Car operating panel
push—button unit

Cab inside

Safety edge

Car door track

7 in to 9 in

Car door

Car sill

Running clearance = 1¼ in.

Hoistway door cam

Door lock actuating rollers

Door space 5 in center —opening and single—slide doors 6½ in two —speed doors

Door lock mechanism

Sight guard (antivision)

Hoistway door sill

Sill overhang = 1 in.

Hoistway door

Clear hoistway line

Front hoistway wall

¼ in.

Door frame

Hoistway door track

(b)

Detail Area *C*

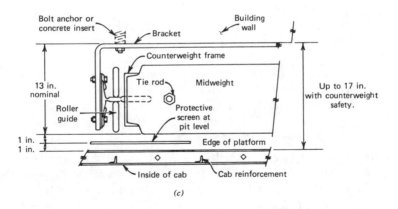

Bolt anchor or concrete insert

Bracket

Building wall

Counterweight frame

13 in. nominal

Tie rod

Midweight

Up to 17 in. with counterweight safety.

Roller guide

Protective screen at pit level

1 in.
1 in.

Edge of platform

Inside of cab

Cab reinforcement

(c)

Figure 8.3. (*b*) Detail of Area *B*; (*c*) Detail of Area *C*.

1. Add 16 in. (400 mm) to the platform width for the total clear hoistway width required [8 in. (200 mm) on each side of the platform for rails] (see Figure 8.3*a*).

2. Add 15 in. (310 mm) for the counterweight at the back [17 in. (430 mm) if counterweight safeties are required] plus 2¼ in. (57 mm) in front of the platform for the clear front-to-back hoistway space required. The 2¼ in. (57 mm) includes 1¼-in. (32-mm) running clearance and 1-in. (25-mm) sill overhang (see Figure 8.3*c*).

3. Door space in front of the hoistway consisting of the total sill and space for doors will vary with the door type and is calculated as follows:

    a. From edge of sill to back of door 2¼ in. (57 mm).

    b. Each door panel is 1¼ in. (32 mm) thick.

    c. One-quarter inch (6 mm) is allowed between each door panel and the door frame. To this is added 1¼ in. (32 mm) from frame construction for a total of 1½-in. (38-mm) door running and frame space.

Therefore, with single-panel center-opening or single-slide doors a total door space of 5 in. (127 mm) is required and with two-speed doors the space is 6½ in. (165 mm) (see Figure 8.3*b*).

To estimate the clear inside dimensions of a given platform size, the following steps are required:

1. Reduce the width of the platform by 2 in. (50 mm) on each side for a total of 4 in. (100 mm). The 2-in. (50-mm) space is needed for structure and fastening as shown in Figure 8.2.

2. Reduce the front-to-back dimension of the platform by 2 in. (50 mm) on the back and by 6 in. (150 mm) if single-slide or center-opening doors are used or 8 in. (200 mm) if two-speed doors are used for a total of 8 or 10 in. (200 or 250 mm) to obtain the gross front-to-back dimension.

The A17.1 Elevator Code allows a net area related to the elevator capacity in pounds as shown in Figure 8.2*c*. This area is measured at 36 in. (915 mm) above the floor and does not include the space required by the car doors nor the framing space around the doors. This framing space can be 2 to 4 in. (50 to 100 mm) deep and to obtain the net front-to-back dimension, the gross depth is further reduced by 2 or more inches (50 mm), depending upon cab design. The inside area can vary with a tolerance of +5% as allowed by the A17.1 Elevator Code for nonconventional shaped cabs.

Example 8.1

Inch-pound: Given a 7-ft-wide by 5-ft-deep platform, what is the elevator capacity? Use center-opening doors.

Width: 7ft   = 84 in. − 4 in. = 80 in.
Depth: 5 ft   = 60 in. − 9 in. = 51 in.
Area: 80 × 51 = 4080 in² ÷ 144 = 28.4 ft²
Allowable A17.1 code area 2500 lb = 29.1 ft²
(Figure 8.2*b*)
Therefore, rate elevator 2500 lb

Metric: The EN81 inside area requirements are shown in Figure 8.2*a*. Given a 2140-mm-wide by 1525-mm-deep platform, what is the elevator capacity? Use center-opening doors.

Width: 2140 mm − 100 mm   = 2040 mm
Depth: 1525 mm − 180 mm   = 1345 mm
Area: 2040 mm × 1345 mm = 2.74 m²

Add back space between doorjamb inside car and depth to face of door
28 mm × 1200 mm = 0.03 m²
Total inside area: 2.77 m²
Allowable area for 1200 kg per EN81 code 2.8 m² (see Figure 8.2*b*)
Therefore, rate elevator 1200 kg

The A17.1 code does not include the area between the inside doorjambs to the face of the doors whereas the EN81 code does, therefore the difference in calculation.

OVERHEAD AND PIT REQUIREMENTS

Doors

The width of the clear opening of hoistway doors is directly related to the available clear width of the hoistway. Single-slide and center-opening door space requirements are established by multiplying the desired width of the opening by 2 and adding 6 to 8 in. (150 to 200 mm) for structural requirements behind the open doors.

For example, if 4 ft 0 in. (1200 mm) center-opening doors are desired, the clear hoistway must be two times 4 ft (2400 mm), plus 8 in. (200 mm), or 8 ft 8 in. (2600 mm) wide.

For two-speed doors, the hoistway width required is 1.5 times the desired width of opening, plus 4 to 6 in. (100 to 150 mm) and 11 in (275 mm) for the jamb. Therefore, the clear hoistway width for a 4-ft (1200-mm) two-speed door is 1.5 × 4-ft plus 15 in. or 7 ft 3 in. (2200 mm), preferably 7 ft 5 in. (2260 mm). The strike jamb side of the two-speed opening must be at least 3 in. (75 mm) in from the platform edge. Two-speed center opening doors are established similar to center opening; however, the door space is 1.5 times the opening width.

In Example 8.1, the 7 ft 0 in. (2140 mm) wide platform given can accommodate, under ideal conditions, a 4 ft 0 in. (1200 mm) wide center-opening door. More practically 3 ft 10 in. (1150 mm) wide center-opening entrance should be employed since building tolerances must be considered. If the platform width is increased to 7 ft 2 in. (2200 mm) per Table 4.4, the 4 ft 0 in. (1200 mm) center opening entrance can be easily accommodated.

Care must always be taken that architectural or structural infringements such as mail chutes or columns are located to avoid interference with the door opening.

As previously discussed, the front-to-back space required for various doors from the edge of the hoistway doorsill to back of the hoistway wall is 5 in. (127 mm) for center-opening doors and 6½ in. (165 mm) for two-speed doors. If flush transom panels are used with center-opening doors above the moving doors, this space must be increased about 1½ in. (38 mm).

The front wall may be of masonry or of gypsum board construction and must preserve the fire integrity of the entrance assembly as well as the fire rating of the hoistway. Entrance assemblies are required to be tested under fire conditions and certified by an independent laboratory. Details of such a test can be found in the American Society for Testing Material publication ASTM 152.

Once the plan of the various elevators in the building is established, attention must be given to providing the necessary vertical space. We discuss conventional traction elevators at this point.

Later in the chapter details about hydraulics, basement machines, and underslung elevators will be discussed.

Pit Space

When the elevator stops at a floor of a building certain parts of the structure of the elevator are either above or below the cab as it is seen at that floor. Below the cab floor is the platform, a structural base composed of either wood and angle iron or all steel. This platform should be cushioned on elastomer pads, on a sound isolation subframe. A present method of weighing the load in the elevator car is to measure the deflection of these elastomer pads and actuate microswitches or a transducer for the various degrees of loading in the car (Figure 8.4).

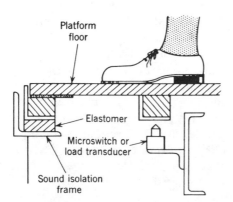

Platform floor

Elastomer

Microswitch or load transducer

Sound isolation frame

Figure 8.4. One form of load weighing by measuring platform deflection.

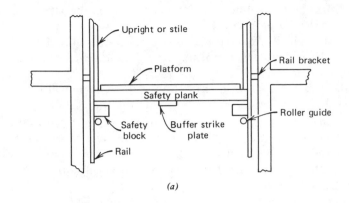

Upright or stile

Platform

Rail bracket

Safety plank

Safety block

Buffer strike plate

Roller guide

Rail

(a)

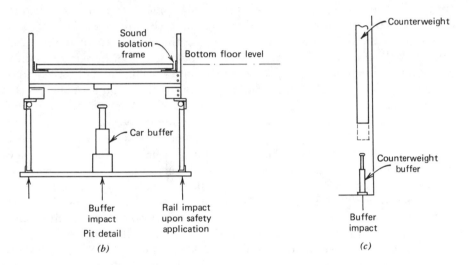

Sound isolation frame

Bottom floor level

Car buffer

Counterweight

Counterweight buffer

Buffer impact

Rail impact upon safety application

Pit detail

(b)

Buffer impact

(c)

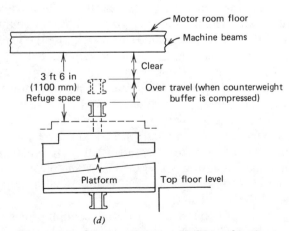

Motor room floor

Machine beams

Clear

3 ft 6 in (1100 mm) Refuge space

Over travel (when counterweight buffer is compressed)

Platform

Top floor level

(d)

Figure 8.5. Elevator layout procedure—elevation.

159

The platform and its sound isolation frame rest on the safety plank, which also supports the elevator safety device, clamps that will stop and hold the elevator on the rails if it overspeeds in the down direction. The safety plank must also be designed to absorb the impact from the elevator buffers located in the pit (Figure 8.5*a*).

When the elevator is stopped at the lowest floor, there will be a few inches overtravel before it strikes the buffer. The buffer is designed to stop the elevator if, for some reason, it travels at its operating speed past the lowest floor. The buffer absorbs the kinetic energy of the moving car and brings it to a stop within the limit of the buffer stroke. The stop is not gentle but it is within safe limits. The buffer is not required by safety codes to bring a free-falling elevator to a stop; this is the function of the car safety.

The depth of the pit must include the depth of the platform and support space required for the elevator car (Figure 8.5*b*) plus operating clearance in addition to the buffer standing and stroke space (Figure 8.5*c*). For elevator speeds up to and including 200 fpm (1.0 mps) spring buffers are used. Over 200 fpm gradual buffers must be used which are usually hydraulic buffers commonly referred to as "oil buffers." The stroke of the buffer is a function of speed and is defined by the A17.1 Elevator Code and is shown in Table 8.1. The depths required for typical pits range from 5 to 16 ft (1.5 to 5 m) or more for very high speeds. Table 8.1 shows depths for the more common speeds.

For any special elevators the buffer stroke may be reduced if certain other precautions are taken. One of these is a means to ensure that the elevator brake is applied if the car is traveling at high speed when it is within a determined distance of the pit. If, for example, it normally takes about 20 ft (6 m) to slow down and stop a 700-fpm (3.5-mps) elevator, and the elevator is traveling that fast within, say, 15 ft (4.5 m) of the pit, applying the brake at that point should slow the car down sufficiently so that a shorter buffer will stop the car.

When the elevator car lands on the buffer and the buffer is fully compressed, the counterweight is at its highest point. The counterweight will also jump a short distance of less than one-half the buffer stroke if it is traveling up at full speed and abruptly stopped. Sufficient clearance above the counterweight must be maintained so that there is no danger of its striking the overhead.

The counterweight will land in the pit if the car passes the topmost landing. The counterweight will have a buffer of equal stroke to the car buffer. In addition space under the counterweight must be provided for normal running clearance (Figure 8.6*b*, Ⓑ) plus sufficient additional space for the normal stretching that is experienced with elevator ropes and to avoid causing the counterweights to land. (Figure 8.6*b*, Ⓒ). This rope stretching can result from temperature changes, aging, or rope strands settling after manufacture because of the addition of load or various other causes. About 0.25% of rope length is a nominal allowance for this contingency after the initial installation stretch of from 0.5 to 1% of the rope length.

In addition to the buffers, the pit must often contain provisions for guiding of compensating ropes. These are ropes attached to the bottom of the car and to

Table. 8.1. Pit Depths. Traction Elevators—Overhead Machines

Speed (fpm)	100	200	300	400	500	600	700	800
(mps)	0.5	1	1.5	2	2.5	3	3.5	4
Depths								
a. With restrained rope compensation	—	—	—	8 ft 0 in. 1.6 m	8 ft 6 in. 2.6 m	9 ft 2 in. 2.8 m	9 ft 10 in. 3.0 m	10 ft 6 in. 3.2 m
b. With chain, free rope or traveling cable compensation	5 ft 0 in. 1.5 m	5 ft 0 in. 1.5 m	5 ft 4 in. 1.6 m	7 ft 10 in. 2.4 m	8 ft 4 in. 2.5 m	—	—	—
c. With reduced stroke buffer and either restrained rope chain, traveling cable or free rope compensation	—	—	5 ft 0 in. 1.5 m	5 ft 4 in. 1.6 m	8 ft 0 in. 2.4 m	8 ft 6 in. 2.6 m	8 ft 6 in. 2.6 m	9 ft 2 in. 2.8 m
Buffer type	Spring	Spring	Oil	Oil	Oil	Oil	Oil	Oil

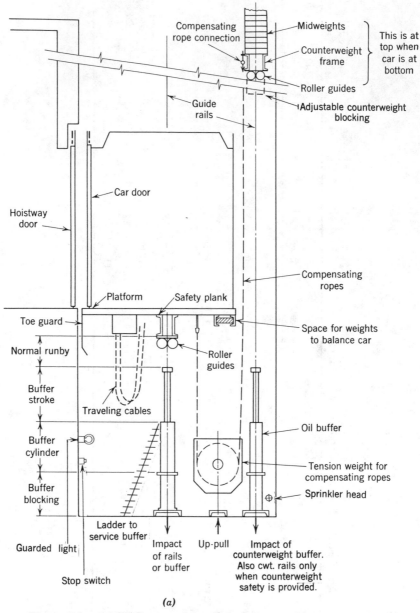

Figure 8.6. (a) Pit Construction and equipment—with rope compensation.

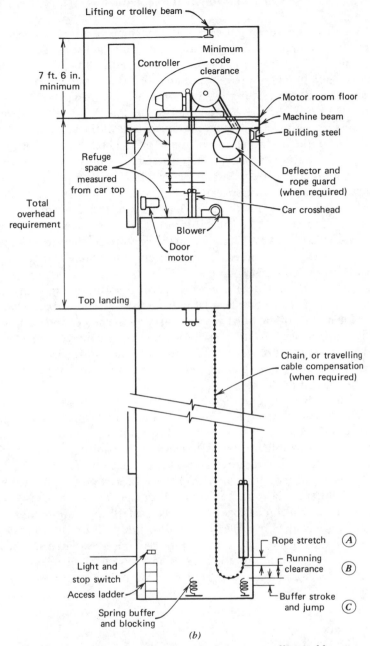

Figure 8.6. (*b*) Pit and overhead space with chain or travelling cable compensation.

163

the bottom of the counterweight side of the machine as the car travels up and down. In taller buildings this compensating rope guide will consist of a weighted sheave riding on short tracks in the pit. In very tall buildings where elevator speeds exceed 500 fpm (2.5 mps), if seismic shock is a possibility or if building sway is anticipated, the guiding mechanism must be arranged, often by code requirements, to link the car and counterweight together in such a way that when the car safety applies it acts on the entire mass of the system. This avoids excessive jump of the up-traveling weights when the high kinetic energy of the down-traveling elevator is arrested. Such a system is known as tied-down compensation or monomass safety and creates a substantial up-pull in the pit.

For slower-speed elevators, special ropes, travelling cable or chain interwoven with sash cord or coated with plastic may be substituted for guided rope as compensation. The sash cord of plastic is used to overcome the normal rattling noise a moving chain makes. Chain is usually confined to elevators of 500 fpm (2.5 mps) or less where an increased noise level is acceptable. Travelling cable may be used up to 700 fpm (3.5 mps).

In determining the pit depth, consideration of other equipment often is necessary. The electrical cables that hang from the bottom of the elevator to a point near the middle of the hoistway and carry all the power and signals to the elevator car must travel into the pit without coming in contact with equipment that would cause premature wear and failure. A tensioning device for the elevator governor or safety rope, which controls the application of the car safety, must also be located in the pit.

The total design of the pit (Figure 8.6) must include provisions for all the necessary elevator equipment, lighting, an elevator stop switch, access ladders, and, in some jurisdictions, sprinklers to guard against possible fire, and sufficient depth for the possible run-by of the elevator. In addition a safety clearance imposed by elevator codes requires at least 2 ft (0.6 m) between the bottom of the car platform and the pit floor when the elevator is on a fully compressed buffer. Sufficient support must be provided for the various impacts that may occur. These impacts include those of the rail on safety application, those on the buffers if they are called on to perform, and the up-pulls on the compensating arrangement in the event of safety application. All these reactions are shown on the elevator layout provided by the consulting engineer or the elevator manufacturer. Architects and structural engineers must provide for them in their total plans.

Overhead Space

The overhead space required by the elevator when it is stopped at the top landing must be allocated in a manner similar to pit space. On top of the elevator car is the mechanism that operates the elevator doors and the elevator lifting beam or crosshead. In addition space must be allocated for the car blower, various elevator operating devices, plus refuge space for a person on the car top to the closest overhead point in the hoistway. This close point may be the bottom of the beams that support the elevator machinery or a sheave used to deflect the hoisting ropes back toward the counterweight.

If the elevator passes the top landing a number of events occur. First the counterweight lands and starts to compress the counterweight buffer. If the elevator is traveling at any speed the car may have a slight jump. If someone is working on top of the car, a safe clearance to crouch will be required. The A17.1 code requires a refuge space for a person who may be on top of an elevator car. All this adds up to the space required for run-by. The space required from the top landing to the top of the motor room floor may be 20 to 30 or more feet (6 to 9 m).

Figure 8.6 shows this space and how it is allocated. Considerable variation may be expected with various speeds, sizes, and arrangements of elevator equipment.

Calculating Pit and Overhead Space

The interior height of the elevator cab is one of the main factors in calculating the required overhead height of an elevator hoistway as measured from the top floor served to the underside of the machine room floor. If we assume a desired clear inside ceiling height of 8 ft (2400 mm), the height of car lighting which can vary from 12 to 18 in. (300 to 450 mm) depending upon lighting fixtures selected must be added. This establishes the overall height of the cab. On top of the car, space is needed between the bottom of the crosshead and top of the cab for either rope shackles or a 2:1 sheave in the crosshead. This is from 12 to 18 in. (300 to 450 mm). The structural crosshead is 10 to 12 in. (250 to 300 mm) high. The overtravel of the elevator is about 6 in. (150 mm) and the elevator car will move upward the length of the buffer stroke plus one-half of the stroke for possible jump when the counterweight lands on its buffer. When the counterweight is resting on its fully compressed buffer, the elevator code requires a minimum of 2 ft (600 mm) usually from the top of the crosshead to the top of the hoistway or the closest striking point.

Summing up the foregoing, the following factors are required:

Requirement	Example
1. Height of cab ceiling (note)	8 ft 0 in.
2. Space for lighting	1 ft 3 in.
3. Space between top of enclosure and under the crosshead	1 ft 0 in.
4. Depth of crosshead	0 ft 10 in.
5. Overtravel	0 ft 6 in.
6. Counterweight buffer stroke @ 500 fpm	1 ft 5 in.
7. Counterweight buffer jump	0 ft 8 in.
8. Rope stretch and counterweight buffer clearance (assume)	0 ft 8 in.
9. 2-ft clearance	2 ft 0 in.
	13 ft 40 in.
Distance from top landing served to underside of the top of the hoistway (slab or structure)	16 ft 4 in. (5000 mm)

Note: Cab ceiling height is also related to landing door height. For example, if the landing doors are 7 ft 0 in. high, the cab ceiling should be a minimum of 7 ft 6 in. high.

The A17.1 Safety Code for elevators also requires a 3 ft 6 in. (1050 mm) refuge space on top of the elevator car enclosure for a person who may be required to be on top. This is an area of 650 in.^{2} (0.4 m^2) with no side less than 16 in. (400 mm) and the vertical distance from this area to the closest point must be 3 ft 6 in. (1050 mm) when the elevator is at its upper extreme of travel. With large high-speed elevators there is usually sufficient space on the car top and overhead space is generally ample to provide the required refuge space. Refuge space usually requires additional overhead for slower-speed elevators especially with hydraulic elevators where there is little need for extensive structure on top of the elevator car.

Machine Room Space

The preferred usual location for traction elevator machinery is directly above the elevator hoistway itself. For traction elevator applications this location can also be below, at the side or at the rear (front will interfere with doors) adjacent to the hoistway. The machine must be sufficiently below so that the lead from the driving machine sheave to the sheaves in the overhead does not cause sidewise stray of the ropes. The machine room space for hydraulic elevators may be remote, with piping carrying the oil to and from the elevator to the pumping unit. Machine below arrangements require special study and additional discussion may be found later in this chapter.

Equipment in the machine room varies with the load the elevator must carry and its speed. The following is usually included: the hoisting machine and the electric elevator controller; a governor for safety application; a motor generator with any elevator of the generator field control type; a floor-selecting device on the larger elevators or those that serve many stops; and, for a group of three or more elevators, a group electrical controller.

Some elevator manufacturers may mount the group controller as part of each elevator controller. With the office-sized groups of six or eight elevators an extra control cabinet or two is usually required for efficient group control.

In addition to elevator equipment (which, by most elevator safety codes is the only equipment allowed in the elevator machine room) a main power disconnect is required for each elevator. Lighting and ventilation are necessary. The ventilation must be capable of removing the heat released by the elevator equipment to maintain a maximum temperature in the motor room of 104°F (40°C)* Heating should be provided in colder climates to keep the machine room no less than 50°F (10°C).

Reasonable access to and from the machine room should be provided. Mechanics must maintain the equipment and occasionally remove and replace

---

* For some solid state controls, 95°F (35°C) is required.

parts. Trap doors and trolley beams provide for the contingency in gearless installations. The repairman's ingenuity must often suffice in the smaller, single-car installations, but prior consideration to the problem will save the owner time and shutdowns. Space should be provided around each piece of electrical equipment, and requirements have been established by the National Electrical Code, C1 or by local codes. This usually consists of from 36 to 48 in. (900 to 1200 mm) in front of electrical controllers depending upon their location in relation to walls or other equipment. Space to remove the armature of a motor should be provided in front of each hoisting motor or generator (Figure 8.7).

Secondary Levels

All the foregoing items must be considered in the allocation of space for the machine room. On the average, the plan of the motor room for each elevator usually requires the space necessary for the elevator hoistway plus that space repeated in front of the hoistway. In the larger installations, elevator equipment may advantageously be placed on two levels.

With smaller or slower installations all equipment is usually located on one level as shown in Figure 8.7 which reflects clearances as required by the National Electrical Code C1, 1980 edition. The latest edition or local codes should be used as a reference. With the larger installation, a secondary sheave located below the driving machine is often necessary, and because there must be access to this sheave, a submachine room or secondary level is created. This level can be half a floor below the elevator machine room floor and be provided with suitable access for personnel (Figure 8.8). The lobby space between elevators at

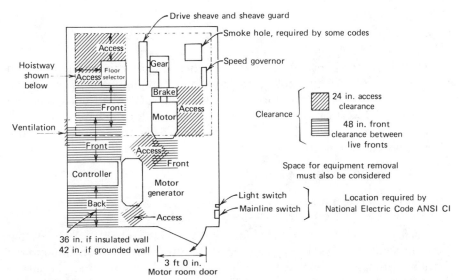

Figure 8.7. National Electric Code (ANSI C1) clearance requirements.

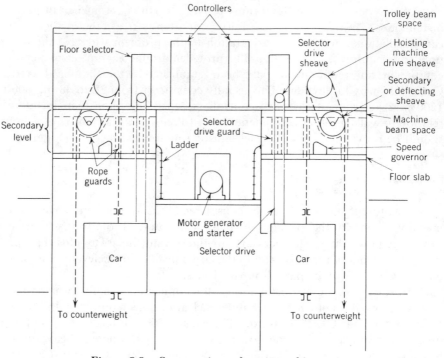

Figure 8.8.   Cross section—elevator machine room.

the floor below the elevator machine room can often be used for additional elevator equipment.

Venting

Elevator hoistways usually extend to the highest point in the building and could become filled with smoke as a result of a building fire. Many building codes require venting of the hoistway to the atmosphere. An open vent can cause wind and weather problems; therefore, most codes allow dampers which will open automatically in the event a smoke detector in the hoistway operates. Local requirements and applicable building codes must be consulted for specific requirements.

ELEVATOR REACTIONS

Concurrent with the space allocations in the machine room are the loads that the equipment imposes. The primary load for a traction elevator is due to the elevator-hauling load which is transmitted through the driving machine sheave shaft to the elevator machine beams. These beams are installed by the elevator contractor and the structure must be designed to provide for the loads imposed. Two or three beams are placed across the hoistway and the reactions at the end or support point of each beam are calculated and indicated on the elevator layout. This load includes the suspended load and an allowance for impact on the structure which is usually double the suspended load. In addition the weight and location of each piece of elevator equipment is shown on the layout and the floor or supporting structure must be designed to accommodate these loads. Connecting conduits for electrical wiring can be buried in the floor or the wiring installed in overhead troughs. Trolley beams for moving equipment should be provided. Loading and the location will vary with each installation and are functions of elevator capacity, roping arrangement, and speed.

Typical Layouts and Manufacturers' Standards

Most of the larger elevator manufacturers have preprepared layout information for a wide variety of elevator applications. These layouts are available on request and will often represent that particular manufacturer's best arrangement for the particular elevator size, capacity, and speed. There may be variations among manufacturers for the same situation. One may use a machine with 2:1 roping whereas another may use 1:1 roping for the same load and speed. If competitive bidding is expected, it is necessary to ensure that any bidder's equipment will fit and the space allocated should suit the greatest requirements.

Many typical layouts can be varied in some dimensions and the manufacturer's representative's advice should be sought. With some typical layouts which may be based on a particular manufacturer's "model" elevator, no variation is permitted since all parts have been pre-engineered and are "stock."

In addition to dimensional and structural loading requirements, the typical layout should include information regarding electrical power requirements and the expected heat to be released by the operating elevator. Any additional items such as access ladders, lighting, power for intercommunication systems, life safety systems, or special requirements should also be provided with the layout data.

For the larger, more complex installation, a competent elevator consulting engineer should be engaged since numerous unique problems of space allocation and complex electrical requirements can arise. In the taller building, there are no standard column spacings and the available land area may not allow use of an acceptable standard module. The structure is often integrated into the elevator hoistways, the elevators may be of unique sizes because of space allocations, electrical requirements become complex when split feeders and standby power aspects are considered, and special building operating, security, and life safety requirements may involve unusual elevator interfaces.

Changing a typical layout to suit some particular situation should be attempted only with expert advice. As may be seen from the earlier discussion of space requirements, an assumed minor change in any dimension may have a chain effect and render the entire layout useless. In European countries many eleva-

DOVER SPF 25/20, 25/35, AND 25/40

NOTE:

Dimensional data shown here comply with the American National Standards Safety Code A17.1 State or local codes must be used if they vary from the national code. Contact your local Dover Elevator representative.

CODE REQUIREMENT	STATE-LOCAL
CLEAR OVERHEAD	
PIT DEPTH	

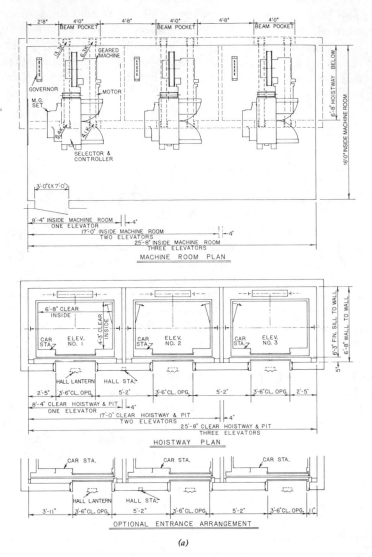

MACHINE ROOM PLAN

HOISTWAY PLAN

OPTIONAL ENTRANCE ARRANGEMENT

(a)

Figure 8.9. Sample forms of standard data as provided by various elevator manufacturers: (a) Dover Elevator Company;

170

WORK NOT INCLUDED IN ELEVATOR CONTRACT*

The following preparatory work is required in order to properly install the elevator equipment. The cost of this work is not included in the elevator proposal, since it is a part of the building construction.

1. A plumb hoistway and pit including lighting, electrical work outlets, and pit ladder. Structure to be adequate to support loads imposed by elevator equipment.
2. Complete electrical service connected to terminals of elevator control.
3. An enclosed elevator equipment room with electrical work outlets, adequate lighting, heating and ventilation. (Minimum machine room temperature 50° F., maximum 100° F.)
4. An emergency power supply source, including necessary transfer switches and auxiliary contacts where elevator operation from an alternate power supply is required.
5. Heat and smoke sensing devices connected to elevator machine room terminals when such devices are required.
6. Temporary barricades around elevator work area during installation.
7. Masonry cutouts and grouting and sill supports where required.
8. Adequate storage facilities for elevator equipment prior to and during installation.
9. Temporary electrical power of same characteristics as permanent supply if required during installation.

*Refer to elevator layout drawings for details of each requirement.

(a)

Figure 8.9. (a) (continued);

medium and low speed traction

Medium and Low Speed Traction Elevators perform efficiently and economically when serving traffic demands in medium and low rise buildings.

HANDICAPPED REQUIREMENTS AVAILABLE TO MEET NEII STANDARDS OR LOCAL CODES

National Elevator Industry, Inc. (NEII) Handicapped Standards are:
Placement of car controls, hall buttons and phone (or intercommunication equipment) for easy access.
Tactile markings for operating switches, buttons and hoistway door jambs.
Handrails in car — dual ray door protection — audible signals in car position indicator and lanterns.

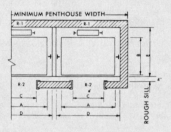

PLAN FOR ONE OR MORE ELEVATORS
CENTER OPENING DOORS SHOWN —
SINGLE SLIDE DOORS OPTIONAL

RECOMMENDED SIZES AND CAPACITIES

TYPE BUILDING	SMALL APARTMENT	SMALL OFFICE	AVERAGE OFFICE HOTEL		LARGE OFFICE OR STORE
CAPACITY	1500#	2000#	2500#	3000#	3500#
A	4'-10''	6'-0''	7'-0''	7'-0''	7'-0''
B	5'-0''	5'-0''	5'-0''	5'-6''	6'-2''
C	2'-8''	3'-0''	3'-6''	3'-6''	3'-6''
D	6'-2''	7'-4''	8'-4''	8'-4''	8'-4''
E	6'-6''	6'-6''	6'-6''	7'-0''	7'-8''

MINIMUM PIT—OVERHEAD & MACHINE ROOM DIMENSIONS

SPEED	100	200	250	300	350	400
L	16'-0''	16'-0''	16'-0''	17'-0''	17'-0''	21'-6''
O	15'-6''	15'-9''	16'-2''	16'-4''	16'-6''	17'-7''
*P (a)	4'-0''					
*P (b)	-	4'-0''	4'-6''	4'-6''	5'-1''	5'-7''
*P (c)	-	5'-8''	6'-0''	6'-6''	6'-11''	7'-4''

*P (a) indicates minimum pit required for elevators with type "A" safety.
*P (b) indicates minimum pit required for elevators with type "B" safety.
*P (c) indicates minimum pit required for elevators with type "C" safety.

NOTES:
1. Reactions include allowances for impact but DO NOT include weight of concrete slab.
2. Pit depths, overhead clearance and penthouse sizes are in accordance with ANSI code requirements. Local codes may vary these requirements.
3. Layouts and dimensions shown are for center opening type entrances.
4. Consult your Montgomery Representative for specific recommendations where space is limited or other conditions may necessitate further study.
5. All data is general. Consult your local Montgomery Representative for exact information for your working drawings.

OVERHEAD LOADS/LBS. APPROXIMATE PER ELEVATOR

CAPACITY	SPEED FPM	R-1	R-2
1500#	100	12000	7500
2000#	100	12500	8800
	200	15200	9900
	250	15500	10800
	300	15800	11000
	350	19800	12000
	400	24000	14500
2500#	100	14900	10300
	200	16700	11500
	250	17200	12300
	300	17500	12500
	350	20400	12800
	400	25000	15000
3000#	100	17100	12100
	200	19400	12200
	250	19800	12600
	300	20200	13200
	350	20400	13300
	400	26500	16000
3500#	100	18300	13300
	200	21000	14100
	250	21300	14400
	300	21800	14700
	350	25200	15100
	400	28000	16800

(b)

Figure 8.9. *(continued)* *(b)* **Mongomery Elevator Company;**

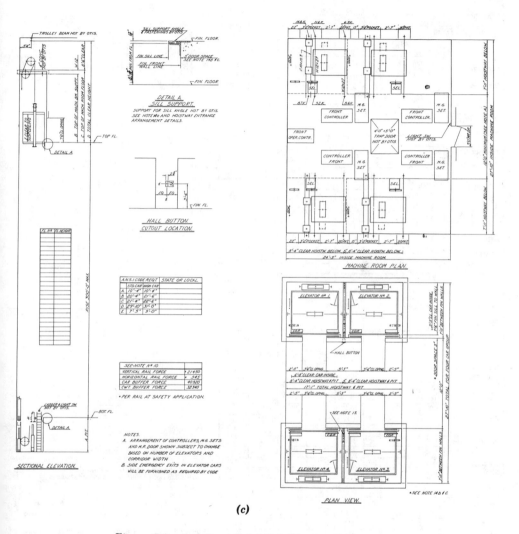

(c)

Figure 8.9. (*continued*) (*c*) Otis Elevator Company;

173

MEDIUM and HIGH RISE
passenger elevators —
geared and gearless
from 2000 to 4000 pound capacity

Geared

			2000¹	2000 ♿	2500¹	2500 ♿	3000 ♿	3500 ♿
Capacities	Pounds		2000¹	2000	2500¹	2500	3000	3500
	Passengers		13	13	16	16	20	23
Rated Speeds Ft./Min.			200 or 350					
Hoistway Dimensions	Clear Hoistway 1 Car Width	W_1	7'8''	7'4''	8'4''	8'4''	8'4''	8'4''
	Clear Hoistway 2 Car Width	W_2	15'8''	15'0''	17'0''	17'0''	17'0''	17'0''
	Clear Hoistway Depth	D_1	5'7''	6'4''	6'2''	6'4''	6'8''	7'4''
	Wall to Wall Depth	D_2	6'0''	6'9''	6'7''	6'9''	7'1''	7'9''
Minimum Overhead			13'11''	13'11''	13'11''	13'11''	13'11''	15'6''
Minimum Pit			5'0''					
Recommended Entrance Sizes²	Width		3'0''	3'0''	3'6''	3'6''	3'6''	3'6''
	Height		7'0''	7'0''	7'0''	7'0''	7'0''	7'0''
Platform Sizes	Width	PW	6'4''	6'0''	7'0''	7'0''	7'0''	7'0''
	Depth	PD	4'5''	5'2''	5'0''	5'2''	5'6''	6'2''
Machine Room Clear Height			8'6''					
Machine Room Dimensions	1 Car Width	W_3	7'8''	7'4''	8'4''	8'4''	8'4''	8'4''
	2 Car Width	W_4	15'8''	15'0''	17'0''	17'0''	17'0''	17'0''
	Depth	D_3	12'6''	13'2''	11'6''	12'3''	12'8''	13'6''

Gearless

			2500¹ ♿	2500 ♿	3000 ♿	3500 ♿	4000
Capacities	Pounds		2500¹	2500	3000	3500	4000
	Passengers		17	17	20	23	27
Rated Speeds Ft./Min.			500, 700, 800 or 1000				
Hoistway Dimensions	Clear Hoistway 1 Car Width³	W_1	8'4''	8'4''	8'4''	8'4''	9'4''
	Clear Hoistway 2 Car Width³	W_2	17'0''	17'0''	17'0''	17'0''	19'0''
	Clear Hoistway Depth⁴	D_1	6'5½''	6'7½''	6'10''	7'6''	7'6''
	Wall to Wall Depth⁴	D_2	6'1½''	7'0½''	7'3''	7'11''	7'11''
Minimum Overhead⁵	500 fpm		15'11''	15'11''	16'8''	16'8''	16'5''
	700 fpm		17'9''	17'9''	17'9''	16'1''	16'1''
	800 fpm		16'10''	16'10''	16'10''	16'10''	16'11''
	1000 fpm		19'10''	19'10''	19'10''	20'0''	20'0''
Minimum Pit	500 fpm		8'9''	8'9''	8'9''	8'9''	8'9''
	700 fpm		10'4''	10'4''	10'4''	10'4''	10'4''
	800 fpm		11'11''	11'11''	11'11''	11'11''	11'11''
	1000 fpm		16'4''	16'4''	16'4''	16'4''	16'4''
Recommended Entrance Sizes²	Width		3'6''	3'6''	3'6''	3'6''	4'0''
	Height		7'0''	7'0''	7'0''	7'0''	7'0''
Platform Sizes	Width	PW	7'0''	7'0''	7'0''	7'0''	8'0''
	Depth	PD	5'0''	5'2''	5'6''	6'2''	6'2''
Machine Room Clear Height			500 fpm - 8'0'' / 700 fpm - 9'0''		800 fpm - 9'6'' / 1000 fpm - 10'6''		
Machine Room Dimensions	2 Car Width	W_3	17'0''	17'0''	17'0''	17'0''	19'0''
	3 Car Width	W_4	25'8''	25'8''	25'8''	25'8''	28'8''
	Depth	D_3	14'3''	14'3''	14'7''	15'3''	15'3''

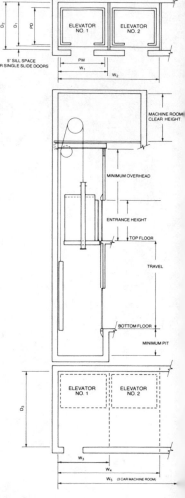

Notes:

1. These elevators do not meet the National Elevator Industry recommendations for handicap requirements.
2. Entrances are available in single slide center opening, or side slide right or left hand configurations.
3. For speeds of 1000 and 1200 fpm increase this dimension by 2'' and 4'', respectively.
4. For speeds over 700 fpm increase this dimension by 2''.
5. If a secondary level is desired add 5'0'' to overhead dimensions.
6. Geared traction elevators are normally used at 350 fpm and below, Gearless above 350 fpm.
7. It is recommended you contact your local Schindler Haughton Representative for speeds above 800 fpm.
8. For preparatory work by other than elevator contractor see page 15.

(d)

Figure 8.9. *(continued)* *(d)* Schindler-Haughton Elevator Corp.;

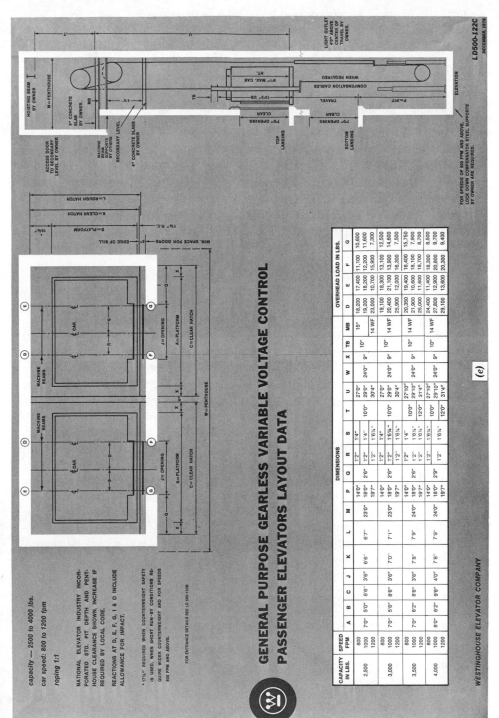

Figure 8.9. *(continued)* *(e)* Westinghouse Elevator Company.

175

tors are completely standardized and sizes established by the International Standards Organization (ISO) so that any deviation may increase the cost of the installation. In the United States if a manufacturer's "model" is intended to be used such changes can impose a substantial price penalty.

Sample typical layouts are shown on in Figure 8.9*a*, *b*, *c*, *d*, and *e*. Note that all the necessary dimensions are provided with a minimum of detail.

ADDITIONAL MACHINE ROOM CONSIDERATIONS

Power and ventilation requirements were mentioned in the discussion of machine room requirements. These items concern the efficient operation of the elevators and particular attention should be given to their provisions.

Energy—Absorbed and Regenerated

Electrical energy is primarily required to start and run an elevator (traction-type) and only partially to lift the load. Except for energy consumed in opening and closing the doors, much of the energy a traction elevator requires to lift loads is returned to the line when such load is lowered. Stated otherwise, the elevators require energy to lift the uncounterbalanced portion of the load when people come to work and those same people, as they leave, theoretically cause the elevator to pump back an equal amount of energy. In practice, friction, impedance, and other factors introduce losses and some energy will always be used by an elevator in excess of energy returned. As an approximate guide, regeneration caused by a fully loaded elevator traveling down is about 75% of the energy required to lift that same load.

Efficiency of traction elevator machinery varies from 50 to 70% for the geared-type elevator with roller guides and from 75 to 85% for gearless elevators. Energy losses are generally changed to heat and must be dissipated by the building ventilation system.

The power required by any fully or partly loaded elevator may be readily derived from the following formula:

Horsepower (hp) =

$$\frac{\text{load in car (lb)} \times \text{percentage of load that is unbalanced} \times \text{velocity (fpm)}}{33,000 \times \text{efficiency of the hoisting machine}}$$

Example: Traction elevator, 2500 lb, 500 fpm, gearless machine, 40% counterbalance.

$$\text{hp} = \frac{2500 \times 0.60 \times 500}{33,000 \times 0.80} = 28 \text{ hp approximate}$$

The energy consumption of a group of elevators operating over a period of time requires a number of calculations, including determination of full-, partial-, and no-load trips, direction, and number of car stops. The task for a particular large building requires the help of a special computer program.

The energy required by an automatic elevator varies with its size, speed, and usage. It can be easily determined by a recording watt-hour meter after the building is completed and has a mature population.

Estimates can be based on known examples. If we take a typical office building as an example and either measure or calculate the energy requirements during the course of a 10-hr day, the following results can be expected.

Example 8.2

Given: low-rise 6 elevators, 3500 lb (1600 kg) at 500 fpm (2.5 mps) serving floors 1 to 10

Average energy required during up-peak traffic	48 kW
Average energy required during heavy two-way traffic	36 kW
Average energy required during light two-way traffic	30 kW
Average energy required during down-peak traffic	30 kW

Assume the following distribution of traffic during a 10-hr day.

Up-peak	1 hr	× 48 kW	=	48 kWh
Heavy two-way	4 hr	× 36 kW	=	144 kWh
Light two-way	4.5 hr	× 30 kW	=	135 kWh
Down-peak	0.5 hr	× 30 kW	=	15 kWh
				342 kWh

Total energy consumption for the day is 342 kWh which divided by 6 is 57 kWh per elevator. At 6 cents a kWh, this amounts to about $21.00 for the 6 elevators per day, or for a 250 working-day year, $5130.00 per year.

Example 8.2 was based on elevators with motor generator sets. If solid-state motor drives are used, the motor generator idling losses are eliminated and the energy requirements are reduced by about 30 to 35%. Therefore, the total energy per elevator would be about 40 kWh per day. The annual savings for the 6 elevators would be $1750.

With higher rises and higher speeds, values as high as 100 kWh per day are possible. As a general rule, elevators in commercial and institutional buildings require from 50 to 100 kWh per day depending upon load, speed, and rise, and elevators in residential buildings from 10 to 25 kWh per day for lower rise (up to about 20 floors) and 25 to 50 kWh per day for higher buildings.

When compared to heating, air conditioning, and lighting, the elevator requires a very small percentage of the total building energy needs, amounting to about 2 or 3% in a large office building.

Table. 8.2. Starting and Running Current; Typical Generator Field Control Elevators @ 460 V, 3φ, 60 Hz

Geared	2500 lb @ 200 fpm 1200 kg @ 1.0 mps	3000 lb @ 300 fpm 1400 kg @ 1.5 mps	3500 lb @ 350 fpm 1600 kg @ 1.6 mps
M.G.[a] starting from rest	40 A	60 M	60 A
Elevator start full load up	45 A	70 A	85 A
Elevator run full load up	30 A	45 A	55 A

Gearless	2500 lb @ 500 fpm 1200 kg @ 2.5 mps	3000 lb @ 700 fpm 1400 kg @ 3.5 mps	3500 lb @ 1000 fpm 1600 kg @ 5.0 mps
M.G. starting from rest	75 A	120 A	250 A
Elevator start full load up	70 A	120 A	180 A
Elevator run full load up	45 A	70 A	110 A

[a] M.G. = Motor generator (values will vary with various manufacturers).
With solid-state motor drive, the M.G. start from rest current is eliminated and the other values remain approximately the same.

The demand an elevator system creates for power is of concern. The feeders that supply the elevator motor must be of sufficient size to serve that demand with a minimum of voltage drop. Elevator motor generators are driven by ac induction motors as are the pumps on hydraulic elevators of about 10 to 75 or more horsepower. When the motor starts from rest the greatest power demand is usually created. An equal or higher-power demand for a generator field control elevator or an elevator with an SCR motion controller may occur when the elevator itself is started up with a full load in the car. Normal running current is generally much lower than that used for either the starting of the motor generator or the starting of the elevator with full load up. Elevator power requirements are usually stated with three values: (a) motor generator starting from rest, (b) elevator starting full load up, and (c) elevator running full load up. The motor current for other conditions is lower.

With any induction motor, resistance starting (or reconnecting motor windings from Y to delta) is a means of reducing the peak starting current of the motor. There is a mathematical probability that all the elevators will not start up with a full load at one time. This is called a diversity factor and can be calculated for any group of two or more cars. Typical values run from 0.87 for two cars to 0.75 for six cars, the value indicating the probable percentage of the total current required to operate all the elevators in the group at one time.

A set of typical values of starting and running current values appears in Table 8.2. The building designer usually is required to confirm with the local electric company the voltage characteristics and availability of power before the elevator manufacturer undertakes the production of equipment.

Electric current requirements for ac resistance traction and hydraulic elevators are easier to calculate. The elevator is either running or stopped, unlike generator field control where a generator is running whether or not the elevator

Table. 8.3. Starting and Running Currents Typical ac Traction Machine (Counterweighted) and Hydraulic Elevators (Uncounterweighted), 460 V, 3φ, 60 Hz (Amperes Starting/Amperes Running)

	2500 lb @ 100 fpm 1200 kg @ 0.5 mps	2500 lb @ 125 fpm 1200 kg @ 0.6 mps	2500 lb @ 150 fpm 1200 kg @ 0.75 mps
ac traction across the line	25/15	35/20	45/25
Hydraulic across the line	175/30	200/35	275/65

	3500 lb @ 100 fpm 1600 kg @ 0.5 mps	3500 lb @ 125 fpm 1600 kg @ 0.6 mps	3500 lb @ 150 fpm 1600 kg @ 0.75 mps
ac traction across the line	50/25	65/30	75/35
Hydraulic across the line	200/35	225/40	300/70

is. With ac traction or hydraulic elevators starting is either across the line or through a resistance step. The horsepower of the elevator motor establishes the electrical power required. Typical values are given in Table 8.3.

Ventilation

The heat released by ac resistance traction or hydraulic elevators is relatively easy to estimate. Most of the energy supplied to the motor will be changed to heat through braking when stops are made, or with a hydraulic elevator, cause the oil to heat. The elevator runs up less than half the time whereas a substantial portion of time is spent stopped. Since most ac resistance traction or hydraulic elevators are found in low-rise buildings, it is fair to estimate a maximum of 20% of time spent will be spent running full load up, the worst condition. Converting this percentage of the elevator motor horsepower to Btu gives a good approximation of the heating expected in the elevator machine room. (One horsepower per hour equals 2544 Btu.) This heat must be dissipated for efficient elevator operation.

Dissipation of machine room heat is important in any elevator installation. If the machine room gets too hot, electrical insulation deteriorates, oil loses its viscosity, and erratic operation such as poor leveling, abrupt starts, and poor brake action can be experienced.

Heat generated by the elevator may be considerable during certain periods of the day. For example, during the morning in-rush in an office building, elevators are expected to be leaving the lobby with full loads, so that energy demands will be at their maximum for a half-hour or so. At that time a typical office building elevator may generate from 25,000 to 40,000 Btu/hr.

Adequate ventilation is required in any elevator machine room to dissipate the heat and provide a temperature of no more than 104° F (40° C)* or no less than 50° F (10° C) and may require cooling or heating. In the smaller, single-car installations a thermostatically controlled exhaust fan may accomplish this. In the larger installations the air conditioning system should be designed to maintain these temperature limits. Table 8.4 shows a sample of typical heat generated by elevator equipment. If the generator is remote from the elevator drive motor, one-half of the heat emanates from the generator or SCR controller and one-half from the hoist motor.

An estimate of machine room heating should be given by the elevator equipment manufacturer. This estimate can be calculated from the number of floors an elevator serves, its rise, the speed, and the duty load, plus information as to its expected use. The percentage of full-load running time is then calculated from probable stop data and the heating losses expressed in Btu or joules. One Btu = 1055 joules.

Table 8.4 gives some typical values for expected heat release.

---

* 95° F (35° C) for solid state controllers.

Table. 8.4.   Sample Heat Release Parameters; Typical Gearless Installations Under Busy Conditions (mj = mega-joule)

1.	2000 lb	@	500 fpm	30-story apartment (250-ft rise)
	900 kg	@	2.5 mps	18,000 Btu/hr per elevator (19 mj)
2.	3000 lb	@	500	10-story office building (100-ft rise)
	1400 kg		2.5 mps	22,000 Btu/hr per elevator (23 mj)
3.	3500 lb	@	700	Serving floors 1, 10 to 20 in an office building
	1600 kg		3.5 mps	35,000 Btu/hr per elevator (37 mj) (240-ft rise)
4.	4000 lb	@	500	10-story hospital (100-ft rise)
	1800 kg		2.5 mps	25,000 Btu/hr per elevator (26 mj)
5.	3500 lb	@	1000 fpm	Serving floors 1, 30 to 45 in an office building
	1600 kg		5.0 mps	40,000 Btu/hr per elevator (42 mj) (550-ft rise)
6.	6500 lb	@	400 fpm	Service elevator in an office building serving floors 1 to 20 (240-ft rise)
	3000 kg		2.0 mps	37,000 Btu/hr per elevator (39 mj)

Emergency Operation

In all hospitals and in many other buildings it is desirable to provide a standby power supply to operate the elevators. In all buildings it is necessary to have a separate power source for the car lights and fan to assure passengers if for any reason car lighting fails. This should be a constantly charging battery pack located on the car top so it can be serviced and energize car lights when needed. A battery light is necessary since standby generators require time to start. Car lighting depends upon traveling cables to the car which may fail. When elevators are essential such as in a hospital, they should be capable of operation during a power failure.

The amount of standby power available determines how elevators can be operated. If the supply is limited the consulting engineer can specify that the elevator equipment manufacturer arrange the installation so that only one elevator operates at one time automatically to evacuate passengers, and then a choice can be made manually by a selector switch for one elevator to remain in service. The electrical contractor provides the wiring and switchover to provide power at the machine room. The elevator manufacturer provides interlocking for each elevator controller to ensure that only one elevator is operated at one time and the selecting switch to determine which elevator will operate after all are evacuated.

The building's standby generator system should supply sufficient power to operate any one elevator (or more if the standby plant is large enough) as well as sufficient provisions to absorb the current regenerated by the elevator if it should travel down with a full load. This absorption of regenerated power is essential to prevent overspeeding of the elevator. The value of absorption needed is given by the elevator equipment manufacturer and must be designed into the standby power system.

Hydraulic elevators have a manual lowering valve that can be used to move a car to a next lower floor or they can be arranged so that the standby power source or a battery pack can be used to cause them to travel to the lowest landing and have the doors open. Smaller geared elevators can be cranked to a landing. Any elevator that stops for other than normal reasons is potentially dangerous. No attempt to move the car or remove the passengers in that car should be made without adequate precautions. The passengers' condition should be ascertained and they should be advised of imminent movement. Intercommunication systems connected to a standby power source are a necessity.

ELEVATOR LAYOUTS

Attention has been given to establishing space, ventilation, and power requirements of passenger overhead traction elevators in particular. Layouts for hydraulic elevators, elevators with basement machines, freight elevators, and other types will follow, in general, the steps that have been outlined. Based on the car size, the hoistway plan is established, the pit and overhead determined, the machine room laid out, supports for equipment and impacts determined, and power and ventilation established. For other than overhead traction elevators the additional following considerations must be made.

Hydraulic Elevators

Hydraulic elevators require a hoistway space for only the car since they usually do not have a counterweight. Sufficient space must be provided for the pump, control equipment, and piping to and from the elevator shaft from a machine room which may be remote. Adequate pit space for the plunger and cylinder supports are necessary and impacts on buffers must be considered. In areas where there is groundwater, consideration must be given to tying down the cylinder and well casing lest it float up as well as protecting it against

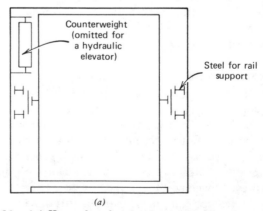

(a)

Figure 8.10. (*a*) Heavy duty freight elevator—schematic;

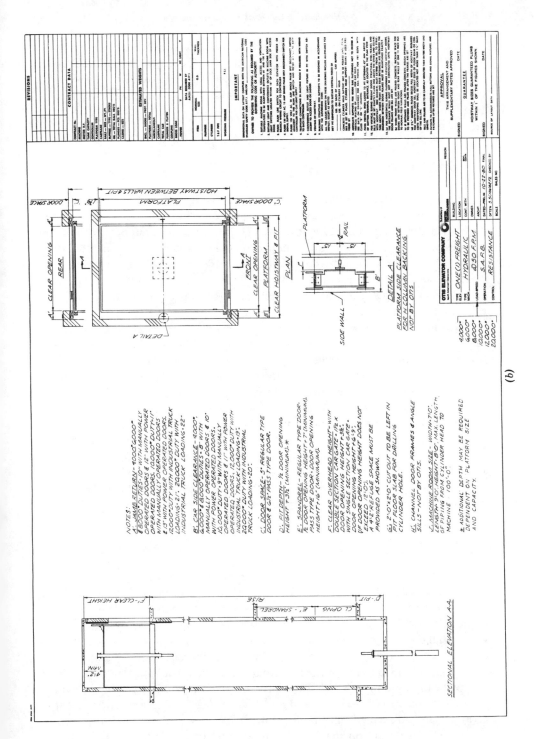

Figure 8.10. (b) Actual layout (Courtesy Otis Elevator Company).

183

corrosion by electrolytic action. In the elevator hoistway adequate run-by space above the top landing as well as sufficient support for fastening rail brackets at established distances (usually at floor levels) must be provided. The vertical loads on the building are minimum which makes the hydraulic elevator attractive for low-rise structures, and, especially, for heavy duty freight and truck elevators.

Freight Elevators

For both electric and hydraulic heavy duty freight elevators, adequate rail supports must be provided, usually at shorter distances than at floor levels depending upon the expected loading on the elevator. These supports can be the building steel at each floor and at intermediate levels, or a solid concrete hoistway into which the elevator contractor can anchor supporting brackets. For elevators loaded by trucks or carrying trucks, fabricated steel supports may be erected at each side of the hoistway (Figure 8.10) to provide for the additional rail brackets required.

The supports for elevator guides and brackets must be designed to absorb the impact of the truck as it loads on the elevator. The impacts consist of the braking load on the truck wheels, the twisting motion of unbalanced loading, and opposing forces on the top and bottom elevator guide shoes transmitted to the rails. Figure 8.11 shows the directions some of these forces will take.

If an elevator is to be loaded by an industrial truck it must be designed to hold both its capacity load and some of the weight of the truck. Depending on the type of load, either all or part of the truck may be on the elevator as the last portion of the load is deposited. The elevator machine (or hydraulic cylinder) brake, car frame, and platform must all be designed to withstand this extra "static" load. The elevator need not lift this load but must be able to level, that is, move the platform level with the floor with the extra load on board.

Basement Traction Machines

It is often desirable to install an elevator without a rooftop penthouse. The hydraulic elevator does not need a penthouse and is limited in speed to about 150 fpm (0.75 mps) and rise to about 70 ft (21 m). Hydraulic elevators have higher electrical horsepower requirements than counterweighted electrical elevators and the high power demand associated with multiple hydraulic elevators may suggest electric traction as the preferred equipment. Installing an electric traction machine in the basement or adjacent to the hoistway at an upper floor below the top landing served has the advantage of the counterweighted elevator with its lower horsepower demand and higher speeds. These traction elevator arrangements are called "machine below."

This class of traction machine below can be of two types, underslung or overslung (also called "direct pickup"). The underslung arrangement consists

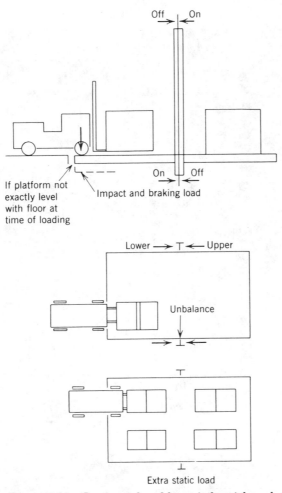

Off | On

On | Off

If platform not
exactly level
with floor at
time of loading

Impact and braking load

Lower → T ← Upper

Unbalance

T

Extra static load

Figure 8.11. Loads produced by an industrial truck.

of lifting sheaves located under the car platform and requires somewhat more pit depth and hoistway space than the overslung (Figure 8.12). The overslung is lifted at the crosshead like any conventional traction elevator (Figure 8.13). Either arrangement requires overhead space for rope sheaves and a machine room located either to the rear or side of the hoistway. With more than one elevator in a line, the rear location is preferred, and with three or more elevators in a line, provides the best layout (Figure 8.14).

With any machine-below arrangement, up-thrust forces on the machine are equal to the entire weight of the elevator and counterweight plus impact allowances which require sufficient tiedown, in the form of anchors, to the building steel or a concrete foundation block.

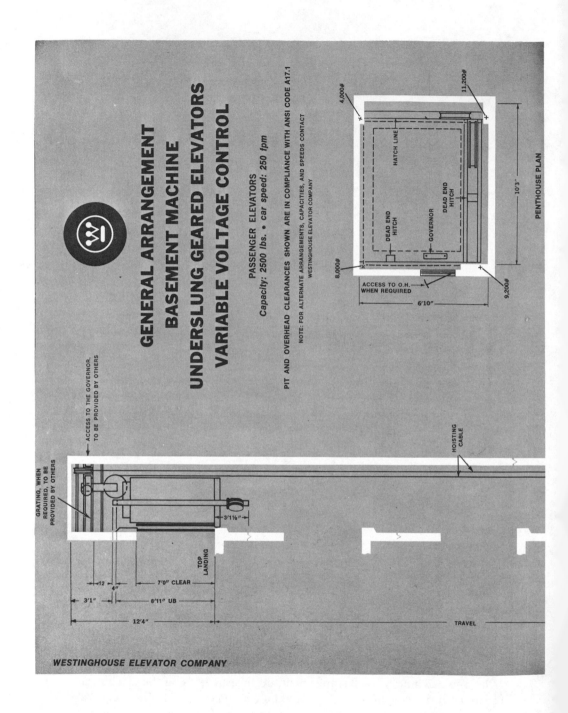

GENERAL ARRANGEMENT
BASEMENT MACHINE
UNDERSLUNG GEARED ELEVATORS
VARIABLE VOLTAGE CONTROL

PASSENGER ELEVATORS
Capacity: 2500 lbs. • car speed: 250 fpm

PIT AND OVERHEAD CLEARANCES SHOWN ARE IN COMPLIANCE WITH ANSI CODE A17.1

NOTE: FOR ALTERNATE ARRANGEMENTS, CAPACITIES, AND SPEEDS CONTACT
WESTINGHOUSE ELEVATOR COMPANY

PENTHOUSE PLAN

HATCH LINE

DEAD END HITCH

DEAD END HITCH

DEAD END HITCH

GOVERNOR

ACCESS TO O.H.
WHEN REQUIRED

4,000#

11,200#

8,000#

9,200#

10'3"

6'10"

HOISTING CABLE

ACCESS TO THE GOVERNOR,
TO BE PROVIDED BY OTHERS

GRATING, WHEN
REQUIRED, TO BE
PROVIDED BY OTHERS

TOP LANDING

3'1½"

7'0" CLEAR

8'11" UB

12

4"

3'1"

12'4"

TRAVEL

WESTINGHOUSE ELEVATOR COMPANY

186

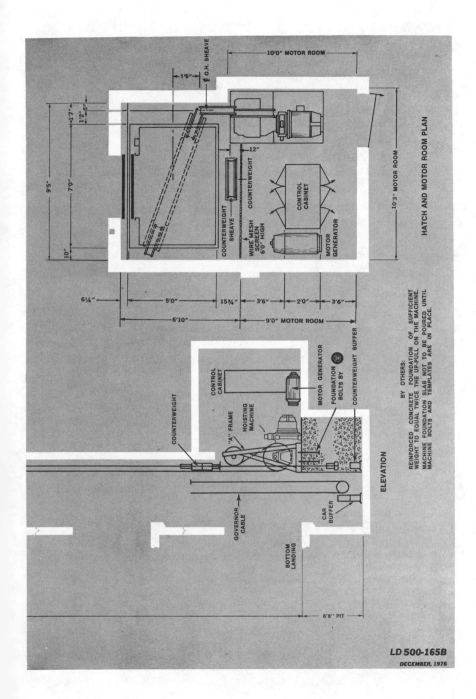

HATCH AND MOTOR ROOM PLAN

ELEVATION

BY OTHERS:

REINFORCED CONCRETE FOUNDATION OF SUFFICIENT
WEIGHT TO EQUAL TWICE THE UP-PULL ON THE MACHINE.
MACHINE FOUNDATION SLAB NOT TO BE POURED UNTIL
MACHINE BOLTS AND TEMPLATES ARE IN PLACE.

LD 500-165B
DECEMBER, 1976

Figure 8.12. Underslung elevator (Courtesy Westinghouse Elevator Company).

187

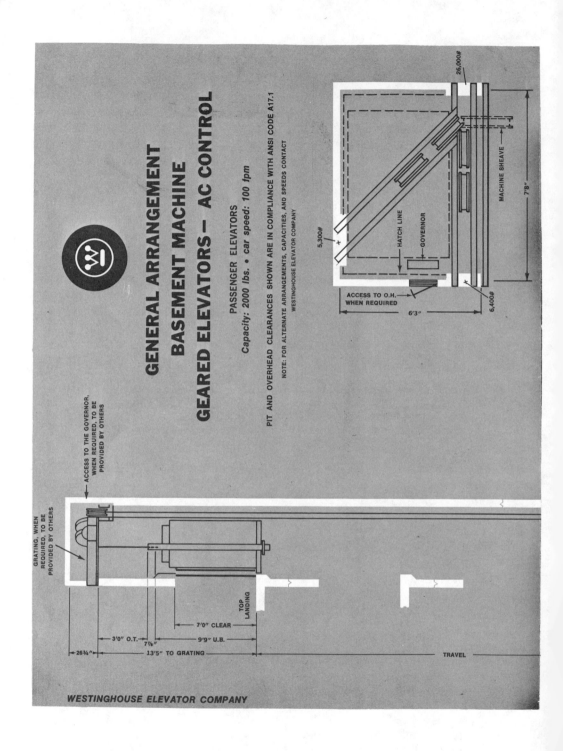

GENERAL ARRANGEMENT
BASEMENT MACHINE
GEARED ELEVATORS — AC CONTROL

PASSENGER ELEVATORS
Capacity: 2000 lbs. • car speed: 100 fpm

PIT AND OVERHEAD CLEARANCES SHOWN ARE IN COMPLIANCE WITH ANSI CODE A17.1

NOTE: FOR ALTERNATE ARRANGEMENTS, CAPACITIES, AND SPEEDS CONTACT
WESTINGHOUSE ELEVATOR COMPANY

26,000#

5,300#

MACHINE SHEAVE

7'8"

HATCH LINE

GOVERNOR

ACCESS TO O.H.
WHEN REQUIRED

6,400#

6'3"

ACCESS TO THE GOVERNOR,
WHEN REQUIRED, TO BE
PROVIDED BY OTHERS

GRATING, WHEN
REQUIRED, TO BE
PROVIDED BY OTHERS

TOP
LANDING

7'0" CLEAR

3'0" O.T.

7⅛"

9'9" U.B.

26¾"

13'5" TO GRATING

TRAVEL

WESTINGHOUSE ELEVATOR COMPANY

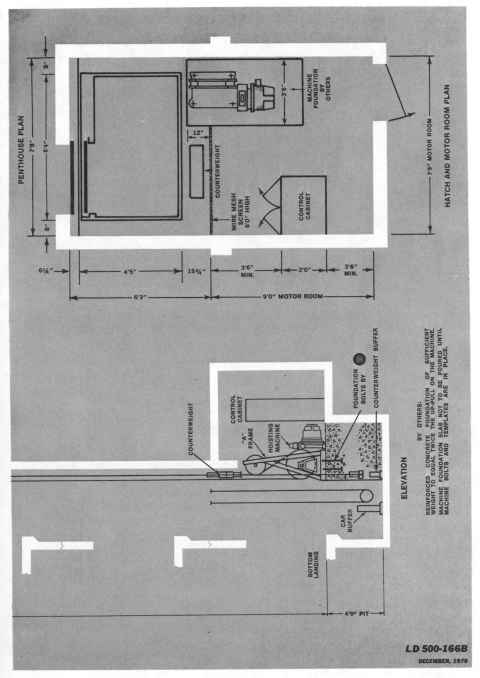

Figure 8.13. Overslung elevator (Courtesy Westinghouse Elevator Company).

189

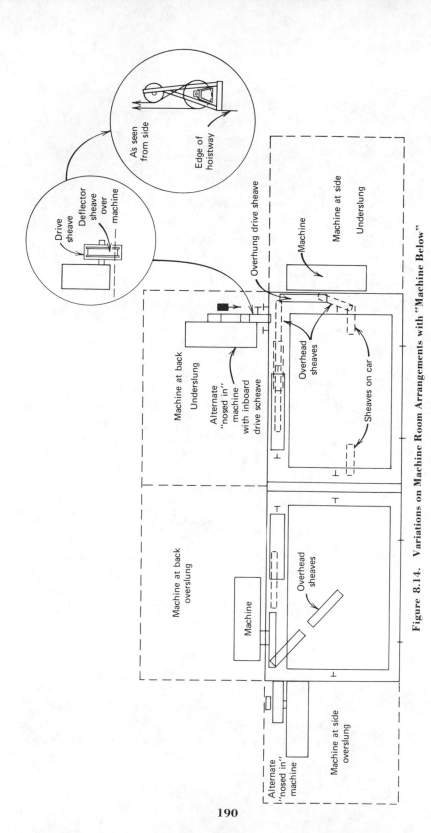

Figure 8.14. Variations on Machine Room Arrangements with "Machine Below"

Underslung or overslung elevators can be designed for a full range of lifting capacity and speed; however, speeds above the geared elevator range are extremely special and costly. Front and rear or front and side entrances on an elevator with a machine-below arrangement are extremely special. Layout space requirements for underslung elevators necessitate space of up to 14 in. (350 mm) on each side of the car for the sheaves under the car. With the overslung arrangement, space requirements are the same as for an overhead traction machine.

Dumbwaiters

Dumbwaiters are either traction or drum type. The size of a dumbwaiter is limited to 9 ft^2 (0.8 m^2) of platform area and a height of no more than 4 ft (1200 mm). Anything over that size must be classified as an elevator or a special lift and comply with codes governing elevators.

Dumbwaiters are operated from the landing and are equipped with doors and electrical locks or contacts to prevent their operation if a door is opened. Typical operations are: (a) call and send for a two-stop dumbwaiter; (b) multifloor buttons at one floor to send the dumbwaiter to any designated floor with return, call, or multifloor buttons at other than the main floor; (c) multifloor button stations at all floors for complete flexibility. Loading can be at counter height, undercounter, or arranged so carts can be rolled on at floor level. The latter requires automatic leveling and carts with wheels of sufficient diameter to bridge the running clearance or the installation of an automatic drawbridge.

Dumbwaiters are manual loading and unloading. Automatic loading and unloading can be provided in conjunction with power operated doors to create an automatic material handling system which will be detailed in a later chapter. Most dumbwaiters are completely standardized and can be arranged with front entrances only, front and rear entrances, or entrances at the front and side. Both power operated doors and manually operated doors are available.

SPECIAL LAYOUT REQUIREMENTS

Counterweight Safeties

Occupied space occasionally must be located under an elevator or groups of elevators as they serve the upper floors in a building or if a garage or other basement space is used under the elevators. In some cities railroad or subway tracks run underneath buildings.

In such cases safety demands provisions against the contingency of a falling elevator or counterweight. The elevator is protected against falling by the required car safety. The counterweight does not normally have a safety but one can be installed and is required if space below the elevator can be occupied. The counterweight rails are made heavy enough to withstand the safety application load and the elevator pit must be designed to withstand this possible impact.

The counterweight with a safety may require up to an additional 2 to 4 in. (50 to 100 mm) of hoistway depth and space for a counterweight governor in the elevator machine room.

Rear Openings

An elevator car with a rear opening in addition to the normal front entrance requires special layout consideration. The elevator counterweight should be located at the side rather than in the rear and the machine room must be designed to locate the machine so that the side counterweight may be accommodated. The common hoistway arrangement for front and rear openings is known in the elevator trade as "no. 4" construction from an early standard layout. This arrangement consists of a support beam for the car rails across the shaft, with counterweight space behind it (Figure 8.15). These provisions require hoistway space for front-and-rear entrances calculated similarly to conventional front-only entrances plus the additional counterweight space. Another use for no. 4 construction is to provide a deeper car if front-to-back space is limited and additional space is available at the side. Extreme care must be taken if two elevators having no. 4 construction are placed side by side. Machine and deflector sheave interference may occur necessitating that one machine be placed at a higher level than the adjacent machine.

Corner-post Arrangement

A variation to provide openings on an elevator at the front and at the side entrances is a corner-post arrangement. The car frame is specially built, the rails are placed in opposite corners of the hoistway, the counterweight is properly

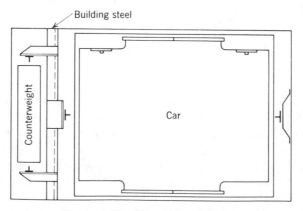

Figure 8.15.   "No. 4" construction.

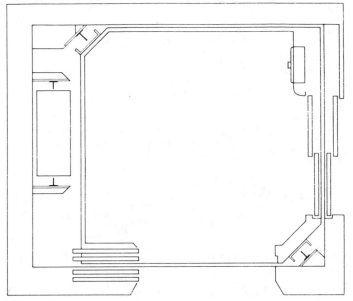

Figure 8.16. Corner-post arrangement (note necessity to use two-speed doors).

located, and the machine room is specially arranged. Space is usually only enough for two-speed doors on both the front and side openings (Figure 8.16).

A corner-post elevator is difficult to lay out and expensive to erect. It should be avoided and elevators with front and rear entrances provided if at all possible.

Special Arrangements

Many other possible elevator layouts include: observation-type elevators on the outside of buildings; wall climbers; explosion-proof elevators in chemical processing plants; shipboard elevators; special elevators to automatically load and unload rolls of paper or pallets; and elevators with revolving platforms. These all require some special considerations and early collaboration with elevator engineers. Similar problems may have been already solved and a specialist's knowledge is needed to aid in the new application. A later chapter on special installations will discuss many unique applications.

Escalators and Moving Walks

ESCALATORS VERSUS MOVING WALKS

In this chapter we emphasize the moving stairway and moving walkway types of vertical and horizontal transportation. Of the two, the escalator (or moving stairs) is the more important since it is used more frequently and has been in use for over 80 years. The moving walkway, either inclined or horizontal, was introduced in its modern form in the 1960s. Earlier versions date back to the Columbian Exposition in Chicago in 1893.

In the later part of 1970, developmental work was begun on an accelerating moving walkway. This device has an entering speed of about 100 fpm (0.5 mps) and will accelerate the pedestrian to about 1500 fpm (7.5 mps). At the end of the travel deceleration will take place and exit speed will be about 100 fpm (0.5 mps). Numerous measures need to be perfected, one of the most important being a handrail capable of matching the acceleration, speed, and position of the steps or pallets. When perfected, the accelerating moving walkway is intended to provide convenient horizontal transportation for distances between the 200 and 2000 ft (60 and 600 m) range. Beyond that distance, an automated moving vehicle seems to be the optimum approach. Such automated moving vehicles will be discussed in a later chapter. Although this chapter emphasizes escalators, the rules and suggestions for escalator application can be extended to include moving walkways.

The Importance of Escalators

There is no better way to guide people in a given path in a building than by providing an escalator. Department store owners discovered this years ago and the most successful stores have their escalators as centers of attraction. The most desirable space is located in line with or next to the escalators. Major expositions have used escalators to direct people to desirable sights and have used moving walkways to keep people moving past exhibits to gain maximum exposure (Figure 9.1).

Transportation terminals, subway stations, and other areas in which large groups of people are required to be moved from one level to another in surges are ideal applications for escalators and moving walks to speed circulation and avoid congestion. Everyone can be moved at a constant speed and people are carried efficiently from one place to another. When people are walking, some are slow, others are fast, some have baggage, others are accompanied by children, so that walking is often slowed to the speed of the slowest pedestrian. With a moving device, the velocity is established.

Escalators provide an effective means to make the second floor or basement space as attractive as street floor space. In a commercial building this increases revenue. In an institutional building service performance is enhanced, horizontal walking distance is shortened, and a greater concentration of service rendered can be attained.

An effective security barrier and checkpoint can be created in a building by using escalators to a second floor lobby and having the elevators start from that floor. All people entering must use the escalator to gain access to the second floor lobby; this expedites checking of identification by a security guard.

Escalators are essential in buildings with double-deck elevators. Since people must separate into those going to the odd numbered floors and those going to the even numbered floors, escalators are the only effective means of providing

Figure 9.1. Modern escalators in a dramatic setting (Courtesy Westinghouse Elevator).

convenience during this separation process. This will be discussed more fully in the chapter on special elevators.

Escalators are found in many places besides their initial field of applications in stores and transportation facilities. Today, schools, hospitals, factories, office buildings and restaurants have escalators. So do hotels, motels, museums, theaters, convention halls, sports arenas and other buildings that must accommodate large groups of people within a short period of time.

A Brief History of Escalators

The modern escalator is a result of two inventions and extensive development. About 1892 Jesse Reno designed and patented a moving inclined ramp featuring cleated triangular shaped platforms on a continually operating belt which were combed at the top and bottom (see Figure 9.2a). An early installation was on the Third Avenue Elevated Line at the 59th Street Station in New York City about 1900 where it was in operation until the line was torn down in 1955. Also, about 1892 G. H. Wheeler invented and patented a flat step moving stairway with a handrail. This invention was developed by C. D. Seeberger and the Otis Elevator Company. The newly developed flat step escalator was exhibited at the Paris Exposition in 1900 (Figure 9.2b). Otis also created the name "escalator" which was a registered trademark of Otis until it was declared in the public domain in the 1930s when it was used as a title in the A17.1 Safety Code.

Seeberger's escalator had flat steps and a triangular diverting baffle at both top and bottom where people had to sidestep on and off the escalator.

Both the Reno type and Seeberger type were manufactured by Otis. In 1922 further developments were made. The result was the forerunner of today's modern escalator which combined Reno's cleats and comb and Seeberger's flat steps. The diverting baffle was eliminated at the landings so people entered and departed in line with the escalator with the help of the flat, cleated steps and combs top and bottom.

Figure 9.2. Early types of escalators: (*a*) Reno type; (*b*) Seeberger flat step escalator with passenger diverter. Passengers stepped on and off escalator to the side as the steps continued under diverter (Courtesy Otis Elevator).

EFFECTIVE APPLICATION

Escalators can be advantageously applied to any building if certain requirements are met. Equipment should be located so that most people entering the building can see it. Access to the escalator must be attractive and in the path of the heaviest expected traffic. Evaluation of expected traffic volume, which could range from a few people continuously to hundreds in a peak 5-min period, depends on the type of building faciity and its use.

Escalators are most effectively used as a continuously running, unidirectional conveyor. They can be started and stopped on demand, but this requires additional special considerations which will be discussed later. Starting, stopping, or reversal is best done by an attendant and with the assurance that no one is riding at the time. Pairs of escalators are necessary for two-way service.

Ample space for people must be provided at the entry and exit landings of an escalator. The escalator can feed people into an area much more rapidly than they can climb a stairway or walk through a restricted opening to leave that area. If an unloading area is restricted, people could be crowded into it with possibly dangerous results since the escalator is an unforgiving conveyance. Such restrictions as doors or gates should be interlocked with the escalator or ramp to ensure that the restriction is removed before the escalator can be run. Where escalators feed a restricted area such as a subway platform, security personnel must be alerted to the possibility of platform overcrowding if subway service is interrupted and instructed to stop the escalators.

In many localities a building can have fewer stairs if fire protective enclosures are provided around the escalator. This enclosure must be equipped with sufficient doors and space at the landings for the doors to swing with the traffic and not impede prompt passenger transfer. When escalator traffic of any magnitude is expected its volume may reach the capacity limit of the unit and ample loading area must be provided. Means should be provided to automatically stop escalators which may travel into an area of the building endangered by a fire. This will be discussed later in this chapter.

One of the attractions of an escalator is its continuous motion, providing service with zero interval in elevator terms. Normally people need not wait but may enjoy service the moment they reach the entry level. If the capacity of the escalator is exceeded, a wait may be necessary; however, the waiting time is readily apparent. If more people are expected to arrive than the escalator system can handle, additional facilities, higher speed, or adequate alternative routes should be offered. Stairs adjacent to pair of escalators are an absolute necessity if the escalators are the primary means of entering or leaving a building lobby. Escalator handling capacity is discussed later in this chapter.

If escalators are the primary means of vertical transportation they need to be supplemented by one or more elevators. Some handicapped persons can usually negotiate an escalator but it is almost impossible for a person in a wheelchair to do so. Similarly, a person with impaired vision would have difficulty picking a proper tread to step on as would many older people with motion difficulties.

Mothers with baby carriages or strollers should be encouraged to use elevators rather than escalators or moving walks. Escalators are almost impossible to negotiate with a carriage and the incline of a walk may make a loaded carriage difficult to hold back. A loaded food cart, which may weigh from 50 to 100 lb (20 to 40 kg) can be dangerous on an inclined walk unless the wheels are locked. Locking must be done automatically since it would be an impossible task to ensure that everyone using a walk will secure their carts. Inclined elevators as an adjunct to escalators will be discussed in a later chapter.

TRAFFIC-HANDLING ABILITY

Escalators and moving walkways are rated by nominal width at approximate hip level in North America and by step width in European countries and by speed in feet per minute. Because escalators and walkways are usually driven by ac induction motors, operating speed is constant under load conditions and rating is at a single speed. The generally recognized escalator speeds are either 90 fpm (0.45 mps) or 120 fpm (0.6 mps) along the incline. Faster escalators have been provided in some areas but their use is not common and a factor of diminishing return can result; the steps may move too fast for people to use them.

Very high-speed escalators are used in the Soviet subway stations in cities such as Moscow, Kiev, and Leningrad. This is due to the exceedingly high rises in many of their stations, up to a record of 214 ft (65 m) in Kiev. Speeds as high as 200 fpm (mps) are used to overcome the extremely long riding time and handling capacity is of secondary importance. Many escalators are equipped for two-speed operation by manual switching; these escalators can be run at 120 fpm (0.6 mps) for the rush period and at 90 fpm (0.45 mps) during the rest of the day with the consequent reduction in operating mileage. The normal angle of incline of an escalator is 30°, give or take a degree or two for particular building conditions and is established by the A17.1 code. In European countries, both 30° and 35° escalators are common, however the 35° escalator is limited to 90 fpm (0.45 mps).

The number of flat steps a passenger encounters upon entering an escalator is extremely important. A flat step is a step that is level with the preceding step prior to the step rising or depressing to form the incline. Observations have shown that the greater the number of flat steps, the more easily passengers adjust to the moving escalator and traffic handling ability is expedited. This is to be expected since a person's stride is about 30 in. (750 mm) and two flat steps are about 32 in. (800 mm). The manufacturer's normal standard is one and one-third flat step. This should be increased to a minimum of 2⅓ when escalators are designed to be used at 120 fpm (0.65 mps). Additional flat steps can also be used for specific building conditions, such as lining up the newels of adjacent escalators.

Table. 9.1.   Escalator Capacities (30° Incline)

Width	Speed	Maximum Capacity Theoretical	Nominal Capacity Observed[a]
32 in. (600 mm)	90 fpm (0.45 mps)	425/5 min, 5000/hr	170/5 min, 2040/hr
	120 fpm (0.6 mps)	566/5 min, 6700/hr	225/5 min, 2700/hr
48 in. (1000 mm)	90 fpm (0.45 mps)	680/5 min, 8000/hr	340/5 min, 4080/hr
	120 fpm (0.6 mps)	891/5 min, 10700/hr	450/5 min, 5400/hr

[a] Based on one person every other step for 32-in. (600-mm) wide and one person per step for 48-in. (1000-mm) wide escalator.

Because moving walkways can be installed at any angle from 0° (a moving walkway) to 15° (inclined walk), operating speed varies with the angle of inclination. At any speed with an inclined walk, the entering and exiting area should move horizontally for boarding or exiting and make a smooth transition to inclined motion. With level boarding and exiting, operating speed can be higher than if the passenger must board at an incline. Operating speeds, angles, and walk widths are established by the A17.1 code, and are briefly shown in Tables 9.1, 9.2, and 9.3. Latest editions of the code and local codes should be consulted.

The nominal hip-high widths of escalators are either 32 or 48 in. For European escalators the corresponding step widths are 600 mm (24 in.) and 1000 mm (40 in.). In addition an intermediate size with an 800-mm (30-in.) wide step is available in countries other than the United States and Canada.

The 32-in. is wide enough for one person per step and the 48-in. allows a person with baggage or two adults to ride side by side. In actual observed practice, one person is on every other step of the 32-in. escalator and one person on each step of a 48-in. escalator. A further advantage of the 48-in. escalator is that people in a hurry may pass a standing rider if all the riders stand to one side. If one were available, an escalator with a 54-in. hip-height width would easily allow passing or two people per step.

Table. 9.2.   Operating Speeds of Moving Walkway [Based on 40-in. (1000-mm) Nominal Trend Width[a]]

Incline of Ramp on Slope	Maximum Speed with Level Entrance and Exit	Maximum Speed with Sloping Entrance and Exit
0 to 3°	180 fpm (0.9 mps)	180 fpm (0.9 mps)
Over 3 to 5°	180 fpm (0.9 mps)	160 fpm (0.8 mps)
Over 5 to 8°	180 fpm (0.9 mps)	140 fpm (0.7 mps)
Over 8 to 12°	140 fpm (0.7 mps)	130 fpm (0.65 mps)
Over 12 to 15°	140 fpm (0.7 mps)	125 fpm (0.63 mps)

[a] Speed, angles, and capacities will vary with width. See A17.1 code.

Table. 9.3.   Moving Walkway Capacities, 40-in. (1000-mm) Nominal Width[a]

	Treadway Speed	Maximum Capacity Theoretical[b]	Nominal Capacity Observed[c]
0° incline	180 fpm (0.9 mps)	1200/5 min, 14,400/hr	600/5 min, 7200/hr
5° incline	140 fpm (0.7 mps)	932/5 min, 11,180/hr	466/5 min, 5600/hr
10° incline	130 fpm (0.65 mps)	867/5 min, 10,400/hr	434/5 min, 5200/hr
15° incline	125 fpm (0.62 mps)	833/5 min, 10,000/hr	416/5 min, 5000/hr

[a] Speed, angles, and capacities will vary with width. See A17.1 code.
[b] 2.5 ft^2 (0.23 m^2) of treadway per person.
[c] 5 ft^2 (0.46 m^2) of treadway per person.

Escalator capacities are generally expressed in passengers per hour. Capacities expressed by escalator manufacturers are theoretical and assume that each step carries either 1¼ or 2 passengers, depending on the width. Such output is never attained. A reasonable estimate of actual output would be about 50% of theoretical output, as shown in Table 9.1.

Walkways are rated in much the same way. Nominal width is expressed as the width at hip height and the A17.1 code limits width depending upon the incline of the walkway. Escalators and moving walkways should have moving handrails at both sides of the steps; however, the A17.1 code allows a moving walkway to have only one handrail if the slope is 3° or less, if speed is 70 fpm (0.35 mps) or less, or if the width is no more than 21 in. A listing of nominal inclines and ratings is shown in Table 9.3.

ARRANGEMENT AND LOCATION

Two general arrangements of escalators are descriptively named *parallel* and *crisscross*. Both arrangements may have up and down equipment side by side or separated by a distance. A third possible arrangement, which could be called *multiple parallel*, consists of more than two escalators side by side between the same exiting and entering levels, primarily to serve more traffic than a single escalator could handle. Flexibility is provided by operating all the units but one in the direction of heavy traffic.

The various arrangements are sketched in Figure 9.3. The crisscross arrangement is the most popular in department stores because it uses floor space effectively, structural requirements are minimized since escalators can be stacked above each other, and it achieves maximum exposure of passengers to merchandise on the various floors. Separating crisscross escalators increases exposure to the various floors and eases the intermingling of riding passengers and people wishing to board. The separated crisscross arrangement is considered the safest by many users because only one escalator is presented to the riding passenger and there is minimum confusion about whether it is going up or down.

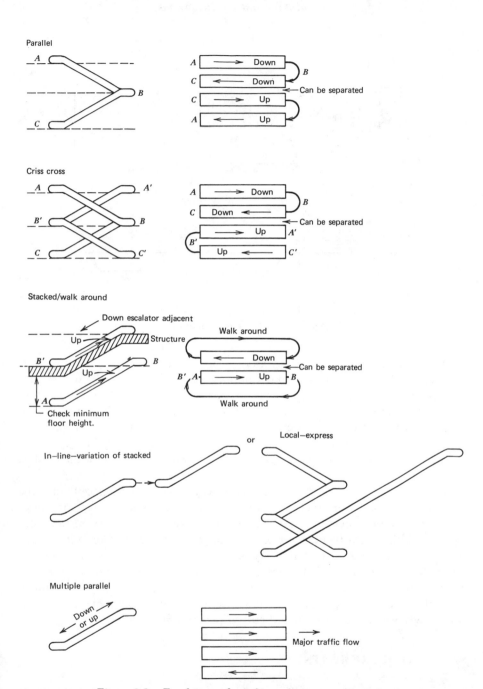

Figure 9.3. Escalator and moving walkway arrangements.

201

Entering group at each floor

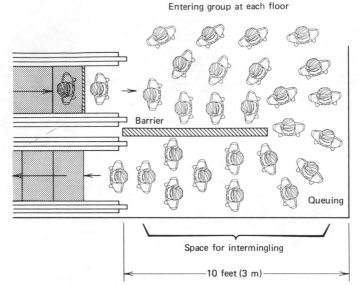

Figure 9.4. Exiting and queuing space.

The parallel arrangement provides the least congested flow with the greatest traffic handling ability and the most impressive appearance to the prospective passenger. Massing the escalator or ramp entrances immediately attracts people to that area. The open appearance at upper landings provides space for decoration or a high-traffic selling area as well as additional open space for intermingling traffic. The arrangement is limited if it is intended to board many people at each floor simultaneously, unless a barrier is created to form a queing line (Figure 9.4).

The multiple parallel arrangement is provided when many people must be transported to another level in minimum time, such as in a busy commuter terminal. Passenger demands may be served by three of a four-unit installation, operated with the traffic morning and evening and reversed to two units in each direction during the rest of the day. Frequent applications are in buildings where the main lobby must be located above the street because of a subway or railroad underneath the building. Here, again, direction of operation is changed to conform with traffic. Units would also be reversed in sports arenas, exhibit halls, and theaters at the start and end of performances.

SPACE REQUIREMENTS

An escalator or moving walk can be thought of as being constructed in three component units: an "upper portion," a "lower portion," and a "midsection."

The lower unit includes the newel (where the handrail is reversed), the lower step return and step tensioning device, and the landing plate and step entry. The upper unit consists of the upper newel, the upper landing, and the driving mechanism (motor and control) for steps and handrail for some manufacturers whereas others have their drive motors in the midsections. The midsection can be of indefinite length (within limits) and consists of balustrading, steps, step tracks, supports, and so on.

Normal support points for escalators and short-run inclined walks are at the top and bottom and are established by a distance from the working points. A working point is where the incline intercepts the horizontal plane related to the combplates; all distances to escalator supports are measured from these points.

Escalator or walk layout procedure begins with determination of the vertical rise and approximate location of the escalator and the upper or lower pedestrian access space, to establish the upper and lower working points. Since escalators are generally inclined at 30° the distance between working points is 1.732 times the rise (Figure 9.5). Once both working points are established the particular manufacturer's standard space requirements must be met to locate the necessary supports that must be built into the building structure. The reactions on those supports depend on the length and width of the escalator.

For the nominal 32-in (600-mm) wide escalator serving a 12-ft (3.6-m) rise, the lower end of the truss imposes a load of about 19.1 kips (8.7 t) and 20.4 kips (9.3 t) at the upper end. These loads include the weight equivalent of plaster facing on the sides of the escalator and normal balustrade treatment. The loads are, for the most part, uniformly distributed over the width of the escalator. Escalators are mounted on the building steel by means of an angle so the load given is for the escalator width. Rises over about 18 ft (5.5. m) for the 48-in. (1000-mm) and over 23 ft (7 m) for the 32-in. (600-mm) escalators as well as for most moving walks usually require intermediate supports. A table of the highest loads imposed by U.S. manufactured escalators is given in Table 9.4.

The loads of moving walkways will vary considerably with the angle of incline and the length of the walkway. When a desired application of a walkway is established, it is best to work with a manufacturer's representative or consulting engineer to determine necessary space requirements and the loads imposed on the structure.

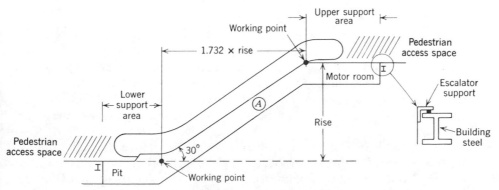

Figure 9.5.   Escalator space requirements. Some escalators have the motor located at A.

Table. 9.4. Maximum Load per Escalator on the Building Structure in kips (1000-lb Units) or metric tons (t) (1000-kg Units)

Rise		10 ft (3 m)	12 ft (3.6 m)	14 ft (4.2 m)
32 in. (600 m)	Upper	19.1 kips (8.7 t)	20.4 kips (9.3 t)	21.7 kips (9.9 t)
	Lower	17.9 kips (8.1 t)	19.1 kips (8.7 t)	20.4 kips (9.3 t)
48 in. (1000 m)	Upper	22.5 kips (9.3 t)	24.1 kips (10.9 t)	25.7 kips (11.7 t)
	Lower	21 kips (9.5 t)	22.6 kips (10.3 t)	24.1 kips (10.9 t)
Rise		16 ft (4.8 m)	18 ft (5.5 m)	20 ft (6.0 m)
32 in. (600 m)	Upper	22.9 kips (10.4 t)	24.2 kips (11 t)	25.6 kips (11.6 t)
	Lower	21.7 kips (9.9 t)	22.9 kips (10.4 t)	24.2 kips (11 t)
48 in. (1000 m)	Upper	27.3 kips (12.4 t)	28.9 kips (13.1 t)	30.5 kips (13.9 t)
	Lower	25.6 kips (11.6 t)	27.1 kips (12.3 t)	28.6 kips (13 t)
Rise		22 ft (6.7 m)		
32 in. (600 m)	Upper	22.5 kips (10.2 t)		
	Lower	14.9 kips (6.7 t)		
	Int.[a]	15.7 kips (7.1 t)		
48 in. (1000 m)	Upper	26.8 kips (12.2 t)		
	Lower	17.6 kips (8 t)		
	Int.[a]	18.7 kips (8.5 t)		

[a] Load at intermediate support.

The plan of an escalator includes the width of the steps, the width of the balustrading plus space for the truss support, and external decoration. A 32-in (600-mm) escalator has an overall width of about 4 ft 4 in. (1320 mm); a 48-in. (1000-mm) escalator, about 5 ft 8 in. (1730 mm) (Figure 9.6). Since most escalators are side by side, the overall width of two is shown. Widths for moving walkways vary with the width of the tread and, to some degree, with the angle of incline.

The overall length of an escalator is much shorter than a moving walkway for the same rise because of the greater angle of inclination of the escalator. An average 32-in. (600-mm) escalator for a 12-ft rise requires about 36 ft (10.9 m) between supports, excluding upper and lower access areas. A 10° walk serving

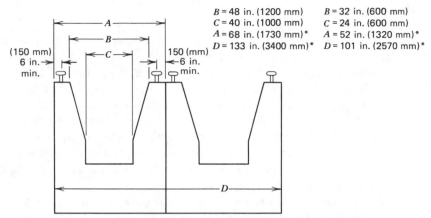

$B = 48$ in. (1200 mm) $B = 32$ in. (600 mm)
$C = 40$ in. (1000 mm) $C = 24$ in. (600 mm)
$A = 68$ in. (1730 mm)* $A = 52$ in. (1320 mm)*
$D = 133$ in. (3400 mm)* $D = 101$ in. (2570 mm)*

* Include a total of 4 in. (100 mm) on both sides for wall finish.

Figure 9.6. Escalator cross section. Note: The 1981 A17.1 code designates escalator width by the width at hip height. The EN81 code designates width by step width.

the same rise will require about 82 ft (25 m), excluding upper and lower access areas. An average escalator truss depth including soffit is about 3 ft 6 in. (1100 mm), whereas the truss for a walk can be of very limited depth, depending on the rise. Escalator trusses are deeper because they must provide space for the returning steps whereas the walk truss need accommodate only returning pallets. A listing of escalator space requirements for various rises is given in Table 9.5.

Table. 9.5. Horizontal Space Required, 30* Escalators.
Note: 32 in. (600 mm) and 48 in. (1000 mm) Are the Same

Rise	Distance Working Point to Working Point	Distance Edge of Support to Edge of Support	
		1⅓ Flat Steps	2⅓ Flat Steps
10 ft. (3.0 m)	17 ft 4 in. (5.3 m)	32 ft 2 in. (9.8 m)	34 ft 10 in. (10.6 m)
12 ft. (3.6 m)	20 ft 10 in. (6.3 m)	35 ft 8 in. (10.9 m)	38 ft 4 in. (11.8 m)
14 ft. (4.2 m)	24 ft 3 in. (7.4 m)	39 ft 1 in. (11.1 m)	41 ft 8 in. (12.7 m)
16 ft. (4.8 m)	27 ft 9 in. (8.5 m)	42 ft 7 in. (12.9 m)	45 ft 3 in. (13.7 m)
18 ft. (5.5 m)	31 ft 2 in. (9.5 m)	46 ft 0 in. (14.0 m)	48 ft 8 in. (14.2 m)
20 ft. (6.0 m)	34 ft 8 in. (10.6 m)	49 ft 6 in. (15.0 m)	53 ft 2 in. (16.2 m)
22 ft.* (6.7 m)	38 ft 2 in. (11.6 m)	52 ft 11 in. (16.2 m)	55 ft 7 in. (16.9 m)
* Distance from lower end support to intermediate support		15 ft 0 in. (4.6 m)	16 ft 4 in. (5.0 m)

Escalator Layouts

The first step in developing an escalator layout is to determine the distance between the working points based on the floor-to-floor height which is the rise of the escalator. From the working points the data given in Table 9.5 and the dimensions shown in Figure 9.7 are used to determine the approximate distance for the upper support point to the lower support point depending upon the number of flat steps required. At this point the designs of various manufacturers should be checked to ensure that sufficient space is allowed for each manufacturer to install the equipment. Alternatively, the greatest dimensional requirement should be developed with the understanding that adjustment can be made once the escalator contractor is chosen.

Once the support points are established, vertical and horizontal spaces for the truss, upper structure, and pit are established. If escalators are to be located one above the other, it is important to have floor-to-floor heights of at least 11 ft 6 in. (3.5 m) to obtain sufficient head room at the entrance to the lower escalator.

After the truss is located, the newels can be established based upon the location of the working points and the number of flat steps. Access area to the escalator is determined from the end of the newel to the nearest obstruction and should be a minimum of 12 ft (3.6 m).

Figure 9.8*a,b,c,* and *d* are typical layouts for Otis, Westinghouse, Montgomery, and Schindler-Haughton escalators representing different approaches

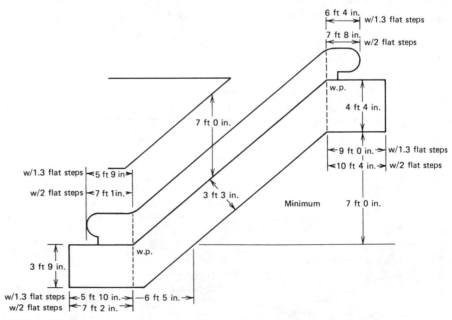

Figure 9.7. Escalator pit and overhead requirements. Note: Dimensions will vary among manufacturers.

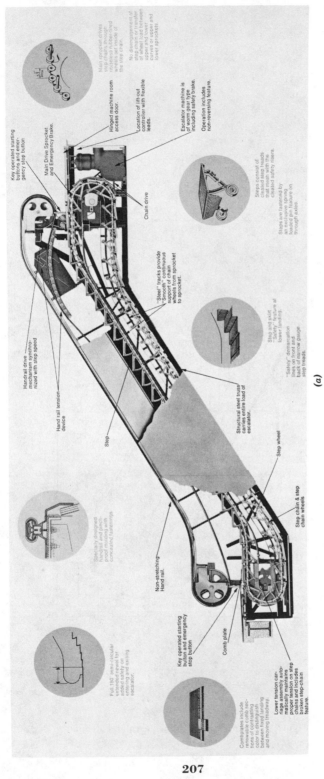

Key operated starting buttons and emergency stop button

Main Drive Sprocket and Emergency Brake.

Chain drive

Handrail drive mechanism synchronized with step speed

Hand rail tension device

Step

Structural steel truss carries entire load of escalator.

Step wheel

Non-stretching Hand rail.

Comb plate

Key operated starting button and emergency stop button

Step chain & step chain wheels

Main sprocket drives step chain through slip-less rubber-lined wheels set inside of the step chain.

No disengagement of step chain or transfer of wheel load between upper and lower curves or inside of upper and lower sprockets.

Hinged machine room access door.

Location of lift-out controller with flexible leads.

Escalator machine is of worm gear type including safety brake.

Operation includes non-reversing feature.

Steps consist of cleated step treads that mesh with the cleated safety risers.

Steps are fastened by an exclusive spring loaded pin feature on through axles.

"Steel" tracks provide "Smooth" continuous support of chain wheels from sprocket to sprocket.

Step and skirt "Safety" feature at lower landing.

"Safety" demarcation lines on front and back of narrow gauge step treads.

Specially designed handrail and pinch-proof molding with concealed fastenings.

Full 180 semi-circular extended newel for added safety on entering and exiting escalator.

Combplates include renewable comb sections of contrasting color to distinguish between fixed landing and moving treadway.

Lower tension carriage assembly automatically maintains proper tension on step chains and includes broken step-chain feature.

(a)

Figure 9.8. (*a*) Otis escalator details;

207

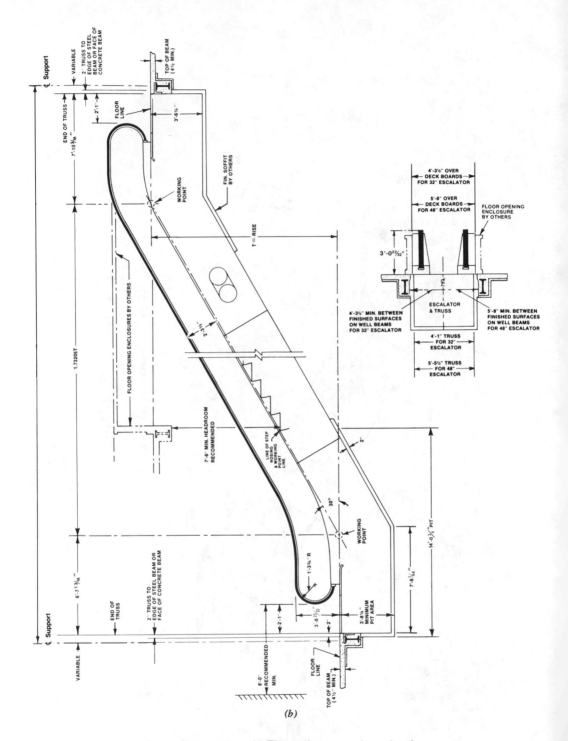

Figure 9.8. (b) Westinghouse escalator details;

208

crystal 3000 glass balustrade

HINGED CAST ACCESS COVER

COMB PLATE

φ WORKING POINT

₵ ESCALATOR & TRUSS

WORKING POINT

COMB PLATE

HINGED CAST ACCESS COVER

FACE OF SUPPORT

FACE OF SUPPORT

L = A + 15-5
A = H x 1.73205
E = (7.6 - D) (1.73205)
7'-6½"
7'-10½"
7'-13¾"
8'-0 REC. MIN.

F

FLOOR OPENING ENCLOSURE

D

5-3 11/16" R
5-6½"
WP
CONTROLLER
2'-11½"
4'-1 13/16"

7-6 RECOMMENDED MINIMUM

2-7 TOP OF HANDRAIL
3½ TOP OF DECK
LINE OF STEP NOSSING
3-2 TRUSS BOTTOM
9"
2'-11"
7'-1 3/4"

SEE FIG. A & B

ELECTRICAL CONNECTION (BY OTHERS)

H
8'-0 REC. MIN.
6'-9⅜
4'-10 5/16" R
1'-4½"

2'-11 1/2"
13¾" R
30
6-4
5-2½
WP
10"
3'-10 MIN. PIT
3'-9 1/8"
6-5
2-6
14-10

DECK DETAIL

TEMPERED GLASS
STAINLESS STEEL DECK
LATH & PLASTER NOT BY MECO
SKIRT PANEL
6
1¼
1½ 3½

SEE FIG. A & B

INTERMEDIATE SUPPORT RECOMMENDED FOR RISES OVER 14-0 FOR 5E
15-6 FOR 4E AND
17-0 FOR 3E ESCALATORS
(consult MECO if deviation is required)

APPROX FILL AFTER SETTING ESCALATOR
FINISHED FLOOR
B POCKET
ACCESS COVER (Non Skid Cast Aluminum by Montgomery)
1-9/16
4½
Beam To Be Sized and Provided By Others
ESCALATOR TRUSS
F P BY OTHERS ON ESCALATOR SIDE AFTER TRUSS IS SET

Fig. A

B POCKET
ACCESS COVER (Non Skid Cast Aluminum by Montgomery)
APPROX FILL AFTER SETTING ESCALATOR
FINISHED FLOOR
1-9/16
ESCALATOR TRUSS
4½
3/4 x 8 CONTINUOUS STEEL PLATE AND ANCHORS

Fig B

₵ ESCALATOR & TRUSS
B
3½
R
2'-11½"
S
7½
Floor Opening Enclosure

Finish By Others. If Heavier Enclosure than Plaster & Wire Lath is Used Refer to MECO for Consideration

SHADED AREAS BY OTHERS

LAYOUT NOTE: The following information, when available, must be shown on all layouts for use of the balustrade manufacturer.
D—Dimension from finished floor to the finished plaster ceiling or bottom of smoke guard.
E-F-G—Detail and kind of wellway railings or fire shutter enclosures which are not furnished by the balustrade manufacturer.

REACTION FORMULAE*

R_U
H
R_L
7'-6½" (1.73205) (H)
7'-10½"

32' ESCALATOR	
RL = (550)H + 10,000	RU = (550)H + 11,100
40' ESCALATOR	
RL = (660)H + 10,570	RU = (660)H + 11,670
48' ESCALATOR	
RL = (660)H + 11,650	RU = (660)H + 12,750

Consult MECO for reactions if intermediate support is used.

(c)

Figure 9.8. (c) Montgomery escalator details;

End view dimensions

Nominal Width	32"	48"
Step Width "W"	24"	40"
Finish Width "Y"	4'0¼"	5'4"
Hand Rail ℄ To ℄ "X"	2'11⅞"	4'3½"
Truss Width "T"	3'9½"	5'1½"
Rough Opening "Z"	4'2½"	5'6½"

Rated capacities per hour

Unit	90 fpm	120 fpm
32"	5000	6666
48"	8000	10,665

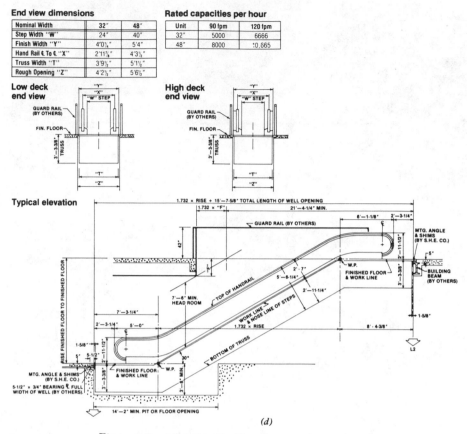

Low deck end view

High deck end view

Typical elevation

Figure 9.8. (*d*) Schindler Haughton escalator details.

to escalator design. Otis' and Schindler-Haughton's driving machine is contained in the upper portion and is located under the landing floor plate. Westinghouse's machine is located in the truss about one-third of the way down. Montgomery's machine is located in the truss below the top working point. The details shown are for escalators with opaque balustrades and will vary for escalators with transparent or translucent balustrades.

Additional Requirements

Equipment access is required to properly service an escalator or walk. The normal access includes the removal of the top and bottom landing plates. The bottom landing plate covers the step or belt tensioning device, and the top landing plate covers the escalator or walkway machine and upper drive for some escalators, and on others the machine is located in the truss. With this latter design, a side access door may be desirable. Alternatively, steps must be removed to service the machine.

Power must be provided to the escalator or walkway controller. The power supply must meet the requirements of the particular manufacturer and usually the respective motor horsepowers of escalators are 10 hp for the 32-in. (600-m) and 15 hp for the 48-in. (1000-mm) for rises up to about 22 ft (6.7 m). Walkway power requirements vary with different widths and angles of incline and equipment suppliers should be consulted.

To dissipate the heat released when an escalator or ramp is operated for long periods under load conditions, ample ventilation for the motor area should be provided. This can be grillwork if the machine room is in an open area or forced ventilation if the machine room is confined. The approximate heat release of escalators will be about 10,000 Btu/hr for the 10-hp, 32-in. (600-mm) escalator and 15,000 Btu/hr for the 15-hp, 48-in. (1000-mm) escalator. The heat dissipation requirements of moving walkway machinery will vary with the horsepower.

FEATURES OF ESCALATORS AND MOVING WALKS

Because escalators and walks should serve people of all ages and abilities, they must be inherently safe. It has been determined that escalators are generally safer than stairs and comparable statistics for moving walkways have not been compiled. Escalator safety has been the result of experience, with much research and development. Early escalators, for example, had wide step cleats with grooves of about ½ in. (13 mm). Well-designed modern escalators have cleats no more than ¼ in. (6 mm) wide and grooves not less than ⅜ in. (9 mm) deep as required by the A17.1 code. These cleats and grooves are "combed" as they move at the top or lower landing to dislodge soft shoe soles and debris to avoid accidents (Figure 9.9).

The A17.1 code requires the same type of combing action on moving walks. Escalator step risers are also cleated so that people who ride with their toes against the riser will not have their soft shoe soles drawn between the steps as they straighten out (Figure 9.10). The footwear shown in the illustration may be dated but the hazard is amply illustrated. As the lower step rises to meet the

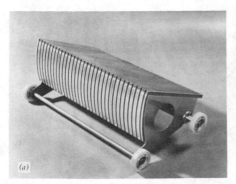

Figure 9.9.   Escalator step with cleated risers.

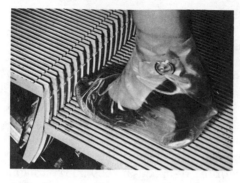

Figure 9.10.   Combing action of cleated risers.

upper step at the exit of an up escalator, the friction of the material against the moving step riser will cause the shoe material to be drawn into the gap between the steps. The combing action will oppose this downward motion. The same hazard exists as the side of the steps travels along the skirt panel.

Some people ride with their feet pressed against the side of an escalator. As the steps flatten there is a possibility of a soft shoe such as a sneaker or rainwear being drawn between the step and the side. If this occurs, a switch actuated by the deflection of the side should stop the escalator (Figure 9.11). Escalator and walk brakes must be designed to stop the fully loaded treadway as quickly as possible to minimize personal injury. A promising development of raised treads adjacent to the skirts discourages riders from pressing their feet against the skirt. The treads need only be raised about ⅛ in. (3 mm). In addition the treads adjacent to the skirt and at the leading and trailing edges of each step can be marked in a high visibility color such as yellow or orange so that riders will avoid these hazardous areas.

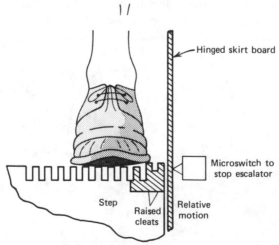

Figure 9.11.   Skirt panel and step arrangement to avoid pinching soft shoes.

All escalators and walkways should be reversible so that their capacity can be utilized in either direction. Reversing switches are key operated and generally located at the top or bottom landing in the newel post. An emergency stop switch is required for escalators and walkways and its location is prescribed by A17.1. To discourage mischief, a hinged plastic cover can be used over the stop switch. Lifting the cover sets off an alarm and the stop switch can be then operated. The alarm can also be monitored at a security station so emergency personnel can be alerted.

The loading and exiting levels of escalators and walks should be extended so that passengers can grasp the handrail and become adjusted to the speed of the steps as they board. Two and one-third flat steps should be provided as a minimum before the incline begins. 2⅓ flat steps are not a manufacturer's standard and need to be specified. Illumination under the steps which shines through the gap between steps is an effective safety feature to provide tread delineation of the points of embarcation.

Balustrade treatments have undergone radical changes since the early days of escalators. A wide variety of materials is available for the balustrading of both escalators and walkways, including stainless steel, bronze, glass (also called crystal balustrading), aluminum in various colors, laminated plastics, tinted glass, lighted glass panels, and fiberglass material. Handrails are now being made in a variety of colors. The basic oval section of the handrail has remained unchanged, as this shape seems to provide the firmest and most comfortable gripping surface. The entry of the handrail into the newel is made as inaccessible as possible to minimize hazard, especially to curious children (Figure 9.12).

Escalator Operation

Escalators are generally manually started and stopped by an attendant using a keyed switch located at the top or bottom of an escalator. The attendant can ensure that no one is on the escalator and start it in the desired direction. Stopping is also done by an attendant who can ensure that no one is riding at the time. The stop can be somewhat harsh since this stop is the same as that used for emergency stopping if the various safety switches are actuated.

Figure 9.12. Extended newels (Courtesy Montgomery Elevator).

It is possible to equip escalators with two levels of stopping deceleration, the emergency stop and a controlled stop which extends the stopping distance of the steps and is somewhat gradual. The controlled stop is limited to a deceleration of about 1.5 fps^2 (0.46 mps^2), whereas an emergency stop deceleration is about 3 fps^2 (0.9 mps^2). The controlled stop was introduced in the United States by the consulting engineering firm of Jaros, Baum & Bolles. William S. Lewis, the partner in charge of vertical transportation, became concerned about escalators in an office building during a fire emergency whereby the escalators, if not stopped, could bring people into a smoke involved floor. By specifying and encouraging the development of a controlled stop, it is now possible to stop the escalators if a smoke detector is actuated, or stop them from a remote control if a fire emergency occurs. The controlled stop has been applied on a number of escalators in various buildings with excellent results.

With a controlled stop and with future development of a gradual, controlled start, it becomes feasible to consider escalators that can be automatically started and stopped and to have a single escalator serve traffic in both the up and down directions. This is presently done without controlled starting and stopping and is accomplished by means of floor treadle or mat switches at the top and bottom landings of escalators and may not be in compliance with safety codes.

The operation of a stopped escalator is initiated, for example, by a person stepping on the top treadle and starting the escalator in the down direction. When the escalator starts, the passenger can travel down and once he or she steps off and passes the lower treadle, the escalator will stop a preestablished time after the lower treadle switch is actuated. Similarly, a passenger wishing to go up steps on the lower treadle and starts the escalator in the up direction, and it will stop a short time after the upper treadle switch is actuated.

The hazard is that any person who may have bypassed one of the treadles may be walking on an escalator and the abrupt start or stop will upset that person. With both a controlled start and stop, that hazard could be minimized.

A single reversible escalator can provide a cost effective vertical transportation solution to an application where heavy traffic in one direction is expected at any time and in the opposite direction at another time, such as in a train station or a commuter parking garage.

Special Applications

Escalators and walkways can be installed practically anywhere. Outdoor applications should be designed to be weatherproofed and include heaters to prevent icing or excessive water accumulation. Operation of escalators under exposed conditions is not recommended since wet or icy steps can be hazardous. Suitable shelter over the escalator or ramp should be provided.

Escalators have been installed aboard ships to operate while the ship is experiencing motion, roll, pitch and yaw. Shipboard applications have included access to and from the flight deck on aircraft carriers, between the restaurant and the kitchen on cruise liners (for waiters carrying trays), and for general passenger use.

Many industrial plants have escalators to serve their employees during shift changes and in the normal course of their duties. The highest and fastest escalators in the world can be found in the USSR subway system and have a vertical rise of up to 214 ft (65 m) and operate at 200 fpm (1 mps).

ESCALATORS VERSUS ELEVATORS

Many office buildings and schools have escalators as the primary means of vertical transportation. Escalators can often be more cost effective than elevators for a given building condition or for solving a particular vertical transportation problem.

One of the best examples is a high school or college in a single building. The major transportation demand occurs when classes change and may be as high as 40 to 50% of the student population moving during a 5-min period. For a six-story school of about 2500 students, 8 to 10 elevators would be required to give everyone floor-to-floor service. This number of elevators is seldom provided and the accepted approach is to have the elevators stop at every other floor and have about half the students walk a floor. With the skip-stop arrangements, only about six elevators are required. With a pair of escalators servicing each floor everyone can ride and the average trip requires no more time than the average elevator trip. The cost of the escalator arrangement plus two elevators for handicapped and freight service may amount to about the same as or less than equivalent elevator service.

Another example of the effectiveness of escalators was shown in an eight-story industrial research building with a population of about 2200 persons. Parking was in four areas on the ground each with good access to main highways so the arrival rate at the building amounts to about 25% of the population in a 5-min period. Sufficient vertical transportation would have required a group of seven elevators with the elevators operated so that one elevator was designated to serve each floor. People would have been directed to line up in front of the elevator for their floor and wait. Once on their floor, they would either have to travel back to the lobby to get to another floor or wait until the peak was over.

A pair of 32-in. escalators, both operated in the same direction during the incoming and outgoing periods provided the necessary short-time handling capacity. Evaluating all the factors, including the usable space required by the elevators versus the escalators, capital investment, maintenance cost, and cost of elevator pit and penthouse structures, is required to make the best recommendation.

Escalators in combination with elevators can solve some intricate vertical transportation problems. An example is a merchandise mart where buyers converge during show times once or twice a year. Various suppliers hire space to display their products on the various floors. The average buyer will start at the top of the building and move down from floor to floor. For this traffic, elevators can be used for travel solely between the top and bottom and escalators used in the down direction only, to take the buyers from floor to floor. Since it may take a buyer more than a day to travel the entire building, elevators could be arranged to stop at some middle floor, in addition to the top, to serve people who may want to return to the lobby during the noon time.

Escalators can be used to serve extra heavy incoming, outgoing, and two-way traffic. Travel time should be considered, and based on average floor heights of 12 ft and an escalator speed of 120 fpm, each floor would take about 22 sec to traverse including the time required for a passenger to turn around and board the next up escalator. This time consists of 6 sec for entering and exiting, 10 sec traveling, and 6 sec to walk around to the next rise of escalators. If maximum riding time of about 3 min is desired, effective application of escalators would be limited to buildings of no more than nine floors high.

CALCULATING ESCALATOR OR MOVING WALK REQUIREMENTS

As with any vertical transportation application the initial step is to qualify the peak traffic demand and the characteristics of that demand. In office buildings, the peak demands may be incoming and outgoing traffic. In a department store, peak demand may be two-way traffic during holiday seasons. In a transportation terminal, peak escalator travel requirements may occur as trains or planes arrive and discharge their passengers. In sports arenas, peak demand can be related to how quickly people can enter the facilities or, perhaps, how quickly the parking lots can be emptied after an event is over.

Examples 9.1 and 9.2 give an example of escalators applied to an office building which has large floor areas and is of limited height. Before the escalators are considered, a number of additional considerations should be made. A building that can provide space for 500 persons per floor will cover an area of at least 60 to 75,000 ft² based on a density of 100 ft² of net area per person. This size building could be 650 ft long and 100 ft deep or 300 ft long by 225 ft deep. If only a single pair of escalators is provided, they should be in the center of the square building so that internal walking distance is minimized. If the building is long and narrow a number of groups of escalators are necessary at points that limit maximum walking distance to a maximum of 200 ft. This latter consideration may change the entire economic aspect of the solution to favor elevators. With any vertical transportation arrangement where more than one group of elevators or escalators serve the same floors, each group must have about 20% excess capacity to compensate for unequal demand.

Assuming that the building is square and that a single center core of escalators can be used, the next step is to estimate the peak demand. If the discharge points of the parking or local transportation system are equally accessible to the entrances of the building, a high rate of arrival can be expected. Assuming further that this will be a single-purpose office building with all employees expected to start and quit at the same time, an arrival rate of up to 20% of the population in 5 min can be expected. The calculations would be as follows:

Example 9.1.   Incoming Traffic

Given: 5-story building, 500 people per floor, arrival rate 20% of population in 5 min
Assume: equal attraction per floor, therefore 100 people must be carried to each floor

Floor	Population	5-min Demand	Escalator Must Carry
5	500	100	100
4	500	100	200
3	500	100	300
2	500	100	400
1	500	100	0

Choose escalator to carry 400 persons in 5 min from Table 9.1.

Floors 1 to 2   48 in. @ 120 fpm Nominal capacity 450/5 min vs. 400 demand
Floors 2 to 3   48 in. @ 120 fpm Nominal capacity 450/5 min vs. 300 demand
Floors 3 to 4   32 in. @ 120 fpm Nominal capacity 225/5 min vs. 200 demand
Floors 4 to 5   32 in. @ 120 fpm Nominal capacity 225/5 min vs. 100 demand

It can be noted that a 48-in. (1000-mm) escalator at 120 fpm (0.65 mps) is required to meet the expected demand from floors 1 to 2 (400 people in 5 min versus an escalator capacity of 450 people in 5 min) and that escalators at 90 fpm (0.45 mps) would meet the requirements for floors 2 to 3, to 4, and 4 to 5. The 120-fpm (0.65-mps) escalators were chosen throughout since escalators of two different speeds should not be mixed because of the desire to keep people moving and the possibility of congestion at the landing. If we wished to determine how quickly people could leave the building, an additional calculation to determine the outgoing traffic capacity of the escalators has to be made as follows:

Example 9.2.   Outgoing Traffic

Floors 4 to 5   32 in. @ 120 fpm Nominal capacity 225/5 min, 45/min
Floors 3 to 4   32 in. @ 120 fpm Nominal capacity 225/5 min, 45/min
Floors 2 to 3   48 in. @ 120 fpm Nominal capacity 450/5 min, 90/min
Floors 1 to 2   48 in. @ 120 fpm Nominal capacity 450/5 min, 90/min

500 persons on fifth floor will require
    11.1 min + 1.3 min (riding time at 22 sec per floor) = 12.4 min
500 persons on fourth floor will require
    11.1 min + 1 min (riding time at 22 sec per floor) = 12.1 min
500 persons on third floor will require
    5.6 min + 0.6 min (riding time at 22 sec per floor) = 6.2 min
500 persons on second floor will require
    5.6 min + 0.3 min (riding time at 22 sec per floor) = <u>5.9 min</u>
                            Maximum time to evacuate = 36.6 min

If it is quitting time, full capacity will probably be used plus stairways, so the time will probably be less than 30 minutes. Since the example is based on a single rise of escalators, both up and down escalators should be provided. Using both up and down escalators operating down at 120 fpm, the time will be less than 15 min.

If the building is evacuated in 15 min, it is highly unlikely that local transportation or the exits from the parking lots can accommodate that type of traffic peak. On that basis, only the single down rise of escalators needs to be used.

In addition to the escalators, service elevators for vehicular traffic, moving furniture, deliveries, mail carts, and other movements necessary for office activity must be provided. In some areas, passenger elevators must be provided based on rulings legislated to provide vertical transportation service for handicapped people qualitatively equal to that provided for all others. Service elevators for materials movement must be in addition to the passenger elevators.

Example 9.3 is a typical store situation in which the interest is in providing transportation to turn over the customer attendance on each floor within some given time period. A customer density of 20 ft² of net selling area per person was used. This will, of course, vary with different types of stores and for different floors within the same store. More expensive shops have lower densities and bargain basements, much greater densities. The requirement is to provide sufficient vertical transportation so that people can be carried to a floor to replace those already there who must have transportation exiting. This continuous replacement of patrons is called turnover.

Example 9.3. Two-way Traffic

Given: 6-story store building, 20,000 ft² net selling area per floor, turnover 1 person per 20 ft² per floor per hour

Demand: each floor $\dfrac{20,000}{20}$ = 1000 persons per floor per hour × 2 (up and down)
                    = 2000 per floor per hour

Floor	Demand	Demand on Escalators per Hour		Floors	Equipment Required	Capacity
		Up	Down			
6	2000	1000	1000			
				5 to 6	One pair 32 in @ 90 fpm	2040/hr/escalator
5	2000	2000	2000			
				4 to 5	One pair 32 in @ 90 fpm	2040/hr/escalator
4	2000	3000	3000			
				3 to 4	Two pair 32 in @ 90 fpm	2040/hr/escalator
3	2000	4000	4000			
				4 to 5	Two pair 32 in @ 90 fpm	2040/hr/escalator
2	2000	5000	5000			
				1 to 2	Two pairs 48 in @ 90 fpm	4080/hr/escalator
1	2000	—	—			

The demand from the first to second and from second to third floors requires two pairs of escalators from floor to floor. These would be put in different locations in relation to the entry to the store and the merchandise featured on each floor. A judgment must be made if 32-in. (600-mm) or 48-in. (1000-mm) escalators should be provided above the second floor. Because of the merchandising plan of the store, it may be desirable to have the wider escalators to encourage greater patronage of the floors above the second floor and to accomodate two "socially related" people per step.

The complete vertical transportation system of any store must include passenger elevators for the one-stop shoppers, shoppers with strollers and for handicapped people. In fact any building with escalators as its primary means of vertical transportation should have an elevator for this contingency. Stocking the various floors in a department store requires service elevators which can combine passenger and freight functions.

Further discussion of stores are given in the chapter on commercial buildings.

APPLICATION OF ESCALATORS AND MOVING WALKS

Proper application of escalators or moving walkways requires determination of the expected demand and the nature of the demand on the system. This is true of any vertical transportation system and must be part of the study undertaken when the facility is planned. An estimate as to how many people will be seeking vertical transportation in a period of time must be made. This is an operational and management problem and may depend on many considerations in addition to those listed in the examples. In a sports arena the rate at which tickets can be sold or collected influences vertical transportation requirements. In a store the availability of parking space, whether mass transportation is a factor, the nature of the business, and the price of the merchandise must all be considered.

If both elevators and escalators serve the same floors of a building and people are given their choice to use either, it has been demonstrated that the following approximate division of traffic will occur; (assuming the elevators provide reasonably good service).

	Division of Traffic	
Floors Traveled	Escalator	Elevators
1	90%	10%
2	75%	25%
3	50%	50%
4	25%	75%
5	10%	90%

Obviously, if minimum elevator service is provided, the percentages will favor the escalators. At about six floors, with both elevator and escalator service available to the sixth floor, people will ride elevators to the sixth floor and travel one or two floors down to their destination by escalator.

Location is all-important in the application of escalators and walkways. They have a known through-put of passengers which, once begun, must be accommodated and suitable provisions must be made. By proper location the use of escalator or moving walkway may be enhanced. If the entrance area is restricted fewer people will be able to use the escalator or walkway than if the entrance is wide open. Convenience is another factor; if the facility will save people walking or climbing they will go out of their way to use it. If a building is on a hillside and the escalator provides ready transportation from one level to the other, all the people in the neighborhood will use that escalator if allowed to do so. Providing community transportation can contribute to the success of commercial facilities. People will use the escalators if it will save them time and stores can obtain exposure to potential patrons.

Application opportunities vary with each location and each building. The foregoing discussion is designed to create the awareness of what a walkway or escalator can do. The successful application will depend on its placement and its convenience as it is being used.

CHAPTER TEN

Elevatoring
Commercial Buildings

DEFINITION

Commercial buildings are buildings in which space is rented or used for a definite commercial purpose. This would include all types of businesses, professional office buildings, stores, industrial buildings, self-parking garages, and so forth. Apartments and hotels will be considered as residential buildings and schools and hospitals as institutional buildings.

Commercial buildings have definite vertical transportation requirements since the arrival and departure of population is usually concentrated within certain periods of the working day. Traffic patterns will vary with the use and location of the building and some of the major variations will be discussed.

Although commercial buildings can be located anywhere and a trend to decentralization has been evident, most of them are still in the central business districts of cities and are usually concentrated in areas with reasonably good horizontal transportation. The efficiency of transportation to and from buildings will greatly influence pedestrian circulation patterns within any building and affect its elevator traffic.

High-density horizontal transportation may be provided by a transit terminal near the building. As the trains arrive, groups of people enter the building, most people timing their transit trip to arrive almost at the time they must start to work. With a train capable of discharging hundreds of people in a short time, the building's vertical transportation system is subject to severe incoming peak demands.

The other extreme of horizontal traffic affecting a building's elevators is a suburban office building with remote parking. Demand on the building's elevators may be directly related to the time required for people to park their cars and walk to the building. If there are local coffee shops, the potential elevator passenger may arrive early, have breakfast, then enter the building. If eating facilities are provided in the building, the people may go to their desks and immediately return to the coffee shop.

How people arrive, when they go to lunch, what they do when they are at work, and how they leave are all factors in elevatoring any commercial building. A systematic consideration of these factors will follow.

POPULATION

As important as the arrival rate of traffic is the number of people who will occupy the building. To some extent, tradition governs allocating space within a building. To perform a job a person requires a certain minimum space, which can be as low as 10 or 15 ft^2 (1 or 1.5 m^2). People must get to and from their desks, which requires another 10 ft^2 (1 m^2). If files or records are used, about another 10 ft^2 (1 m^2) must be allocated for that function. The minimum space per person is therefore about 30 to 40 ft^2 (3 to 4 m^2). If dealing with visitors is necessary, additional room to transact business is required which is about 25 ft^2 (2.5 m^2). Size is also status and office workers expect a minimum of about 100 ft^2 (10 m^2) to perform their tasks. The average manager gets about 200 ft^2 (20 m^2) and the executive may have over 300 ft^2 (30 m^2).

The nature of the task to be performed greatly influences commercial building population. A law office, with its necessary reference files and library, requires more square feet per occupant than a drafting office, in which total working area is within reach of the drawing board. Computer departments with their large machines must have more space per occupant than the word processing department, where a person can operate at a single console.

In professional buildings the examining room may require specialized equipment and an area of 200 ft^2 (20 m^2) or more. Technicians' shops—eyeglass fitting, denture making, photo laboratories, and so on—are compact and may average less than 50 ft^2 (0.5 m^2).

Where there are many tenants on a single floor, a considerable amount of space is required to provide the necessary anterooms and passageways. With single large tenants, more of the space is usable since passageways can be minimized and public space is often limited to elevator lobbies and a corridor around the building core.

This latter aspect of larger open floor areas is being refined with a trend toward open space planning and landscaped offices. The traditional partitions are eliminated and modularized work stations consisting of desks, tables, and files integrated as a semimovable unit are utilized. In this way, grouping of people with associated functions can be accomplished and aisle space minimized. The open space plan is more applicable to larger areas of 5000 ft^2 (500 m^2) or more and may result in population densities approaching 100 ft^2 (10 m^2) or less without the sensation of crowding. The vertical transportation engineer must pay close attention to space utilization, experience, and trends in the area where a new building is planned so that accurate population estimates can be made.

Office buildings can be classed as diversified, single-purpose, or as combined single-purpose-diversified. The completely diversified office building can be one is which no more than one tenant occupies more than a single floor and less than 25% of the tenants are in a similar line of work. This last qualification is important because if all tenants are in the same business competition may cause them to start work at the same time, have the same luncheon habits, and have similar patterns of visitor traffic. The tenants may be diversified but their impact on the building will be the same as that of one large firm. We shall refer to this as single-purpose diversified occupancy.

The single-purpose building is exactly that, one firm occupying the entire building or a substantial portion of the building. The notable difference between a pure single-purpose building and a single-purpose diversified building is that the first provides opportunities to control traffic by staggering employee working times. In buildings with multiple groups of elevators, i.e., low-rise, mid-rise, high-rise, it is quite possible for one section to be single-purpose and another diversified if one tenant occupies the floors served by a single group of elevators.

In determining how many elevators are required for a given building, it is necessary to quantify and qualify its population. It is seldom known at the time the building is planned exactly who will occupy each floor, and the quantity of population must be averaged for each floor based on the type of tenancy expected. Table 10.1 gives typical values of population related to the net usable square feet in each building. This net usable square feet should not be confused with net rentable area which often includes columns, toilets, elevator lobbies, radiator or convector space, and a portion of an air conditioning equipment room which may be on another floor. If net rentable area is used as a basis for population, an allowance must be made.

The thoroughness with which the building program is planned has a great influence on the population of the building. If a building is being built on an investment basis, tenant rental to follow, population should be established conservatively, based on the experience of comparable buildings (Tables 10.2 and 10.3). If the building is for a specific tenant who has planned and allocated space requirements, advantage should be taken of that planning in establishing population, with allowance for expansion.

Because elevators must be planned to serve the needs of the building population, the importance of correctly estimating population cannot be overemphasized. It can be costly to underestimate population or make undue allowances for absent employees. These variables can depend on current business conditions which may change and affect the entire basis of elevatoring. The essential

Table. 10.1. Population Factors, Commercial Buildings

Diversified	150 ft^2 (15 m^2)	net usable area per person[a]
Diversified single-purpose	135 ft^2 (13.5 m^2)	net usable area per person
Single-purpose	120 ft^2 (12 m^2)	net usable area per person

[a] Net usable area = gross area less elevator shaft and lobby space, mechanical space stairways, janitorial, columns, toilets, corridor around core, and convector space.

Table. 10.2.   Approximate Net Usable Area,[a] Various Height Buildings [15 to 20,000 Gross ft^2 (1500 to 2000 m^2) per Floor]

0 to 10 floors	Approximately 80% gross
0 to 20 floors	Floors 1 to 10 approximately 75% gross
	11 to 20 approximately 80% gross
0 to 30 floors	Floors 1 to 10 approximately 70% gross
	11 to 20 approximately 75% gross
	21 to 30 approximately 80% gross
0 to 40 floors	Floors 1 to 10 approximately 70% gross
	11 to 20 approximately 75% gross
	21 to 30 approximately 80% gross
	31 to 40 approximately 85% gross

[a] Net usable area is approximately 85% of full-floor standard rentable area.

consideration is that space is available which can be used by personnel whether or not it is occupied at any particular time. It is also extremely important to estimate both resident and visitor traffic.

This latter point can be emphasized when the arrival traffic in an existing building is observed. Traffic will arrive at the building over about a 1- to 2-hr time period and the total count should be close to the building population on

Table. 10.3.   Suggested Building Population Factors Related to Building Height—Based on Net Usable Area

0 to 10 floors	125 ft^2 (12.5 m^2) per person
0 to 20 floors	Floors 1 to 10 125 ft^2 (12.5 m^2) per person
	11 to 20 130 ft^2 (13 m^2) per person
0 to 30 floors	Floors 1 to 10 125 ft^2 (12.5 m^2) per person
	11 to 20 130 ft^2 (13 m^2) per person
	21 to 30 140 ft^2 (14 m^2) per person
0 to 40 floors	Floors 1 to 10 125 ft^2 (12.5 m^2) per person
	11 to 20 130 ft^2 (13 m^2) per person
	21 to 30 140 ft^2 (14 m^2) per person
	31 to 40 150 ft^2 (15 m^2) per person
Other Commercial Space	
Professional buildings	200 ft^2 (20 m^2) per doctor's office
Self-parking garages	300 ft^2 (30 m^2)[a] 1.2 persons per auto
Stores	Customer density of 10 to 40 ft^2 (1 to 4 m^2) of net selling area[b]
Industrial buildings	
Factories	Depends on manufacturing layout and product
Drafting	80 to 100 ft^2 (8 to 10 m^2) per draftsman

[a] 300 ft^2 is an average for small and large autos.
[b] Net selling area is area open to the public.

that particular day. During that time period, a certain 5 min will represent the peak of the arriving traffic. In some buildings there may be two or three five-min periods where the peak traffic is almost identical due to the variances in horizontal external traffic to the building or starting times of firms occupying the space.

If the traffic were to be observed over many days, a certain pattern would emerge and the average peak 5-min traffic based on those many observations is what the elevators should be designed to serve. The peak 5 min also has a direct relation to the resident population. The percentage may remain the same but the quantity of people will vary as the population changes. The following section describes the characteristics of peak traffic.

ELEVATOR TRAFFIC IN COMMERCIAL BUILDINGS

Once the population that requires vertical transportation is established, the next step is to determine the quantity and characteristics of vertical transportation required. As discussed in the chapter on elevator traffic, we are seeking a critical 5-min traffic period on which to base handling capacity and against which to check all other active traffic periods.

The up-peak or start-to-work period has traditionally been the basis for establishing the quantity of elevator-handling capacity for an office building. This is usually a critical period as surface transportation and subways discharge passengers at the building and the arrivals are at their highest. It is essential to clear the lobby and get people to their desks so they can begin work.

During recent times, starting about 1970, a number of changes have been evolving. Many firms have recognized the difficulties of commuting and are allowing employees flexibility in starting and quitting times. In some localities, local governments, in an effort to reduce traffic congestion, have encouraged staggered starting and quitting times. Progressive firms have recognized that there is no need for all departments to start work at a fixed time and have instituted a staggered work schedule. The net result of these efforts has been a remarkable change in elevator peak traffic wherein a considerable down traffic exists during heavy periods of up traffic. No longer is the up-peak a pure up-peak, but rather an up-peak with approximately 10% of the total up traffic traveling down—up-peak with 10% down traffic.

The traffic will usually peak during a 5-min period before the time most people start work. The intensity of this peak consisting of primarily up traffic with some down traffic is stated as a percentage of building population and forms a means of relating one building to another and the basis of establishing the elevatoring of each (Figure 10.1).

The other periods of the day when elevator traffic may be critical are at lunchtime and quitting time. Because it is general practice to stagger lunchtimes, these traffic peaks are usually not as severe as the incoming peak. This is not always true since lunchtime may also be a very critical traffic period for a single tenanted building if a full-service cafeteria is provided. Quitting time is an

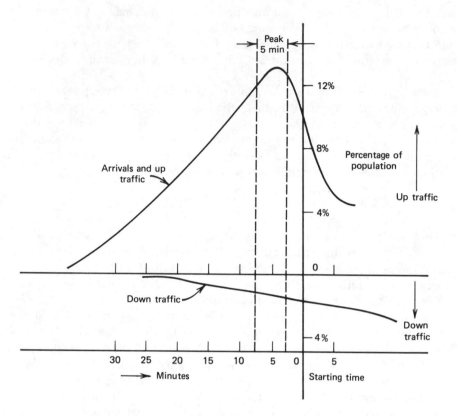

Figure 10.1. Typical arrival rate and elevator traffic distribution in an office building.

intense period of elevator traffic; however, elevator efficiency is greater during an outgoing rush. People will crowd elevators more than they will during arrival periods since they are usually more anxious to leave work than they are to get to work. With this crowding, and because the passengers are distributed over many floors rather than waiting in the lobby as during up peak, the elevators tend to make fewer stops, hence more trips in a given period of time. The net result is that, for a given number of elevators, the outgoing capacity is a substantial percentage greater than for incoming capacity.

If good horizontal transportation exists, the incoming peak traffic with or without opposing down traffic will remain the critical traffic period. If an extended arrival rate is expected, the noontime period of two-way traffic may become critical. In professional office buildings, stores, and industrial buildings with shift changes the two-way traffic period is the most critical.

Sample percentages of the population that must be served during a critical 5-min period in various building types are given in Table 10.4.

Table. 10.4.   Expected Peak Traffic Periods—Various Commercial Buildings

	Percent of Population in a 5-min Period		
	Peak Arrivals	Up-Peak with 10% Down Traffic	Noontime or Two-way
Office Buildings			
Diversified offices	10 to 11%	11 to 12%	10 to 12%
Diversified single-purpose	11 to 13%	12 to 15%	12 to 15%
Single-purpose	12 to 18%	13 to 20%	13 to 17%
Other Building Types			
Characteristic Peak Traffic:			
Professional buildings	Peak traffic	Two-way, based on 1 to 2 visitors per doctor each 15 min coming and going	
Garages—self-parking (assume sufficient ramps to fill or	Peak traffic commuter garage	10 to 15%, one-way traffic	
empty garage in 1 hr)	Peak traffic, store, transpor-tation terminal, garage	10 to 15%, two-way traffic	
Stores	Population to be turned over, i.e., up and down (two-way) in 1 hr		
Industrial buildings	Peak traffic, 15 to 20% (up-peak or two-way).		

Interval

The quality of service given by any elevator system is also reflected by the interval or frequency of that service. Because the reputation of commercial buildings and the rentals they can command are based on the quality of service the building offers, the best quality of elevator service is a necessity. A building may offer the best space layout and services, but if people must wait too long for elevator service the value of other advantages can be lost. Excessive waiting times have been determined by analyzing service in buildings where complaints were minimal, and experience has shown that an up-peak without down traffic loading interval of between 25 and 30 sec produces excellent elevator service in any office building. Loading intervals of from 30 to 35 sec are considered good where some degree of down traffic exists. Loading intervals of over 35 sec are certain to lead to complaints in office buildings of any type and should be avoided.

Two-way traffic operating intervals of from 30 to 40 sec are generally accept-able provided that the elevator operating system is designed to maintain that range or better. With a two-way operating interval of 40 sec designed and poor

Table. 10.5. Suggested Intervals

Office Buildings	Up Peak[a]	Up Peak[a] with 10% Down Traffic	Two-way[b]
Diversified	25 to 30 sec	30 to 35 sec	35 to 45 sec
Diversified single-purpose	23 to 28 sec	28 to 33 sec	33 to 43 sec
Single-purpose	20 to 25 sec	25 to 30 sec	30 to 40 sec
Professional buildings	—	—	30 to 50 sec
Self-parking garages	40 to 50 sec	—	40 to 60 sec
Stores	—	—	30 to 50 sec
Industrial buildings	25 to 30 sec	—	30 to 40 sec

[a] Loading interval—time between elevators departing from a main terminal.
[b] Operating interval—frequency of elevators passing an upper floor.

elevator operation, waits over 40 sec will be frequent and complaints will be received. In general 30 sec is the maximum wait an average person will accept without complaint in a busy commercial atmosphere.*

In professional buildings, stores, or industrial buildings, the maximum two-way operating interval should never exceed 50 sec. The preferable safe maximum should be 40 sec for quality service. Again the elevator operating system should be designed to maintain that maximum or better.

The use of Table 10.5, on suggested intervals, is predicated on providing sufficient traffic-handling capacity as shown in Table 10.4. Multiple entrances to the building, upper-floor cafeterias, roof or basements stops on the elevators, and any odd openings will adversely affect the operating interval and must be considered in calculations.

Capacity and Speed

The combination of capacity and speed (often referred to as duty) of elevators for commercial buildings should be selected to provide service of the highest quality. The typical office building tenant is continually using elevators in his or her own and other buildings and soon learns what to expect in the way of service standards.

The minimum recommended size of an elevator for any building is the 2500-lb (1200-kg), 82 in. (2100 mm) wide by 51 in. (1300 mm) deep car inside. This size allows the architect to take advantge of superior inside decoration and allows the use of the 48-in. (1200-mm) center-opening doors.

For most office buildings the 3000-lb (1400-kg) car should be the minimum. Office buildings that are prominent and heavily traveled should have 3500-lb (1600-kg) and 4000-lb (1800-kg) elevators and monumental buildings such as

---

* Refer to Chapter 5 for a discussion of interval related to waiting time.

the headquarters of large corporations should always have 4000-lb (1800-kg) or larger elevators.

For commercial buildings other than offices the minimum recommended size of the elevator varies with the traffic demand and use of the building as shown in Table 10.6.

Elevator speed may be of secondary importance versus prompt floor-to-floor travel. If an elevator cannot attain a floor-to-floor performance time of 9 to 10 sec including prompt and efficient door-opening and closing speed, it is going to seem to be a slow, sluggish elevator. This performance is a function of the elevator motion control system and should be of the highest quality in any prestigious building. Speed becomes very important where express runs are provided such as in the high-rise portions of an office building of 15 or more floors (see Table 10.8). Speed is also of considerable importance if the elevator must make many stops such as in a tall, slim building of 20 or more floors where only one group of elevators may be provided to serve many floors.

For the low-rise, suburban-type office building of three or four floors, hydraulic elevators with speeds of up to 150 fpm (0.75 mps) are often acceptable. Consideration should always be given to underslung traction elevators which

Table. 10.6. Suggested Elevator Capacity (pounds)—Commercial Buildings

Type of Building	Class of Building		
	Small	Average	Large or Prestige
Offices, suburban	2500	3000	3500
Service elevator[b]	4000	4000	4000
Offices, downtown	3000	3500	4000[f]
Service elevator[b]	4000	4500	6500
Professional offices			
Passenger[a]	2500	3500	4000
Service[b]	4000	4000	5000
Stores			
Passenger[c]	3500	3500	4000
Service[d]	4000	4000 to 6000	6000 to 8000
Garages	2500	3000	3500
Industrial[e]	4000	4000	4000

[a] As a practical consideration, 4-ft center-opening doors on a 3500-lb car will allow a mobile stretcher to enter.
[b] A hospital-shaped car should be provided.
[c] Wide, shallow cars with widest possible center-opening door desirable.
[d] Capacity is primarily to obtain largest size elevator possible.
[e] Consideration should be given to combination passenger-freight elevators.
[f] Special single purpose occupancy may require 5000 lbs.

Metric equivalent, 2500 lb = 1200 kg   3000 lb = 1400 kg
                        3500 lb = 1600 kg   4000 lb = 1800 kg
                        6500 lb = 3000 kg   5000 lb = 2300 kg

Table. 10.7. Suggested Elevator Speeds

	Class of Building			
	Small	Average	Large or Prestige	Service
Office buildings (including professional offices)				
Up to 5 floors	200 fpm[a]	300 to 400 fpm	400 fpm	200 fpm[a]
5 to 10 floors	400 fpm	400 fpm	500 fpm	300 fpm
10 to 15 floors	400 fpm	500 fpm	500 fpm	400 fpm
15 to 25 floors	500 fpm	700 fpm	700 fpm	500 fpm
25 to 35 floors	—	1000 fpm	1000 fpm	500 fpm
35 to 45 floors[b]	—	1000 to 1200 fpm	1200 fpm	700 fpm
45 to 60 floors[b]	—	1200 to 1400 fpm	1400 to 1600 fpm	800 fpm
over 60 floors[b]	—	—	1800 fpm	800 fpm
Stores				
Up 2 to 5 floors	150 fpm[a]	200 fpm	300 fpm	200 fpm[a]
5 to 10 floors	400 fpm	400 fpm	500 fpm	400 fpm
10 to 15 floors	500 fpm	500 fpm	500 to 700 fpm	400 fpm
Garages				
2 to 5 floors	200 fpm[a]	200 fpm[a]	200 fpm[a]	
5 to 10 floors	200 fpm	300 fpm	400 fpm	
10 to 15 floors	300 fpm	400 fpm	500 fpm	

[a] 150-fpm hydraulic acceptable.
[b] Skylobby design should be considered for this height.
Metric equivalent, 150 fpm = 0.75 mps, 200 fpm = 1 mps
                    300 fpm = 1.5 mps, etc.
                    mps (approx.) = fpm/200 = mps

Table. 10.8. Time and Distance Required to Attain Full Speed or to Slow Down from Full Speed (Approximate)

Ultimate Elevator Speed, fpm (mps)	Time (sec)	Distance, ft (m)	Minimum Number of Floors[a] Required for Express Run
500 (2.5)	2.9	12 (3.6)	2
700 (3.5)	3.6	20 (6)	4
1000 (5.0)	4.7	38 (11.5)	6
1200 (6.0)	5.5	53 (16)	10
1400 (7.0)	6.2	70 (21.2)	12
1600 (8.0)	6.9	80 (24.2)	14

[a] Based on 12-ft (3.6-m) floor heights, 1 m = 3.3 ft (approx.).

230

are capable of higher speed without the necessity of extensive overhead structure. Table 10.7 shows recommended elevator speeds for various types of commercial buildings.

When buildings have high- and low-rise elevators, the high-speed elevators are required to traverse the express run in the shortest possible time. The limitation is the rate of change of acceleration as described in Chapter 7. This results in a certain minimum number of floors being required for the express run with various speeds of elevators. This variation is shown in Table 10.8. An example of low-rise–high-rise elevatoring is given later in this chapter.

LAYOUT AND GROUPING OF ELEVATORS

In any multistory commercial building the vertical transportation system should visually dominate the lobby. Since the system is, in effect, the main entrance to the upper floors, people should be directed to the elevators or escalators both physically and visually. Signs, clearly visible from each building entrance, should plainly indicate each system and the floors it serves.

Vertical transportation should be grouped in one area, either the central "utility core" of the building or a service tower along one wall. Long corridors from the main entrance to the elevator lobbies should be avoided. The main entrance will be the one closest to the main horizontal transportation. People will not take an indirect route to a main entrance if there is a secondary entrance next to the transit station.

Elevators in office buildings are commonly installed in groups of either four, six, or eight cars. In many buildings the high-rise shaft space should be conserved for service elevators, building services, stairs or smoke shafts. The highest-rise elevators may be in a five- or seven-car group. This allows the sixth or eighth shaft of a six- or eight-car core to be used for a service elevator.

As pointed out in the section on grouping of elevators, the six- and eight-car groups should have open-ended lobbies to leave space for people to wait before they board elevators. The space adjacent to the elevator lobby is a necessary reservoir for people entering at the main entrance and its extent is based on the ability of the elevators to serve the incoming traffic. It should be equal, as a minimum, to the elevator lobby area.

Some typical core arrangements for high-rise buildings are shown in Figure 10.2. The elements included in the core are usually two stairways, electrical closets, telephone closets, air distribution shafts, toilets, pipe shafts, passenger elevators, and a service elevator. The balancing of required core space and gross area per floor is often the result of repeated alternative arrangements until the most favorable net to gross area is accomplished.

Once determined for a particular building size, the layout and space requirements of elevators is relatively inflexible. The design of the building should always proceed by designing the elevators and core first if a satisfactory elevator plant is expected and the cost of redesigning is to be avoided. The elevator

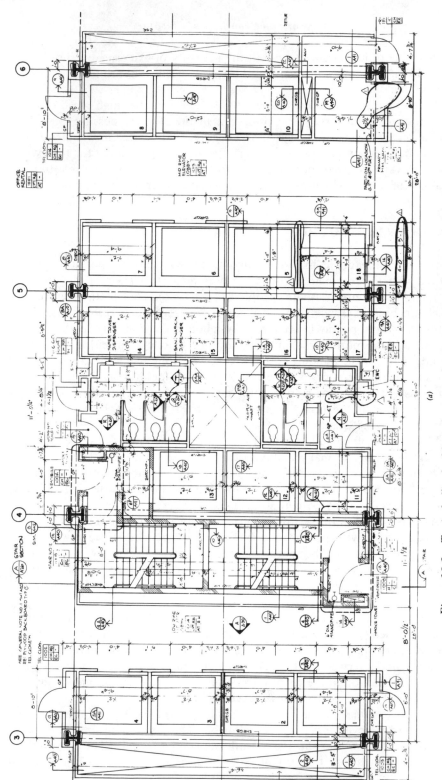

Figure 10.2. Typical office building elevator and utility cores: (*a*) three groups of elevators in a 40-story building (Courtesy Skidmore, Owings, and Merrill);

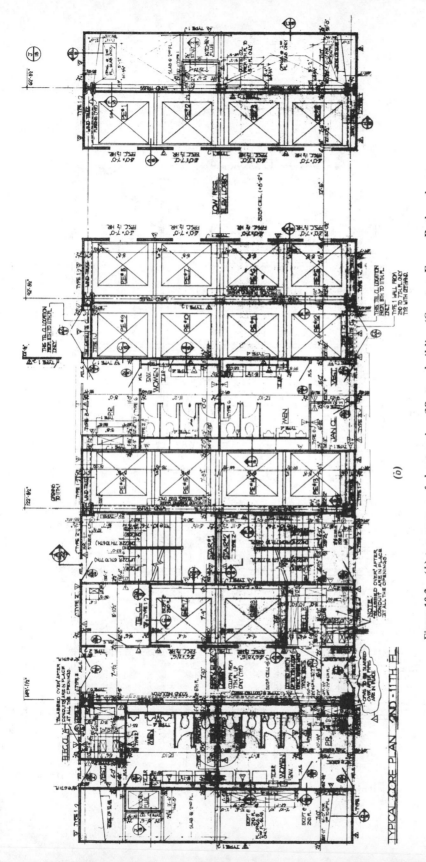

Figure 10.2. (*b*) two groups of elevators in a 28-story building (Courtesy Emery Roth and Sons);

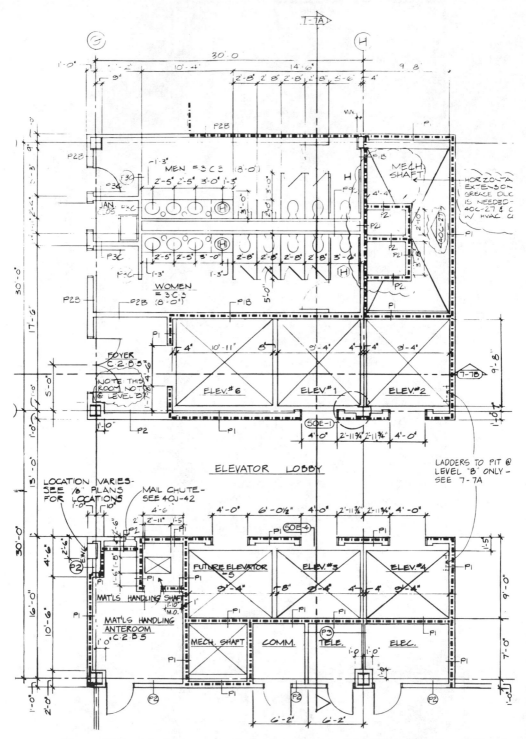

Figure 10.2. (*c*) single group of elevators in a low-rise building (Courtesy Hellmuth, Obata, and Kasselbaum).

design must be resolved early because most other aspects of the building design depend on the elevator design, and it is often necessary to contract for elevators long before contracts are awarded for the other aspects of the building.

TRANSFER FLOORS

Local and express or multiple rises of elevators in a building should always be provided with a transfer floor. Any tenant who occupies floors served by more than one group of elevators requires a transfer floor. Otherwise, people would have to travel to the lobby and change to the next group of elevators. At the transfer floor people will also have to change elevators but will not be backtracking. The time saved can be substantial and depends on the relative interval in each group of elevators and the traffic at that moment.

At the transfer floor the elevator operating system should be arranged to allow the higher-rise elevator to stop only for an up landing call or down car call. This avoids the possibility of a tenant using the high-rise group to reach the transfer floor in the morning, when it is the first stop after the express run, so employees do not have to wait for intermediate stops, and the lower-rise elevators in the evening, when the tenant floor is the first stop in the down direction. If the employees were to fill the car, they would get priority down service. Response only to up landing calls and down car calls at the higher-rise transfer stop avoids this possibility. This operation is referred to as limited transfer floor service.

A second alternative to transfer floor and elevator flexibility requirements is to provide for future entrances on the high-rise elevators at overlapping floors. The entrances could be blocked closed door frames, and as the building matures and changes in elevator requirements are apparent, these entrances could be put into use and necessary adjustments made to the elevator operation.

Transfer between a high-, intermediate-, and low-rise elevator can be a requirement in a single-purpose office building. Normally a single transfer floor on each group of elevators should be sufficient; however, if the nature of the business is such that a great deal of interfloor traffic will take place, openings can be installed in a high-rise elevator hoistways at the low-rise transfer floor. The opening that serves the low rise should be designed to operate only in response to an up landing call or down car call. Such arrangement is often referred to as a "crossover" floor.

ELEVATOR OPERATION IN COMMERCIAL BUILDINGS

Based on an estimate of the expected traffic in a building, the elevators proposed, and the handling capacity and interval calculated for various periods of the day, a traffic flowchart can be prepared. This flowchart will indicate the complexity of the elevator traffic and the varieties of elevator operations required to handle that traffic. A rough chart based on observations made in similar existing buildings can be used as a start and additional refinements added as requirements are analyzed in greater detail. An example of such planning is shown in Figure 10.3.

Figure 10.3 shows the expected traffic in a typical diversified office building, a pattern characteristic of most such buildings. The intensity will vary inversely with the degree of tenant diversity.

In planning elevator operations the service expected of the elevators must be considered. As an example, if much visitor traffic is expected during the day, the system may have to provide for varying degrees of two-way traffic in the heavy up, heavy down, and balanced directions. If considerable traffic between floors is expected, additional provisions for traffic of that nature may be needed.

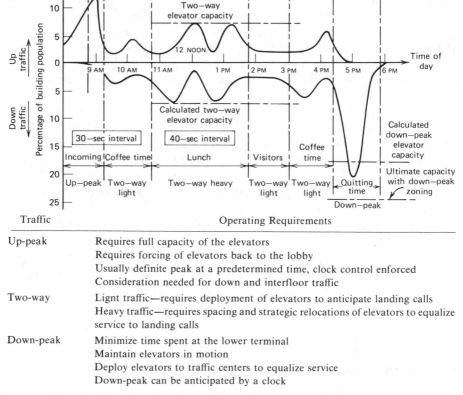

Traffic	Operating Requirements
Up-peak	Requires full capacity of the elevators
	Requires forcing of elevators back to the lobby
	Usually definite peak at a predetermined time, clock control enforced
	Consideration needed for down and interfloor traffic
Two-way	Lignt traffic—requires deployment of elevators to anticipate landing calls
	Heavy traffic—requires spacing and strategic relocations of elevators to equalize service to landing calls
Down-peak	Minimize time spent at the lower terminal
	Maintain elevators in motion
	Deploy elevators to traffic centers to equalize service
	Down-peak can be anticipated by a clock

Figure 10.3. Typical traffic flow—diversified office building traffic observed on and off elevators at the lobby.

Each building has its unique requirements depending on location, expected tenancy, the nature of the tenants' businesses, and visitor traffic, as well as external transportation to and from the building. Typical traffic flow diagrams for other types of commercial buildings (see Figure 10.4), supplemented by information on operation given in Chapter Seven, may be used as a guide for a particular building in selecting a suitable operating system.

Special Requirements

Commercial buildings are, in general, public buildings and are expected to have many visitors. The elevator layout and design should be such that it will be attractive to the visitor and offer service. Tenants, as well as visitors, appreciate attractive elevator design in addition to prompt and efficient service with minimum effort.

To fulfill this requirement the architect must employ an informational system that is clear and distinct. Signs that indicate the floors served by each elevator group must be clearly visible from anywhere in the building lobby. Landing lanterns for each elevator should be easily discerned from every point in the elevator lobby. The elevator call button must also be plainly identifiable and readily accessible from routes of passenger approach and located so that passengers will wait near the elevators. Floor numerals applied to the edge of hoistway doors at each floor help the rider identify his or her floor. Codes enacted to provide service to the handicapped require such an approach.

Communication between elevator pasesngers and building staff through a central telephone system or intercom is required in many localities and a rule in the A17.1 elevator code. If the building has a reccptionist or lobby floor attendant, the attendant should be able to communicate with the elevators so that action can be taken if necessary. In many buildings a lobby indicator panel shows the position of the various elevators and can be equipped with flickering indicators to call attention to a car delayed beyond a predetermined time. Communication by intercom can establish the reason and necessary help can be summoned. Building personnel can also observe possible elevator system failures at the lobby station by means of indicator lights and take appropriate action. Such lobby stations are often located at a security desk together with other building monitoring functions. In some localities, a fire command center is developed so that emergency operations including elevator operations can be directed from a central location.

In smaller buildings the indication and communication equipment can be located in the building office, security room, or at any place where surveillance is available. In buildings where only part-time help is in attendance an emergency call system to a central protective agency can be provided. Many local safety codes as well as the A17.1 Elevator Code require communication outside the building if the building is not manned 24 hr each day.

Operations to call a particular car to the lobby for cleaning or some other purpose are desirable. One such arrangement is a number of key switches, one

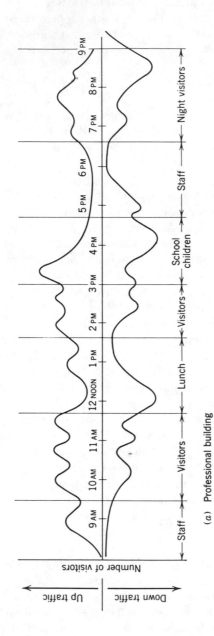

(a) Professional building

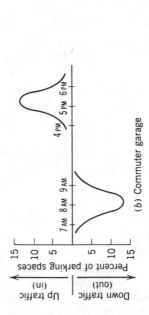

(b) Commuter garage

Figure 10.4. (a) Professional building—elevator traffic in and out at lobby. (Scale will depend upon the number of professional offices in the facility.) (b) Commuter garage.

238

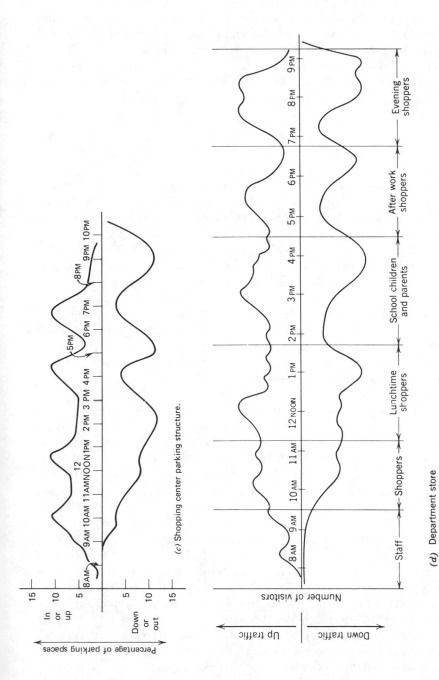

(c) Shopping center parking structure.

(d) Department store

Figure 10.4. (c) shopping center parking structure. (Traffic is based on automobiles and occupants in and out of garage entrance. If garage is a basement or above-ground facility, location of pedestrian exit and entrance floor assumed at ground level.) (d) Department store or shopping mall. (Scale will depend on the total traffic which will be divided approximately 90% escalators and 10% elevators.)

for each car, in a control panel. Operating the switch can call the car so it may be placed on special service such as operating from its car buttons independently of the group or so it may be shut down for necessary cleaning or maintenance. The switch should also serve to close elevator doors when a car is locked out of service for, say, nighttime security.

Emergency requirements such as fire safety, earthquake, riots, and power failures require additional operational considerations. These will be discussed fully in a later chapter.

Service Elevators

A separate service elevator or elevators is a necessity in any building of over 250,000 ft^2 (2500 m^2) of net usable area (300,000 ft^2 gross) or where the passenger elevators are in full use during most of the day. In smaller buildings, the need for freight service is light enough so that one of the passenger elevators may be diverted to this purpose during off-peak periods. This is not feasible if much visitor traffic is expected and a separate service elevator should be provided. One service elevator for each 300,000 to 500,000 ft^2 gross is suggested.

Transportation of myriad items used for tenant activities and building operation requires service elevators. These items may range from mail and office supplies to new furniture, soft drinks, coffee carts, lunch wagons, wall partitions during renovations, or masonry supplies. Increasing cost of service personnel makes it economical to schedule renovation of repairs during the day and, for tenant convenience, management wants to avoid passage of building material or maintenance workers through the passenger lobby.

In many buildings simple freight traffic can attain volumes amounting, in the form of mail, trash, office material, and supplies, to about 150 to 200 lb per employee per week. A 250,000-ft^2 net usable area building may have as many as 2000 employees and will require about 250 tons of material to be moved in and out each week. With a service elevator that has a 4000-lb capacity minimum, and averaging trips every 10 min at half-capacity, handling 250 tons will require a full 40 hr.

Most organizations depend on mail and insist on its prompt delivery. Concurrent mail and passenger peaks will pose a dilemma for management. An unsatisfactory solution is to spread the mail on the lobby floor to be picked up by tenants as they enter the building. A separate service elevator can expedite mail delivery and provide the additional handling capacity required for service needs during the rest of the day. Various mail distribution systems are available and utilize automatic unloading dumbwaiters or conveyors. These specialized systems increase the utilization of the service elevators especially if mail is a priority demand as would occur in a single-purpose building. Automated mail handling systems will be discussed in a later chapter.

The sizes recommended for service elevators will vary with building use. The 4000-lb hospital-shaped platform as discussed in Chapter Twelve on institutional buildings is the most satisfactory minimum. The dimensions of the entrance should consider the largest-sized equipment expected to be in the building and the normal size of the building doors. If wider elevator entrances are required, two-speed sliding or two-speed center-opening doors should be provided. The use of vertical biparting doors is generally not recommended for commercial buildings and nonattendant operation.

A larger elevator with a capacity of 6500 lb is usually required in the larger buildings. The elevator should have a cab interior height of 12 to 14 ft (4300 mm) and doors 8 ft (2400 mm) or higher. A hoist beam at the top of the elevator cab allows rolls of carpet which are 12 to 15 ft (4000 mm) long to be moved and carried. The extra height is also an aid in building completion and renovation since it allows long pieces to be fabricated off site and easily transported up into the building. The additional capacity and size allows the transportation of air conditioning equipment, transformers, large office machines or anything of a larger weight or size than can be carried in a passenger elevator.

In professional buildings, especially medical office buildings where people may arrive on stretchers, a service elevator is necessary to handle the stretcher or any unusual loads. For medical office buildings of any size the 4000-lb hospital-shaped car is usually the most satisfactory service elevator. For a smaller medical building, a 2500-lb car with 42-in. side-sliding doors or a 3500-lb elevator with a 48-in. center-opening door may be used for mobile stretchers.

Easy access to the service elevator at the loading dock and at the upper floors is of prime importance in locating the service elevator. If one of the passenger elevators is to be used for part-time service, traffic flow between that elevator and the building service entrance must be considered. A rear entrance on the elevator may solve the access problem. If a rear entrance is to be used it should be designed to be operative during controlled periods and with the use of an attendant. A separate landing call button is desirable which will operate an annunciator in the elevator.

Seperate service elevators are of inestimable advantage during the building construction periods. Once the building is topped out, the service elevator can be quickly completed sufficiently for temporary operation and allow early removal of an outside hoist. The cost advantage may pay for the investment in the elevator many times over.

Cafeterias and Restaurants

A club, restaurant, cafeteria, or any eating facility in a commercial building requires a separate service elevator. It is difficult to carry foodstuff, carts, garbage, or bottles on an elevator without damaging the sides, spilling, or leaving an odor. Protective pads can be used, but strict management control is required to ensure their use. A facility usually requires considerable time to transfer its supplies from a loading dock area and such supplies will often arrive at the height of the morning rush.

Eating facilities in a building also place extra demands on the passenger elevators. The size of the facility, its expected turnover, and its mode of operation must be considered in the initial elevatoring of the building. In general any facility designed to serve about 300 people which is open to the public requires an additional passenger elevator, plus use of the building's service elevator, over the normal requirements of the building. An additional passenger elevator would be required for each multiple of 300. Because public restaurants are often located on the top floor and generally cater to people outside of the building, the extra elevator(s) can be the shuttle type and travel only between the restaurant and the ground floor. Many successful top-floor restaurants use outside, glass-enclosed elevators to attract and accommodate their clientele. This type of equipment is discussed in a later chapter.

Cafeterias that are to be used by the building's tenants are preferably located in the basement or on the second floor. The main building elevators should not stop at the cafeteria floor and the cafeteria should be served separately by escalators or by separate shuttle-type elevators. Cloakrooms on the cafeteria floor will reduce elevator use by employees who eat and leave the building. Reducing this traffic lessens interference with the next group of employees going to lunch. Separate kitchen-supply elevators should be located near a loading dock and are desirable for the basement or second-floor cafeteria.

Locating a cafeteria on an upper floor such as the transfer floor between a high-rise and low-rise elevator group or on any upper floor will create a constant stop on the elevators during the noontime traffic period. Elevatoring calculations must consider this extra stop as well as the amount of traffic expected as a result of the cafeteria location. An example is given in Example 10.2D. Any upper-floor eating facility requires that a separate service elevator be provided in the building. Cafeterias should never be located at an intermediate floor within a group of elevators unless it is exclusively for the residents within that group and ample elevator service is available.

TYPICAL OFFICE BUILDING ELEVATORING

Before starting the planning of an office building of any size, the architect or builder should investigate existing buildings of a similar size and nature. The real estate boards can suggest comparable examples. The plot size and local zoning laws usually establish the per-floor area and height of the new building as well as required use and other economic factors. Initial elevator requirements are usually based on very meager information that includes the expected height and approximate floor area of the building and its intended use. With more specific information a better-elevatored building can result.

Sample calculations for elevatoring a diversified office building appear in Example 10.1, where the type and size of the building is indicated and its elevatoring is developed. Introduction of a garage facility in the basement changes an acceptable elevator solution to a marginal one. By providing a separate shuttle elevator between the lobby and garage floor, the four elevators are acceptable. In addition to enabling the four main elevators to serve the building, the use of a shuttle facilitates use of the garage. The shuttle elevator allows the garage to be operated independently of the main elevators and possibly used as a public facility on weekends and evenings without impairing the security of the building.

Solving elevatoring problems often takes the form of considerable compromise between the optimum size of the building and the number of elevators to be provided. Each elevator represents a cost factor in both initial equipment investment and rentable area sacrificed. With land selling as high as $1500/ft^2 in some areas of large cities the greatest return on total investment is necessary. Commanding rents to yield an adequate return requires the best possible elevator service. Space is necessary, however, so for tall buildings the initial cost of elevators relative to all the other aspects of a building is generally about 12% of total construction cost. The difference between the best and poorest elevators from a speed and capacity standpoint is only about 10 or 15% of their cost—less than 2% of the entire project—but the return in terms of satisfied tenants can be far greater.

Example 10.1

A. Given: diversified offices, suburban location, investment-type building; 12,000 ft^2 gross per floor, 10 floors, 12-ft floor heights. Population: 12,000 × 0.80 = 9600 ft^2 net per floor @ 125 ft^2 per person = 77 people per floor. Total population floors 2–10 = 693 people.

Assume: 10 passengers up per trip, elevators travel at 400 fpm, 48-in. center-opening doors, floors 1 to 10, 12 ft floor-to-floor, 9 × 12 = 108 ft, elevator rise

Probable stops: 10 passengers, 9 stops = 6.2, no highest call return

Time to run up, per stop: $\dfrac{108}{6.2}$ = 17.4 ft. rise per stop

  17.4 ft at 400 fpm = 6.1 sec

Time to run down: $\dfrac{108}{1+1}$ = 54 ft $\dfrac{(54 - 17.4) \times 60}{400}$ + 6.1 = 11.6 sec

Elevator performance calculations:

Standing time

Lobby time 10 + 0.8	=	10.8 sec
Transfer time up stops 6.2 × 2	=	12.4
Door time, up stops (6.2 + 1) × 5.3	=	38.2
Transfer time, down stops 1 × 4	=	4.
Door time, down stops 1 × 5.3	=	5.3
Total standing time		70.7 sec
Inefficiency, 10%	=	7.1
	Total	77.8 sec

Running time

Run up 6.2 × 6.1	=	37.8
Run down 2 × 11.6	=	23.2
Total round-trip time		138.8 sec

$$HC = \frac{(10 + 1) \times 300}{138.8} = 23.8 \times 4 = 95 \text{ people}$$

Percent HC: $77 \times 9 = 693$, $95/693 = 13.7\%$

Interval: $138.8/4 = 35$ sec

Four 2500-lb elevators @ 400 fpm required as a minimum.

B. If a single basement garage for 200 cars is added: 200 automobiles at 1.5 people per auto = 300 people = about 50% of building population. Therefore it is likely that elevators will travel to the basement every second trip. Recalculate round-trip time: building B, floors 1 to 10

Additional time required for trip to basement garage

Stop at lobby—part of original round trip

Run to basement, 10 ft	=	5.1 sec
Transfer at B	=	4.0 sec
Add door time B and 1, 2×5.3	=	10.6 sec
Run up to 1, 10 ft	=	5.1
		24.8 sec
Assume every second trip to B, 24.8/2	=	12.4 sec

New round-trip time $138.8 + 12.4 = 151.2$ sec

$$\text{New HC} = \frac{(10 + 1) \times 300}{151.2} \times 4 = 87 \text{ people}$$

New percent HC: $87/693 = 12.6\%$

New interval: $151.2/4 = 38$ sec

Example 10.2   Example of Low-rise-High-rise Elevators in an Office Building

Given: single-purpose office building, downtown location, 20,000 ft^2 gross per floor, 20 floors, 12 ft floor to floor, 20-ft lobby height. Population factor average 125 ft^2 per person based on net usable area.

A. Population calculations—assumptions

Floors 2 to 11 $20,000 \times 0.75 = 15,000 \div 125 = 120$ people per floor

Floors 12 to 20 $20,000 \times 0.80 = 16,000 \div 125 = 128$ people per floor

B. Criteria: 5-min elevator capacity—15 to 20%, interval—30- to 35-sec

C. Elevator calculations

1. Low rise: assume 6 3500-lb elevators @ 500 fpm for floors 1 to 11, 48-in. center-opening doors, Rise: $20 \text{ ft} + (9 \times 12) = 128 \text{ ft}$

Probable stops: 16 passengers, 10 stops = 8.2

Up floor-to-floor time $\dfrac{128}{8.2} = 15.6 = 4.7$ sec

Down floor-to-floor time $\dfrac{128}{1 + 1} = 64 = \dfrac{(64 - 15.6) \times 60}{500} + 4.7 = 10.5$ sec

Elevator performance calculations:
Standing time

Lobby time 16 up + 2 down = 14 + 1.6	=	15.6 sec
Transfer time, up stops 8.2 × 2	=	16.4
Door time, up stops (8.2 + 1) × 5.3	=	48.8
Transfer time, down stops 1 × 4	=	4.0
Door time, down stops 1 × 5.3	=	5.3
Total standing time		90.1 sec
Inefficiency, 10%	=	9.0
	Total	99.1 sec

Running time

Run up 8.2 × 4.7	=	38.5
Run down 2 × 10.5	=	21.0
Total round-trip time		158.6 sec

Population: $120 \times 10 = 1200$

$$HC = \frac{(16 + 2) \times 300}{158.6} \times 6 = 204 \text{ people}$$

Percent HC: $204/1200 = 17\%$

Interval: $158.6/6 = 26.4$ sec

2. High Rise: assume 6 3500-lb elevators @ 700 fpm for floors 1, 12 to 20
Rise: $20 \text{ ft} + 18 \times 12 = 236 \text{ ft}$
Probable stops: 16 passengers, 9 stops = 7.6
Up floor-to-floor time: local run 12 to 20 = $8 \times 12 = 96/7.6 = 12.6 = 4.5$ sec

Down floor-to-floor time: $\dfrac{96}{1 + 1} = 48 \text{ ft}$

$$\frac{(48 - 12) \times 60}{700} + 4.5 = 7.6 \text{ sec}$$

Express run: $\dfrac{(236 - 96 - 12) \times 60}{700} + 4.5 = 15.5 \text{ sec}$

Elevator performance calculations:

Standing time			Up Transit Time
Lobby time 16 up + 2 down = 14 + 1.6	=	15.6 sec	15.6
Transfer time, up stops 7.6 × 2 = 15.2 use	=	16.0	16.0
Door time, up stops (7.6 + 1) 5.3	=	45.6	45.6
Transfer time, down stops 1 × 4	=	4	—
Door time, down stops 1 × 5.3	=	5.3	—
Total standing time		86.5 sec	77.2 sec
Inefficiency, 10%	=	8.7	7.7
	Total	95.2 sec	84.9 sec

Running time

Run up express	=	15.5	15.5
Run up local 7.6 × 4.5	=	34.2	34.2
Run down local 2 × 7.6	=	15.2	—
Run down express	=	15.5	—
Total round-trip time		175.6 sec	134.6 sec

Population = 9 × 128 = 1152

$$HC = \frac{(16 + 2) \times 300}{175.6} = 31 \times 6 = 185 \text{ people}$$

Percent HC: 185/1152 = 16%

Interval: 175.6/6 = 29.3 sec

The foregoing calculations show that the six elevators low rise and six elevators high rise will serve the incoming traffic to the building. When the building is designed a transfer floor should be provided at the eleventh floor on the high-rise group so people from the low rise can readily travel to the high rise and vice versa. This transfer floor does not affect the elevator calculations since the eleventh floor population will continue to be served by the low rise and the eleventh floor high rise stop as a transfer floor will be served by only allowing the elevators to stop for up landing calls and down car calls. The eleventh floor car button in the elevator should only operate in the down direction.

To elaborate on the building, a further assumption can be made that it is desired to locate a cafeteria on a middle floor such as the eleventh floor and have both the low-rise and high-rise elevators stop there. To determine if this is feasible, a two-way traffic calculation must be made.

D. Two-way traffic

1. Assume: Cafeteria serves 500 meals per seating, 3 seatings. Usual design for a cafeteria facility in a single-purpose building is to serve about 60% of the population.

Total population 2352 − 1500 = 832 eat out

2. Assume cafeteria turnover ½ hr

$$\text{Cafeteria traffic } \frac{500 \text{ in} + 500 \text{ out}}{6(30 \text{ min})} = \frac{1000}{6} = 167 \text{ in and out per 5 min}$$

People not eating, interfloor ⅓(2352) = 784 × 10% = 78 per 5 min

People in and out of building

$$\frac{2}{3} \times \frac{1}{3}(832) = 185, \frac{185 \text{ in} + 185 \text{ out}}{6} = 62$$

= 62 in and out per 5 min

Total lunch time traffic (167 + 78 + 62) = 307 people on both elevator groups

$$\text{Prorate low rise } \frac{1200}{2352} \times 307 = 157 \text{ people, two-way demand on low rise}$$

$$\text{high rise } \frac{1152}{2352} \times 307 = 150 \text{ people, two-way demand on high rise}$$

3. Calculate two-way traffic on the low-rise elevator.
   Probable stops: assume 8 people up, 8 people down; 5.9 stops up, (.7 × 5.9) = 4 stops down

   Time to run up, per stop $\dfrac{128}{5.9}$ = 21.7-ft rise per stop

   $$21.7 \text{ ft} = 5.4 \text{ sec}$$

   Time to run down $\dfrac{128}{4+1}$ = 25.6 ft = 5.8 sec

   Elevator performance calculations:
   Standing time

Lobby time 8 in + 8 out	= 16.0 sec
Transfer time, up stops 5.9 × 2	= 11.8
Door time, up stops (5.9 + 1) × 5.3	= 36.6
Transfer time, down stops 4 × 4	= 16.0
Door time, down stops 4 × 5.3	= 21.2
Total standing time	101.6 sec
Inefficiency, 10%	= 10.2
Total	111.8 sec

   Running time

Run up 5.9 × 5.4	= 31.9
Run down 5 × 5.8	= 29.0
Total round-trip time	172.7 sec

   $$HC = \frac{(8+8) \times 300}{172.7} = 28 \times 6 = 168 \text{ vs } 157 \text{ people, demand.}$$

   Interval: 172.7/6 = 28.8 sec

4. Calculate two-way traffic on the high-rise elevators.
   Probable stops: assume 10 people up, 10 people down; 9 upper floors + cafeteria, up stops = 6.5, down stops (6.5 × 0.7) = 4.6

   Time to run up, per stop $\dfrac{(96 + 12)}{6.5}$ = 16.6-ft rise per stop

   $$16.6 \text{ ft} = 4.7 \text{ sec}$$

   Time to run down, $\dfrac{(96 + 12)}{4.6 + 1}$ = 19.3 ft = 5.2 sec

   Express run time $\dfrac{(236 - 108 - 19.3) \times 60}{700}$ + 5.2 = 14.6

   Elevator performance calculations:
   Standing time

Lobby time 10 in + 10 out	= 18.0 sec
Transfer time, up stops 6.5 × 2	= 13.0
Door time, up stops (6.5 + 1) × 5.3	= 39.8
Transfer time, down stops 4.6 × 4	= 18.4
Door time, down stops 4.6 × 5.3	= 24.4
Total standing time	113.6 sec

Inefficiency, 10%   = 11.4

                                                         Total 125.0 sec

Running time

Run up express   = 14.6

Run up 6.5 × 4.7   = 30.6

Run down express   = 14.6

Run down 5.6 × 5.2   = 29.1

Total round-trip time   213.9 sec

$$\text{HC} = \frac{(10 + 10) \times 300}{213.9} = 28 \times 6 = 168 \text{ vs } 150 \text{ people, demand.}$$

Interval: 213.9/6 = 35.7 sec

The foregoing elevator arrangement will support the personnel requirements for a cafeteria on the eleventh floor based on the assumptions made. An essential consideration is that sufficient service elevator capacity be provided to serve both the normal needs of the building and the food service requirements. A separate service elevator is a necessity for a building of this size and essential for a single tenant operation. For a single tenant, an automated mail handling system, independent of the service elevator, should be provided.

The final recommendation for the building would be six low-rise elevators, 4000 lb (1800 kg) at 500 fpm (2.5 mps), serving floors 1 to 11, six high-rise elevators, 4000 lb (1800 kg) at 700 fpm (3.5 mps), serving floors 1, 11 (transfer), 12 to 20, and one service elevator, 6500 lb (3000 kg) at 400 fpm (2 mps) serving all floors. If a basement is provided an additional shuttle passenger elevator, 3500 lb (1600 kg) at 125 fpm (0.6 mps), should be provided to serve from floor 1 to the basement.

RULES OF THUMB

Rules of thumb for evaluating the elevatoring of a planned office building quickly are helpful if requirements are relatively clear-cut and elevators need *not* serve more than one entrance, a garage, or a cafeteria, nor serve as part-time service elevators. In other words the rules are only valid when the elevators serve direct traffic between the office floors and the lobby.

Another rule of thumb is to provide at least one elevator for each 225 to 250 building occupants. This rule is related to the 35,000-ft^2 rule on the basis of population density of 125 ft^2 per person. Again, the rule fails for buildings over 20 floors or with floor areas of less than about 10,000 ft^2.

A better guide is given in Figure 10.5. By referring to the figure, the approximate elevatoring may be established for a diversified office building requiring the percentage of handling capacity and having the density of population shown. This chart is based on simple elevatoring without a garage floor, no double lobbies, no odd stops, and no other complexities. The size and speed of the elevators are to be established from the guides shown earlier in this chapter.

A detailed elevator study should be made where the building is near the upper extremes of the curves.

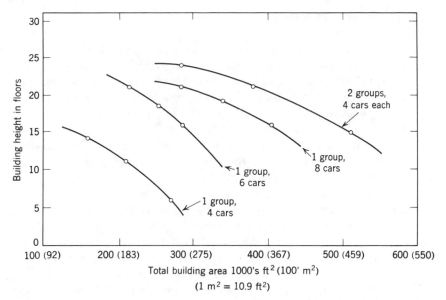

Figure 10.5. Office building elevatoring, diversified office building based on: 1, no base-ments, no upper-floor garages or eating facilities; usable office floors only. 2, 30-sec loading interval. 3, 200-sec maximum round-trip time. 4, 125 ft² (11.5 m²) per person. 5, 12 ft (3.5 m) floor-to-floor height. 6, 12% up traffic per 5 min (percentage of population). Note: One separate service elevator needed for each 300,000 ft² (27,500 m²) of office space.

MAXIMUM ELEVATOR SIZE IN OFFICE BUILDINGS

Elevators of more than 4000-lb capacity should not be used in office build-ings without special provisions for loading and operation. The 4000-lb car is about the maximum that people will fill to capacity when left on their own. Its nominal capacity is 19 people, which is quite a large group and will make the elevator appear crowded even though it is not. Other people will usually wait for the next car. In addition, if a 4000-lb (1800-kg) car serves more than 12 to 14 stops, the trip becomes too long, people are irritated, and building reputation suffers. Investigation should be made for low- and high-rise elevators for build-ings of 15 floors and over.

Elevators larger than 4000 lb (1800 kg) can be effectively used in some buildings. The floor area per floor should be 25,000 ft² (2500 m²) or greater and each elevator group should be a minimum of six elevators to provide a suitable interval. Lobby space should be wide enough to accommodate the expected heavy loading and extra heavy exiting traffic. Elevators should be operated so that all elevators are available for loading if they are at the main lobby and loading time factors should be automatically adjusted to suit the varying loads expected.

Large insurance and banking operations centers have successfully used ele-vators of 5000- or 6000-lb capacities. Further discussion of larger elevators is given in a later chapter on double-deck elevators and sky lobbies.

ELEVATORING STORES

As noted in the chapter on escalators, escalators are usually the preferred vertical transportation in stores. Modern merchandising depends on people visiting each floor to shop and exposure of patronage to the displays on each floor is by escalator. The escalator is a center of attraction, is constantly operating, and requires very little decision or patience on the part of the user.

Demands on the vertical transportation in a store vary with selling space (net selling area) on each floor, the density at which customers are expected to occupy that space, and the rate the per-floor customer population must be "turned over," that is, carried up and down to and from that floor.

Nominal densities in selling area vary from 10 ft² (1 m²) per person on the "bargain" floors to 40 to 50 ft² (4 to 5 m²) in exclusive departments. The normal time of turnover expected is an hour, about the time a person will spend if he or she is a serious shopper and seeking a particular item of any consequence. In some downtown specialty shops, where much patronage comes from office workers during lunch time, a half-hour turnover may be more realistic.

Escalators should provide transportation for about 90% of the shoppers, with the other 10% expected to use elevators. The elevators should also be capable of serving about 10 to 15% of the staff in a 5-min period during working hours to provide for lunchtime or shift changes. If the store's offices are above the selling levels the entire population of the office floors needs elevator service similar to a single-purpose office building. Store elevators should be located so that the view from the elevator encompasses prime selling area. When a store is large enough to require a group of four or more elevators, multiple groups of four elevators located in different quadrants of a large floor are suggested. The elevators should be in a line of four or less to avoid unnecessary holding of an elevator at a floor. With more than four elevators in a line, the distance between the first and last car tends to be too great and prospective passengers can miss the end cars if they were not waiting near them.

Department store passenger elevators should be large in size to accommodate shoppers with strollers and wheelchairs. If separate service or freight elevators are provided, the passenger elevators of 3500-lb (1600-kg) or 4000-lb (1800-kg) capacity with 48-in. (1200-mm) center-opening doors are generally ample and acceptable. If separate service elevators are not provided, the passenger elevators must be used as combination elevators and larger sizes must be provided.

Sufficient service elevators must be provided in a store for stock handling. A complete department store may require elevators of sufficient size to handle 12-ft (3600-mm) rolls of carpet, for which a 5000-lb (2250-kg) elevator with a 12-ft-high cab height can be provided. The more usual requirement is for racks of dresses or coats which may be from 5 to 8 ft (1600 to 2400 mm) long. In general an 8000-lb (3600-kg) passenger-type elevator with a platform approximately 8 × 8 ft (2400 × 2400 mm) is recommended. This arrangement allows automatic operation and use by both passengers and freight. Providing two-speed center-opening doors 60 in. (1600 mm) wide will allow any large packages to pass with ease. Easy access from loading docks to the service elevator is a necessity.

Restaurants and cafeterias in department stores require the same considerations as in office buildings. If the escalators do not serve the cafeteria floor ample elevator service must be provided.

PASSENGER ELEVATORS IN GARAGES

The number of passenger elevators required for a self-parking garage depends largely on the efficiency of the garage design. That efficiency in turn depends on how quickly the city streets can deliver or absorb autos from the garage.

If we assume that a well-designed garage has sufficient ramps to allow all its spaces to be filled (or emptied) within an hour, we will have some basis for elevator calculations. The nature of the garage is also important. Use by commuters who will park in the morning, leave their cars all day, and drive away in the evening presents a different problem than use by transients who will come and go during the course of a day and park while they shop or attend to business. The commuter garage has the simplest elevator traffic to serve, all out in the morning and all in the evening. The transient garage presents a two-way elevator traffic pattern and shoppers with children usually comprise the greatest traffic volume.

A garage connected to an airport or downtown bus transportation terminal will have a combination of long-term and transient parking. In airport terminals, many people will park their cars and leave on early flights and return in the evening, while others will park to meet incoming flights and leave in an hour or so after they meet their party. Peak traffic is generally established by the transients and the approach should be to determine their peak usage and establish the elevators accordingly.

Actual demand on elevators varies with the two types of traffic and is a function of the number of automobile passengers. The commuter may or may not have a partner and, on the average, each car has 1.1 to 1.4 occupants. The shopper usually is accompanied by other people for an average of 2 to 3 occupants. This varies with the type of parking facility, its location, as well as the area of the country. In airports, traffic counts have indicated about 1.5 occupants per auto.

If the garage can be filled, emptied, or turned over in an hour's time, the number of cars that can be parked times the average number of occupants divided by 12 (for 12 5-min periods) gives the expected demand for elevator service.

Example 10.3 gives two typical situations. Similar calculations would be used for basement or upper-floor garages in buildings where separate garage elevators are provided.

It is reasonable to expect some patrons to use stairs in a garage if they are conveniently located and well marked. With large garages connected with sports arenas or theaters the impact of the crowd exiting from an event is such that escalators may be warranted. For general convenience, in any large multi-level garage, escalators are recommended. If the garage covers a large area multiple locations may be necessary. Multiple locations required additional security and reliability considerations.

For a garage up to about 50 ft of elevator rise, the hydraulic-type elevator operating at 150 fpm usually provides a cost effective elevator especially when garage design warrants single elevators in a number of locations.

Example 10.3.   Garages—Self-parking—Passenger Elevators

A. Commuter garage: 6 levels, 10-ft floor heights, 200 automobiles per parking level. Expected elevator demand: 5 (upper levels) $\times$ 200 = 1000 automobiles $\times$ 1.4 people per auto = 1400 people

$$\frac{1400}{12} = 117 \text{ people (all up or all down) in 5 min}$$

Assume: 16 passengers per trip, 150 fpm, 6 stops, floors G, 1 to 5, 10 ft floor to floor, 50-ft rise

Probable stops: 14 pass up, 5 stops = 5 probable stops

Time to run up, per stop $\dfrac{50}{5}$ = 10-ft rise per stop

$$10 \text{ ft} = 7.1 \text{ sec}$$

Time to run down, $\dfrac{(50 - 10) \times 60}{150} + 7.1 = 23.1 \text{ sec}$

Elevator performance calculations:

Standing time

Lobby time 16 in or out	=	14 sec
Transfer time, up stops 5 $\times$ 2 = 10, use	=	16
Door time, up stops 5.3 $\times$ (5 + 1)	=	31.8
Transfer time, down stops	=	—
Door time, down stops	=	—
Total standing time		61.8 sec
Inefficiency, 10%	=	6.2
	Total	68.0 sec

Running time

Run up 5 $\times$ 7.1	=	35.5
Run down 1 $\times$ 23.1	=	23.1
Total round-trip time		126.6 sec

$$\text{HC} = \frac{16 \times 300}{126.6} = 38 \times 3 = 114 \text{ vs } 117 \text{ people, demand.}$$

Interval: 126.6/3 = 42.2 sec

The calculations as shown for all incoming traffic are essentially the same for outgoing traffic.

If three elevators are in the same location, handling capacity and interval is good. Three 3500-lb elevators should be provided.

Because of size of floor for 200 automobiles (60,000 ft² approximately), at least two locations are required.

Recalculate: demand = 117 + 20% (split-location inefficiency) = 140 or 70 people elevator demand per location

Assume: 12 passengers per trip, 200 fpm, 6 stops, floors G, 1 to 5, 10 ft floor to floor, 50-ft rise

Probable stops: 12 pass up, 5 stops = 4.6 stops

Time to run up, per stop $\dfrac{50}{4.6}$ = 10.9-ft rise per stop

$$10.9 \text{ ft} = 6.4 \text{ sec}$$

Time to run down, $\dfrac{(50 - 10.9) \times 60}{200} + 6.4 = 18.1 \text{ sec}$

Elevator performance calculations:

Standing time

Lobby time 12 pass in or out	=	11.0 sec
Transfer time, up stops 4.6 × 2 = 9.2 use	=	12.0
Door time, up stops (4.6 + 1) × 5.3	=	28.7
Transfer time, down stops	=	—
Door time, down stops	=	—
Total standing time		51.7 sec
Inefficiency, 10%	=	5.2
	Total	56.9 sec

Running time

Run up 4.6 × 6.4	=	29.4
Run down 1 × 18.1	=	18.1
Total round-trip time		104.4 sec

$$HC = \frac{12 \times 300}{104.4} = 34 \times 2 = 68 \text{ vs } 70 \text{ people, demand.}$$

Interval: 104.4/2 = 52.2 sec

Two elevators each in two separate locations will provide sufficient handling capacity at an acceptable interval. Two 3000-lb elevators @ 200 fpm in each location should be provided.

B.   Some garages may have complex traffic patterns and an analysis of the various conditions should be made to determine the most critical. For example, assume a garage connected to a department store: 7 levels, 10-ft floor heights, bridge at fourth level, 75 cars per level; expected elevator demand, turnover entire garage in 1 hr; 5 (upper levels, exclude bridge floor) × 75 = 375 automobiles × 2 people per auto = 750 people.

750 × 2 (up and down per hr) ÷ 12 = 125 up and down per 5 min

Analyzing the possible traffic patterns that may predominate suggests three:

1. Most people enter and leave at the fourth level.
2. Most people enter and leave at the ground level.
3. Equal traffic in and out of both ground and fourth level.

For an example and to show a method, a detailed analysis will be made of the third condition which appears to be the worst one.

In	Floor	Out	In	Floor	Out
	7	1a	1a	7	
	6	1a	1a	6	
	5	2a	2a	5	
4a	4	4a	4a	4	4a
2a	3	2a	2a	3	2a
2a	2	2a	2a	2	2a
4a	G			G	4a
	Up trip			Down trip	

a = groups of 2 people traveling to or leaving one automobile

1. Elevator on the up trip makes 5 intermediate stops and 2 lobby stops.
2. Elevator on the down trip makes 4 intermediate stops and 1 lobby stop at the fourth floor.
3. Assume each lobby stop is for 16 people in and 16 people out = 28 sec.
4. Assume 200-fpm elevator speed, 48-in. center-opening doors.

$$\text{Time to run up, per stop} \frac{60}{6} = 10 \text{ ft rise per stop}$$

$$10 \text{ ft} = 6.1 \text{ sec}$$

$$\text{Time to run down,} \frac{60}{6} = 10 \text{ ft} = 6.1 \text{ sec}$$

Elevator performance calculations:
Standing time

Lobby time 3 × 28	=	84.0 sec
Transfer time, up stops, assume 4 × 5	=	20.0
Door time, up stops (6 + 1) 5.3	=	37.1
Transfer time, down stops 4 × 5	=	20.0
Door time, down stops 5 × 5.3	=	26.5
Total standing time		187.6 sec
Inefficiency, 20%	=	37.5
Total		225.1 sec

Running time
  Run up 6 × 6.1   =   36.6
  Run down 6 × 6.1   =   36.6
  Total round-trip time   298.3 sec

$$\text{HC} = \frac{(16\,\text{up} + 16\,\text{down})\,300}{298.3} = 32\ \text{people,}\ \frac{125}{32} = 4\ \text{elevators.}$$

Interval: 298.3/4 = 74 sec, longer than 60-sec criteria

The interval of 74 sec for four elevators is long but may not be too long since the one possible traffic pattern investigated was the worst condition. Further investigation should be done so it can be judged what the average interval will be between the best and worst. Additional consideration in the final judgment should be given to the fact that 100% usage of the garage will be a peak situation and the 1-hr turnover considered will be highly unlikely with the other traffic congestion which will probably be experienced.

It is recommended and many local codes require that the elevator lobby in underground garages be totally enclosed to avoid the possibility of gasoline fumes, which are heavier than air, accumulating in the elevator pit. It is also a good safety practice to isolate the elevator lobby from the traffic by a suitable enclosure on any garage floor. It is also good commercial practice to provide a warm lobby for waiting patrons and to avoid cold weather elevator operating problems. In some areas, raising the entrance of the elevator above the parking floor level is sufficient to meet the code.

PROFESSIONAL BUILDINGS

Professional buildings are specialized office buildings generally devoted to medical and dental use. Their success depends on services offered and accessibility to a large population center. It is not unusual to find 200 or more doctors, dentists, optometrists, x-ray labs, and other services concentrated in one downtown building.

Traffic in such buildings consists of patients visiting these offices, and its volume is a function of the normal turnover of visitors. The average doctor may take care of about three to four visitors per hour, each visitor usually coming with a companion. Based on the number of professional offices, the elevator traffic will amount to about eight people in and eight people out per office per hour.

As an example, if the building has 200 offices the critical elevator traffic would be about 16 people times 200 offices divided by 12 5-min periods or about 267 people in and out of the building in a 5-min period. Peak elevator traffic usually occurs between 2 and 4 in the afternoon (see Figure 10.4).

If the number of offices is not known at the time the building is being planned an estimate on the basis of 300 ft^2 (30 m^2) of net area per office can be made. A further estimate of two people per office and a 5-min elevator peak traffic of 20% of this population, two-way traffic, leads to an approximate solution of the elevatoring problem.

The suggested speed, capacity, and interval may be found in Tables 10.5, 10.6, and 10.7. In addition, consideration should be given to the possible need in some professional buildings for stretcher service on the elevator. A hospital-shaped service elevator or special provisions such as a wider passenger-shaped platform with wide doors should be considered. The 3500-lb (1600-kg) passenger cars with 48-in. (1200-mm) center-opening doors can accommodate a standard 76 × 22-in. (1930 × 600 mm) mobile stretcher, or any size wheelchair, as well as a number of passengers. For the smaller medical building, a 2500-lb (1200-kg) car, specially arranged with 42-in. (1100-mm) side-sliding doors can also carry mobile stretchers.

MERCHANDISE MARTS

The success of merchandise marts depends in part on the speed with which exhibits of wares can be set up as well as resident displays. A mart caters to people interested only in a restricted line of merchandise. Shows will run from one to three days, the usual buyer trying to visit each display in the building the first day and concentrating on a limited number on succeeding days.

A common pattern for buyers is to start at the top and work their way down, using the elevators, stairs, or escalators if available.

Escalators or elevators on approximately the same basis as a diversified office building of the same area will provide approximately the vertical transportation needed for passengers—4000-lb (1800-kg) elevators are recommended. Ample, large freight elevators are necessary to accommodate the merchandise that must be moved for temporary or permanent display. The number and size of the freight elevators will depend on the requirements of the building and the time allowed to set up for the show. Combination elevators which can be used for merchandise movement and subsequent pedestrian movement may be considered but are usually not practical.

INDUSTRIAL BUILDINGS

Elevatoring of industrial buildings depends on their general function, the specific type of work to be performed, and the personnel practices of the occupying firm. There are no general rules and the extent of each specific traffic problem must be determined and treated accordingly.

In larger plants escalators have proved advantageous in that they can accommodate large numbers of people in a short time and are ideal for shift changes. With a single shift operation, they can be reversed to accommodate the traffic flow. Elevators have the advantage of being able to double as freight elevators during working hours and to serve the employees at incoming, luncheon, and quitting times. Employee demand for service is usually established by strict working rules, and because everyone usually starts or finishes at the same time the peak traffic is exceedingly heavy.

SUMMARY

It is obviously impossible to cover all the vertical transportation situations that may arise in the elevatoring of commercial and industrial buildings. Particular attention must be paid to avoiding situations that will reduce the efficiency of the elevator plant, such as two entrances to the building, odd floors served, unnecessary special operations, too many priority services, and upper-floor cafeterias. The best elevator or escalator installation is usually the simplest and most direct, with one clear, unobstructed path for everyone.

CHAPTER ELEVEN

Elevatoring Residential Buildings

Residential buildings are buildings where a number of people live either permanently or temporarily. They include hotels, motels, apartments, senior citizen housing, dormitories, and other residence halls. Elevator traffic in such buildings is generally not as intense as in commercial or institutional buildings and a greater tolerance of waiting is found. Availability and capacity as well as a prompt trip and furniture moving capability are the more important criteria.

POPULATION

Each occupant is allotted a certain minimum space to live and sleep, as little as 100 ft^2 (10 m^2) in some low-cost housing projects and considerably more in luxury apartments. The average, when the layout and room utilization of a residential floor is unknown, is about 200 ft^2 (20 m^2) of net area per person. Design and use of a building will alter this average. For example, in hotels and motels the average occupancy per room is relevant. In apartments, because room count is often distorted by assigning half- and quarter-room values to such areas as foyers and closets, the number of bedrooms and the average occupancy per bedroom is the criteria. In dormitories, which may start as large single suites and change to two- or three-person bedrooms, the 200 ft^2 (20 m^2) per person average is best to use.

In most apartments and residences people have a more relaxed attitude toward vertical transportation than in commercial buildings and will tolerate a longer average wait for service and a longer trip time.

In hotels and motels, because people are paying for service, they expect more prompt and efficient vertical transportation. There is a paradox, however; during a convention or gathering when crowds are coming and going, the average patron is tolerant of delays and accepts them in the carnival atmosphere that prevails. Waiting for a minute or two may be acceptable but beyond that irritation may become apparent.

CHARACTERISTIC TRAFFIC AND INTENSITY

Vertical traffic in residential buildings is predominantly two-way. In downtown apartment buildings and hotels substantial down peaks occur at the start of the working day and up peaks as business people arrive home. In in-town and suburban apartments the peak traffic occurs in the early evening when people are leaving for evening activities and others are returning after shopping or other activities. In a hotel the late afternoon is often marked by a check-in peak, meetings breaking up, and people returning to rooms as others leave to seek refreshment and in the morning, by breakfast and checkout.

In dormitories and residential halls the two-way peak may be in the evening when residents are going to and returning from dinner. This peak is also influenced by whether the dining facilities are cafeterias or dining rooms. In senior citizen housing recreational periods may cause the greatest elevator activity.

The percentage of a building's population that the elevators must serve during the critical 5-min traffic periods varies with the facility. For a hotel that hosts frequent conventions elevators must be able to serve from 12 to 15% of the population during a 5-min peak period. In a luxury apartment where the number of children is expected to be low, the percentage of the population served during a 5-min peak period may be as low as 5%. Required capacity for other building types varies between those extremes. Information about recommended handling capacities, intervals, and population criteria appears in Table 11.1.

CALCULATING ELEVATOR REQUIREMENTS

Elevatoring a residential building can proceed as with any other type of building. The population is determined and the number, speed, and capacity of elevators are assumed and verified for handling capacity and interval. Double entrances to the building, garage stops above or below the main entrance floor, and other services or facilities such as rooftop swimming pools, restaurants, and lounges all require additional elevator service. The extent, use, and capacity of such facilities must be determined and the impact on the elevator situation calculated.

Service requirements may be quite stringent in residential buildings. In hotels and motels there is always someone moving in and out, chambermaids at work, and, in the higher-class establishments, considerable room service for refreshments and food.

Luxury apartments hardly deserve that appellation without a separate service elevator. In a large apartment building the frequency of moves in and out may require use of a service elevator 4 to 6 hr daily just for moving.

A nominal 10% inefficiency factor in standing time was applied to office building elevatoring calculation when 48-in. (1200-mm) center-opening doors were used to reflect the expected promptness and attention of people in a business atmosphere. In residential buildings, this inefficiency will be greater depending upon the nature of the tenancy. Suggested percentages are shown and recommended to be used in calculations.

Each major type of residential structure is reviewed here to show the impact of expected activity on the vertical transportation.

ELEVATOR EQUIPMENT AND LAYOUT

Rules of elevator location and grouping introduced in Chapter 2 apply to residential as well as other buildings. Elevators should be a center of attraction in the lobby and readily accessible on each floor. In hotels and dormitories, where many people come and go at the same time, ample elevator lobby space must be provided at the entrance and on other floors where people are expected to gather.

Service elevators should be located in separate alcoves with ample lobby space at each floor to turn any carts to be carried as well as to accommodate waiting or stored vehicles. This is especially important in hotels, where the room-service tables and carts with their dirty dishes may be stored until picked up by service personnel.

Elevator hoistways should be isolated from sleeping rooms by lobbies, mechanical shaft space, or stairwells. Although elevators are relatively quiet, air noises of an elevator traveling through a shaftway are noticeable when other building noise is low. In addition such mechanical noises as the opening of doors on a floor or the passing of the counterweight within an inch or two of a wall are unavoidable. If a sleeping person's head is on the other side of the wall within 12 in. (300 mm) of the passing car or counterweight [assuming an 8-in. (200-mm) wall], the person will hear the noise and complain. If sleeping rooms must be placed next to the elevator space, ample sound insulation should be provided in the form of dead air space or a mineral wool blanket inside the finished wall and sound isolation between the elevator rail and its bracket to the building structure should be considered.

Similarly, sleeping spaces should not be located next to elevator machine rooms. Electric motors starting and stopping and relays operating can be objectionable, especially in the middle of the night.

HOTELS AND MOTELS

From an operational point of view the distinction between motels and hotels is not always sharp. From the vertical transportation aspect a motel is defined as a lodging in which room service demands are minimal and a hotel as one in which considerable service is expected. The major difference, then, is in the number of service elevators required.

A further distinction is necessary. If the establishment has considerable convention facilities, large dining or meeting rooms, or a ballroom that is not located on the ground floor, it is more a hotel than a motel.

Parking

Today, hotels and motels are expected to have ample parking facilities within or near the structure. Even if there are in-building facilities they should be separated from guest rooms in the interest of security. This may require attendant parking with checking in and checking out at the entrance to the garage. With self-parking one of the best security arrangements is a separate shuttle elevator for the garage area. Not only can the desk clerk see who is coming and going, garage floor stops for the main elevators are eliminated which improves their efficiency, and may minimize the number required.

In spite of impaired security and the inefficiency created by garage stops many hotel and motel operators insist that the main elevators serve both the garage and guest floors. If this is a building design criteria, a constant lobby stop plus additional stops both up and down in the garage area must be considered in elevator calculations.

Meeting Rooms

Another critical area in elevatoring a hotel or motel are the meeting rooms or ballroom floors. When meetings are starting and breaking up these floors will be constant elevator stops so that adequate allowance in elevator trip time must be made. The best arrangement is to locate the meeting floors where escalators can connect them to the street or lobby floor. (It is presumed that the lobby floor is at street level or connected with it by escalators.) Separate escalator service to the meeting room floor will permit public use of the facility with minimum interference with the hotel guests. When large meetings are breaking up the elevators may be overwhelmed by many people wishing to get back to their rooms.

Kitchens

Kitchen facilities should be at the lower levels. A kitchen service elevator should connect the kitchen to the loading dock as well as to the restaurant level or ballroom level. Dumbwaiters can be used to connect the kitchen to the various food shops usually located at the main level. The kitchen service elevator should be a service type with wide horizontally sliding doors that are operated automatically.

SERVICE ELEVATORS

Elevators for room service should be large enough to handle carts or portable tables. 4000-lb (1800-kg) cars with 48-in. (1200-mm) wide center-opening doors are recommended. The number of service elevators depends on many factors, and experience has shown that a minimum of one service elevator for each 200 to 300 rooms in a hotel is necessary. This minimum number requires that schedules, deliveries, and movement of linen and other supplies to each floor be restricted to other than peak dining hours and that special functions be held to a minimum. If a number of special facilities such as rooftop restaurants or lounges must be serviced, or if there are a considerable number of hospitality rooms, additional service elevators are necessary.

A complete study of service requirements can be made by determining the average time per delivery and relating that to the number of deliveries in a given period of time and the average time required per elevator. Such factors as the number of room-service meals the kitchen can prepare in a given time, the number of service employees who must be transported, their shift changes, or local labor requirements, and frequency of special parties must be considered.

During the 1920s, the era of luxury hotel room service when many larger hotels were built, the rule of thumb for service elevators was one for each passenger elevator. Today, with swifter intercity transportation and a greater turnover in hotel guests plus considerably less room service, an approximate ratio of service elevators to passenger elevators should be from 50 to 60%.

SAMPLE HOTEL ELEVATORING

Sample calculations for elevatoring a typical hotel or motel are shown in Example 11.1, using factors from Table 11.1. Note the impact of the garage, although minimum use was assumed. Also note that a restaurant on the top floor of a hotel or any other building requires considerable additional service. This impact cannot be minimized because the restaurant will be used concurrently with other activities and by visitors as well as guests of the hotel. The traffic restaurant patrons create will often be opposed to other traffic in the hotel.

Example 11.1.   Hotel-Motel

Given: 25-story + 3-basement hotel; first and second floors 20 ft, typical floors 10 ft; self-parking for 300 cars in basements; first floor, lobby and restaurants (kitchen in basement); second floor, meeting rooms and 1200-person ballroom; floors 3 to 25 guest rooms, 17 rooms per floor; located near college football stadium; will be used for conventions, may have 300-person restaurant on twenty-sixth floor.

Required: Number of passenger elevators and service elevators—recommended sizes

Calculations: 23 floors of rooms $\times$ 17 $\times$ 1.9 (convention occupancy) = 391 $\times$ 1.9 = 743. 743 $\times$ 12.5% = 93 people, 5 min, two-way peak demand.

Assume: 3500-lb passenger elevators @ 500-fpm, 48-in. center-opening doors
Assume: 10 people up, 10 people down, average travel to twentieth floor (20% highest call and return)

Probable stops: 19 upper floors = 7.9 up $\times$ 0.7 = 5.5 down

Time to run up, per stop: $\dfrac{(17 \times 10) + (2 \times 20)}{7.9} = \dfrac{210}{7.9}$

$$= \ 26.6\text{-ft rise per stop}$$
$$26.6 \text{ ft} = \ 6.1 \text{ sec}$$

Time to run down, per stop: $\dfrac{210}{5.5} = 38.2$ ft = 7.4 sec

Elevator performance calculations:

Standing time

Lobby time 10 in + 10 out	= 18.0 sec
Transfer time, up stops 7.9 $\times$ 2	= 15.8
Door time, up stops (7.9 + 1) $\times$ 5.3	= 47.2
Transfer time, down stops 5.5 $\times$ 4	= 22.0
Door time, down stops 5.5 $\times$ 5.3	= 29.2
Total standing time	132.2 sec
Inefficiency, 10%	= 13.2
	Total 145.4 sec

Running time

Run up 6.1 $\times$ 7.9	= 48.2
Run down 5.5 $\times$ 7.4	= 40.7
Total round-trip time	234.3 sec

$\text{HC} = \dfrac{(10 + 10) \times 300}{234.3} = 25.6$ people, $\dfrac{93}{25.6} = 3.6 = 4$ elevators

Interval: 234.3/4 = 58.6 sec

The four elevators will provide the necessary handling capacity but the interval is approaching the upper limit of 60 sec which is recommended as a maximum for good elevator service. Adding a garage stop will further increase the interval as follows:

Assume: 1 garage stop per trip

Travel to the garage below lobby floor 10 ft

10 ft	= 4.3 sec
Passenger transfer at garage	= 4.0 sec
Door operation at garage and at lobby (2 $\times$ 5.3)	= 10.6
Travel to and from garage (2 $\times$ 4.3)	= 8.6
Round-trip time without garage	= 234.3
New round-trip time	257.5 sec

New interval: 257.5/4 = 64.4 sec

The interval will exceed the criteria and become unacceptable. Therefore separate elevator service to the garage must be considered.

Similarly, adding a large, 300-patron restaurant on the top floor will have a pronounced effect on the elevators. A new calculation needs to be developed since the original study included a substantial reduction in stops and travel for expected travel to the highest average floor call and return. With the upper-floor restaurant, frequent trips to the top landing will be experienced and shown as follows:

Restaurant: 300 seats, assume turnover each 1.5 hr

$$\frac{300 \text{ in } + \text{ 300 out}}{90 \text{ min}} \times 5 \text{ min} \qquad\qquad = \quad 33 \text{ people}$$

Hotel demand, per 5 min   93 people

Total demand   126 people

33 is about ⅓ of 93 so each third elevator trip will be to the top. Therefore highest call is reduced by one-third.

Travel 270 ft total, 210 ft used, (60 × ⅓) = 20, now 230 ft.

Upper floors 25 total, 19 used (5 × ½) = 2, now 21

Recalculate:

Assume: 10 people up, 10 people down

Probable stops: 21 upper floors = 8.1 up × 0.7 = 5.7 down

Time to run up, per stop: $\dfrac{230}{8.1}$ = 28.4-ft rise per stop

$$28.4 \text{ ft} = 6.2 \text{ sec}$$

Time to run down: $\dfrac{230}{5.7}$ = 40.4 ft = 7.6 sec

Elevator performance calculations:

Standing time

Lobby time 10 in + 10 out	= 18.0 sec
Transfer time, up stops 3.1 × 2	= 16.2
Door time, up stops (8.1 + 1) × 5.3	= 48.2
Transfer time, down stops 5.7 × 4	= 22.8
Door time, down stops 5.7 × 5.3	= 30.2
Total standing time	135.4 sec
Inefficiency, 10%	= 13.5
Total	148.9 sec

Running time

Run up 8.1 × 6.2	= 50.2
Run down 5.7 × 7.6	= 43.3
Total round-trip time	242.4 sec

$$\text{HC} = \frac{(10 + 10) \times 300}{242.4} = 24.7 \text{ people}, \quad \frac{126}{24.7} = 5.1 = 5 \text{ elevators}$$

Interval: 242.4/5 = 48.5 sec

The addition of the top-floor restaurant requires a fifth elevator which has the further benefit of improving the interval to 48 sec, which is good. With the fifth elevator, a garage stop could also be served.

Additional considerations will include the necessary service to the ballroom and kitchen and the final recommendations should include the following, assuming a top-floor restaurant:

Passenger elevators: Five 3500 lb (1600 kg) @ 500 fpm (2.5 mps) serving B, 1 to 26 with 48-in. (1200-mm) center-opening doors.

Service elevators: Three 4000 lb (1800 kg) @ 400 fpm (2 mps) serving B, 1 to 26 with 48-in. (1200-mm) center-opening doors.

Kitchen service: one 4000 lb (1800 kg) @ 125 fpm (0.6 mps) serving B, 1 and 2.

Ballroom: Two 48-in. (1200-mm) wide escalators 1 to 2.

Ballroom freight (for displays): One 12,000 lb (5500 kg) @ 50 fpm (0.25 mps) freight elevator with 10-ft (3-m) wide by 22-ft (6.7-m) long platform (for automobile displays), 10-ft (3-m) wide by 10-ft (3-m) high vertical biparting doors stopping at the ballroom and a loading dock.

The ballroom freight elevator recommendation is based on the fact that the 1200-person ballroom will accommodate many more people than the hotel and is expected to be used for conventions and exhibits which will attract many outsiders and may include large exhibits.

APARTMENTS

During the 1970s a minor revolution took place in the design of low-rise apartment houses. This was the inclusion of an elevator in apartments of three or more floors. This was made possible by the intensive development of the hydraulic elevator which provides a low-cost installation for a limited rise of up to 70 ft (21 m). Legislation for the handicapped has required developers to include elevators in any project. For the higher rise, downtown apartments, the need for intensive land development has increased the height of such buildings to 30 or more floors which encourages the use of local and express elevators in such buildings. In some areas, combination buildings, featuring stores on the lower three or four floors, office space above, and apartments above the offices, connected to the street with a sky lobby arrangement, are being developed. This sky lobby will be discussed in a later chapter.

Many apartment buildings are now sold as condominiums or cooperatives representing the ownership of the building as compared to a landlord-tenant relationship. The clientele of the higher-priced condominium will be the same as that of the high-rental apartment and the lower-priced cooperatives will be similar to lower rental and development apartments. This section will use the term "rental."

The demands for vertical transportation in apartments follow the normal day of the building's residents. Outgoing traffic is heavy in the morning as people leave for work, with corresponding incoming traffic in the early evening as they return. These are the average peak traffic periods of apartments that house primarily business people.

In apartments and housing projects with a predominantly family occupancy the needs of children and homemakers influence the traffic pattern. Traffic reaches a forenoon peak as shopping expeditions take place and a distinct afternoon peak as the children return from school and go out to play.

In both downtown and family-type apartment buildings the critical peak traffic that determines elevator capacity occurs in the late afternoon and early evening when people are returning for meals and others are going out for evening recreation. Studies have shown that this traffic is two-way in nature and amounts to about 5 to 7% of the building's population in a peak 5-min period.

Population

The population of an apartment building is based on the number of bedrooms provided in the building, counting so-called efficiency (one-room) apartments as one bedroom. Occupancy per bedroom varies with the type of apartment building and its rental range. The building housing business people will be almost as densely occupied as the family type. Occupancy of the first may average 1.75 people per bedroom because of the considerable apartment-sharing by working people. With low-rental apartments occupied by families with many children, an average of people per bedroom should be used. When the rents are high and rooms are spacious, an average occupancy estimate of 1.5 per bedroom is acceptable.

Elevator Capacity

The minimum recommended size of an elevator for any apartment building is the 2000-lb (900-kg) car with 68 in. wide by 51 in. deep (1750 × 1300 mm) inside dimensions, which is also the minimum size required to accommodate a wheelchair. Minimum doors are 36-in. (900-mm) single-slide side opening (Figure 11.1).

In some localities, building codes require that at least one of the elevators in a residential building be large enough to accommodate a mobile stretcher. A 2500-lb (1200-kg) elevator with a car having 82 in. wide by 51 in. deep (2100 × 1300 mm) inside dimensions with 42-in. (1100-mm) one- or two-speed side sliding doors will accommodate a 76 × 22 in. (1950 × 560 mm) mobile stretcher. The 2500-lb (1200-kg) elevator can also expedite moving operations which may be quite frequent in some areas. If only one elevator is required for the building, it should be 2500 lb (1200 kg). If more than one is required, all should be the 2500-lb (1200-kg) elevator described above.

Elevators must be able to carry at least 5% of the population for the high-rent apartment buildings and from 6 to 7% for the moderate- and low-rent apartments. The more economical the rent, the more children are likely and the higher the traffic peaks that will occur.

Service elevators are a necessity for a high-rent apartment building and essential for any apartment building with 500 or more apartments. Recent statistics show that one out of every 10 families moves once a year. Assuming that a move out and a move in will tie up one elevator most of a day, 500 families in a building will tie up an elevator one day per week just for moving—in addition to normal deliveries of furniture, rugs, groceries, cleaning, and various services as well as normal building maintenance. If building management imposes any restriction that all deliveries must be made on service elevators, one service elevator for every 300 units should be provided.

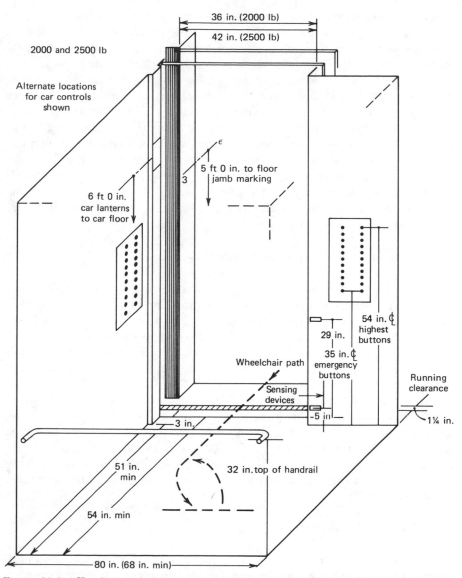

Figure 11.1. Handicapped elevator requirements (courtesy National Elevator Industries Inc.).

A service elevator in any apartment house may pay for itself in improved rental income. This may begin with the opening of the building. If it has only two elevators and the owner is trying to rent before the building is completed, the following situation frequently occurs. To eliminate an outside hoist, the building contractor will take an elevator from the elevator contractor to use temporarily once it is usable and before it is finally completed for use by construction workers and the rental agent to show apartments. If an apartment is rented the tenant usually wants to move in as quickly as possible and the owner is more than willing to accept the rent. Because the completion of the building and moving in cannot take place during the same hours, the owner is faced with overtime moving at premium rates. This can persist for some time because when the building contractor relinquishes the elevator he has been using, it will require considerable finishing to make it ready for tenant use. In addition if both elevators are to operate as a group, their controls must be tied together by the elevator contractor, who must work on both elevators at night, at the owner's expense, to complete the installation. A third car, or a service elevator, would avoid overtime costs for both moving and elevator installation, which have often amounted to a substantial part of the cost of the extra elevator! The tenants enjoy the convenience of a service elevator and rental return per apartment is likely to be higher.

Tenant Garages: Public Use and Penthouse Floors

All floors in an apartment building should be served by all main elevators. Lower-floor garages in apartment houses exact the same elevator trip time penalty as in any other type of building. If an apartment has such garage floors they should be served by all the elevators. If the tenants must wait for, say, only one of two cars to get to and from the garage they are inconvenienced whenever they use such facilities. A separate shuttle elevator should be provided if security is expected to be a problem.

The same consideration of having all elevators serve any public-use floors or the use of a shuttle elevator will also apply to basement laundries, storage areas, recreation areas such as rooftop swimming pools, and terrace floors. It is most important that penthouse floors be served by all the elevators of a group; if not, the highest-rental floors in the building will have substandard elevator service. Alternatively, a shuttle elevator from the top common floor to the penthouse can be provided.

SAMPLE CALCULATIONS

The same calculation procedure for other buildings will be used for apartment houses. Since the 36-in. (900-mm) single-slide door is frequently used, a standing time inefficiency factor of 10% plus the suggested inefficiency of 10 to 15% for apartment houses should be used. Frequent elevator reversal below the top floor is expected and should be estimated at the floor which is the top of 75% of the expected population.

An apartment house elevator is seldom filled to capacity. Unlike office buildings, where elevator passengers seldom have parcels, someone always has something such as a baby stroller, food parcels, luggage, or laundry in an apartment elevator.

The 2000-lb (900-kg) elevator has a nominal full capacity of 8 people and the 2500-lb (1200-kg) elevator a nominal full capacity of 10 people. Elevator calculations for apartment houses are based on two-way traffic. The normal probable stop values can be used or probable stops estimated at one stop per person carried to develop a short method of calculation which will be demonstrated.

Example 11.2.   Apartments

A.   Given: 20-story apartment building, moderate rental, 12 bedrooms per floor, 9-ft floor heights

Demand: 12 × 1.75 (people per bedroom) = 21 per floor 21 × 19 = 399 total × 6% = 24 people, 5-min demand.

1.   Probable stops: assume 8 people per trip; 4 up, 4 down; high call floor 400 × 75% = 300 ÷ 21 = 14 floors above, floors 1 to 15, 14 floors, 4 people = 3.6 stops up × 0.7 = 2.5 down

Time to run up, per stop: $\dfrac{14 \times 9}{3.6}$ = 35-ft rise per stop

Assume 300 fpm, 35 ft = 10.4 sec

Time to run down: $\dfrac{14 \times 9}{2.5}$ = 50 ft, 50 ft = 13.4 sec

Elevator performance calculations:

Standing time

Lobby time 4 in + 4 out	=	8.0 sec
Transfer time, up stops 3.6 × 2	=	7.2
Door time, up stops (3.6 + 1) × 6.6	=	23.8
Transfer time, down stops 2.5 × 4	=	10.0
Door time, down stops 2.5 × 6.6	=	16.5
Total standing time		65.5 sec
Inefficiency, 20%	=	13.1
Total		78.6 sec

Running time

Run up 3.6 × 10.4	=	37.4
Run down 2.5 × 13.4	=	33.5
Total round-trip time		149.5 sec

$HC = \dfrac{8 \times 300}{150}$ = 16 × 2 = 32 vs 24 people, demand.

Interval: 150/2 = 75 sec

2.   Short method of calculating apartment house elevators.

   1.   Assume 8 people per round trip = 8 stops.

   2.   Assign a per-stop value, for example: 36-in. (900-mm) single-slide doors.

a.	Door time	6.6 sec
b.	Transfer time	2.0
c.	Acceleration and deceleration	2.0
		10.6 sec

    d.   +20% inefficiency   2.0

    e.   Total (approx.)   12.0 sec per stop

3. Calculate. Per stop 8 × 12   =   96.0 sec

                Lobby time 8 people   =   8.0

Running time $\dfrac{(14 \times 9) \times 2 \times 60}{300}$ = $\dfrac{50.0}{154.0 \text{ sec}}$

Round-trip time, total

$\text{HC} = \dfrac{8 \times 300}{154} = 15.6 \times 2 = 31 \text{ people}$

Interval: $154/2 = 77$ sec

The results compare favorably to the longer calculations. Two 2000-lb (900-kg) elevators @ 300 fpm (1.5 mps), minimum, two 2500-lb (1200-kg) elevators @ 350 fpm (1.65 mps) recommended.

B.   Given: 30-story apartment building downtown, 18 bedrooms per floor, 9-ft floor heights

Demand: $18 \times 1.75 = 32$ per floor, $29 \times 32 = 928$ total $\times 6\% = 56$ people, 5 min demand.

Probable stops: assume 10 people per trip, 5 up, 5 down; high call floor $928 \times 75\% = 696 \div 32 = 22$, 22 floors, 5 people $= 4$ stops $\times 0.7 = 3$ stops down, assume 500 fpm

Time to run up, per stop: $\dfrac{22 \times 9}{4} = 50\text{-ft rise per stop}$ =   8.8 sec

Time to run down: $\dfrac{22 \times 9}{3} = 66$ ft, 66 ft   =   10.7 sec

Elevator performance calculations:

Standing time

   Lobby time 5 in + 5 out   =   10.0 sec

   Transfer time, up stops 4 × 2   =   8.0

   Door time, up stops (4 + 1) × 6.6   =   33.0

   Transfer time, down stops 3 × 4   =   12.0

   Door time, down stops 3 × 6.6   =   19.8

   Total standing time   82.8 sec

Inefficiency, 20%   =   16.6

                        Total 99.4 sec

Running time

   Run up 4 × 8.8   =   35.2

   Run down 3 × 10.7   =   32.1

   Total round-trip time   166.7 sec

$\text{HC} = \dfrac{10 \times 300}{167} = 18 \text{ people}, \dfrac{56}{18} = 3 \text{ elevators required}$

Interval: $167/3 = 56$ sec, good

C.   Given: 30-story apartment building, low rental, 20 bedrooms per floor, 9-ft floor heights

Demand: $20 \times 2 = 40$ people per floor $\times 29 = 1160 \times 7\% = 81$ people 5 min demand.

Calculating the high call floor would give the twenty-third floor (22 above the lobby), the same as Example 11.2B. Further calculation would show that a 10-passenger elevator would have the same round-trip time and handling capacity as Example 11.2B. Therefore $81/18 = 4.5$, 5 elevators would be required.

Further study as follows would indicate that five elevators is neither necessary nor cost effective.

Probable stops: assume 16 people per trip, 8 up and 8 down

High call floor $1160 \times 75\% = 870 \div 40 = 22$, 22 floors, 8 people, 6.8 stops up $0.7 = 4.8$ down; assume 500 fpm, 48-in. center-opening doors

Time to run up, per stop: $\dfrac{22 \times 9}{6.8} = 29$-ft rise per stop

$29 \text{ ft} = 6.3 \text{ sec}$

Time to run down: $\dfrac{22 \times 9}{4.8} = 41 \text{ ft}$, $41 \text{ ft} = 7.6 \text{ sec}$

Elevator performance calculations:

Standing time

Lobby time 8 in + 8 out	=	14.0 sec
Transfer time, up stops 6.8×2	=	16.0
Door time, up stops $(6.8 + 1) \times 5.3$	=	41.3
Transfer time, down stops 4.8×4	=	19.2
Door time, down stops 4.8×5.3	=	25.4
Total standing time		115.9 sec
Inefficiency, 15%	−	17.4
	Total	133.3 sec

Running time

Run up 6.8×6.3	=	42.8
Run down 4.8×7.6	=	36.5
Total round-trip time		212.6 sec

$\text{HC} = \dfrac{16 \times 300}{213} = 22.5$ people, $\dfrac{81}{22.5} = 3.6 = 4$ elevators

Interval: $212.6/4 = 53$ sec, good

Four larger 3500 lb elevators @ 500 fpm (1600 kg @ 2.5 mps) would be required to provide the necessary handling capacity and would provide an excellent interval of 53 sec. These larger elevators are, however, expensive and a further study should be made for the most cost effective solution as follows, using the short form calculation.

Assume low-rise elevators serve floors 1 to 14.

$13 \times 18 \times 1.75 = 410 \times 75\%$ $308 \div 32 = 10$ floor high call

8 people per round trip

Stops 8×12	=	96
Lobby	=	8

Run $\dfrac{(10 \times 9) \times 2 \times 60}{200}$ $= \underline{54}$

Round-trip time $= 158$

$HC = \dfrac{8 \times 300}{158} = 15$ people, use 2 elevators $= \dfrac{30}{410} = 7.3\%$

Interval: $158/2 = 79$ sec

Assume high-rise elevators serve floors 1, 15 to 30.

$16 \times 18 \times 1.75 = 504 \times 75\% = 378 \div 32 = 12$, 26 floor highest call

8 people per round trip

Stops 8×12 $= 96$

Lobby $= 8$

Run $\dfrac{(26 \times 9) \times 2 \times 60}{500}$ $= \underline{56}$

160

$HC = \dfrac{8 \times 300}{160} = 15$ people, use 2 elevators $= \dfrac{30}{504} = 6\%$

The four elevators, two 2500-lb @ 200 fpm (1200 kg @ 1.0 mps) low-rise serving floors 1 to 14 and two 2500-lb @ 500 fpm (1200-kg @ 1.5 mps) high-rise serving floors 1, 15 to 30 will provide an elevator solution to this building. A conservative approach would be to use the low-rise and high-rise arrangement plus a fifth, larger elevator to operate as a full-time service elevator serving all floors, and operate either with the low-rise or high-rise in the event one of the elevators is out of service for any reason.

Variations in apartment building sizes and heights are endless. In addition, certain minimum standards for elevators have been established by various lending and governmental agencies as well as being required by codes for the handicapped. The criteria for elevator service shown here generally comply with these regulations but local requirements may differ and should be ascertained.

DORMITORIES AND RESIDENCE HALLS

Most of the aspects of apartment elevatoring already discussed also apply to residence halls and dormitories. The essential difference is that dormitory residents are on a more fixed schedule as far as working and mealtimes are concerned. They are usually required to attend classes at certain hours and often take meals in minimum time which result in more severe impact on vertical transportation.

Experience has shown that about 10% of a dormitory's population seek elevator service during a critical 5-min peak, which usually occurs before suppertime and varies with the type of dining facility. If a formal dining room is provided where students must be seated at a given hour the peak will approximate 15% of the population, will be a down peak (assuming lower floor dining), and will utilize the full capacity of the elevators. If there is a cafeteria, only a 10% peak is expected.

Elevators of 2500-lb (1200-kg) capacity with 48-in. (1200-mm) center-opening doors are recommended for dormitories, to provide ample size for students as well as for moving operations and servicing the building. Separate service elevators are seldom required, as their function can be performed by the passenger elevators during slow periods of the day.

Dormitory elevators should be as student- and vandal-proof as possible, as should many apartment elevators, especially in development-type housing. Car operating fixtures should be of substantial construction, extraneous switches should be omitted, and operating buttons should be made of solid metal. Landing buttons, too, should be of solid metal, for it is not unusual for students to use their feet to register calls. Such construction is designated as "vandal proof."

The protective edge on the car doors should be of metal to foil attempts to carve it with a pocket knife. Indicators can be picked out with a sharp instrument and thus should be inaccessible. For anything that is fastened with a screw, the screws should have spanner or allen heads or be avoided entirely; young minds are challenged by what may be behind the cover.

Key-operated switches are a challenge and are best omitted. If a floor must be blocked off at certain times it is best to provide a locked elevator lobby. A load-weighing switch that will sound a loud alarm if the car is filled to 125% of capacity is recommended.

Elevator requirements should be calculated on the same basis as for apartment houses. The elevators should be determined for the apartment house percentage of 6 to 7% two-way traffic and a 10 or 15% capacity at mealtimes, usually a down peak. Population should be based on approximately 200 ft^2 (20 m^2) of net area per student as shown in Table 11.1

Floors with special facilities such as lounges, recreational areas, laundries, and cafeterias require special consideration because of their expected impact on vertical traffic.

As discussed in Chapter Two, skip-stopping is never recommended in dormitories or apartments. There is almost no saving in number of elevators and the saving in entrances creates moving and building service problems. With any emphasis on serving handicapped people, skip-stop arrangements increase management and elevator assignment problems.

SENIOR CITIZEN HOUSING

Apartment dwellings serving the needs of elderly and semi-invalid people are receiving considerable emphasis. Each suite may have its own kitchen facilities as in a conventional apartment house, or community-type dining may be provided. The buildings are designed for people of advancing years who can take care of themselves or require only minor aid.

Because these people are totally dependent on elevators for vertical transportation, each building needs a multiplicity of units to ensure continuity of service. In addition the elevators must be designed to wait for the user, whose

Table. 11.1. Residential Buildings

Type of Building	Population Criteria	Recommended 5-min Capacity (%)	Interval Range (sec)	Inefficiency Factor (%)
Hotel	1.5 to 1.9 people per room	12 to 15	40 to 60	10
Motel	1.5 to 1.9 people per room	10 to 12	40 to 60	10
Apartments				
Downtown	1.5 to 1.75 people per bedroom	5 to 7	50 to 70	15
Development	1.75 to 2 people per bedroom	6 to 7	50 to 90	20
Dormitories	200 ft^2 net per person	15	50 to 70	15
Residence halls	Same as dormitories			
Senior citizen housing	1.25 to 1.5 people per bedroom	6	50 to 90	25

Suggested Elevator Size		
Type of Building	Passenger Elevators Size; Door Type and Size	Service Elevators Size; Door Type and Size
Hotel	3500 lb 48 in. center-opening	4000 lb 48 in. center-opening
Motel	2500 to 3000 lb 42 in. center-opening	3500 lb 48 in. center-opening
Apartments	2500 lb 36 in. single-slide	2500 lb 42 in. two-speed
Dormitories	3000 lb 42 in. center-opening	Use passenger elevators at off-peak times
Residence halls		
Senior citizen housing	2500 lb 42 in. two-speed	Suggest 4000 lb hospital type

274

Suggested Elevator Speed (fpm)

Building Height	Hotels–Motels	Apartments and Senior Citizen Housing	Dormitories and Residence Halls
2 to 6 floors	150[a]	150[a]	150[a]
6 to 12 floors	300	200	200
12 to 20 floors	400 to 500	400	400
20 to 25 floors	500	500	500
25 to 30 floors	700	500	700
30 to 40 floors	700 to 1000		
40 to 50 floors	1000 to 1200		

For buildings of this height, local and express elevators should be considered

[a] Use traction or hydraulic elevators.

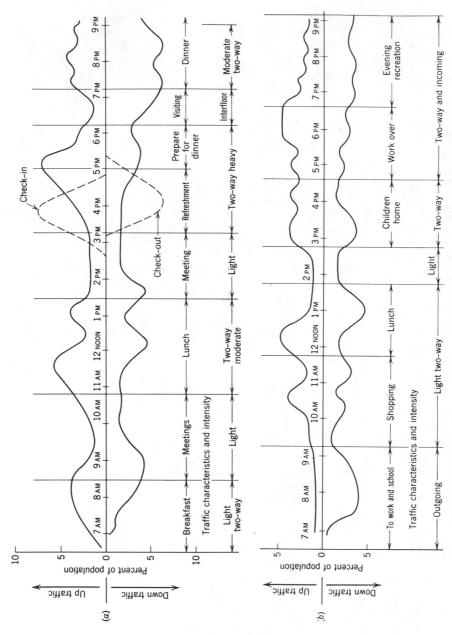

Figure 11.2. Traffic flowcharts—traffic in and out of elevators at a main lobby. (*a*) Motel—hotel convention type; (*b*) apartment building.

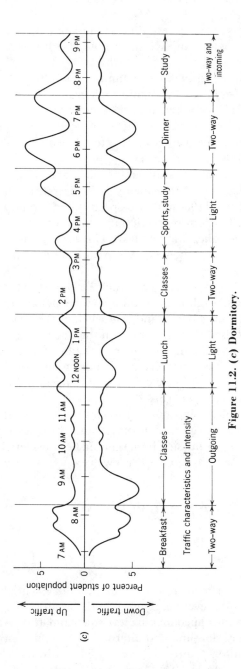

Figure 11.2. (*c*) **Dormitory.**

277

movements and reaction time will be slow. Many passengers will have poor eyesight and use walking aids.

Because occupants are subject to illnesses and may require the use of stretchers for medical care, at least one elevator should be planned for this contingency.

Handling capacity needs will be minimum, no more than about 5 to 6% of the population in a 5-min period. Waiting will be no problem, provided the wait is comfortable; a convenient bench in the elevator lobby on each floor is suggested. Room occupancy is very light; many apartments will have only one person, which indicates an average of 1.25 to 1.5 people per bedroom, depending on the expected use of the facility.

The elevator cars should be wide and shallow and equipped with handrails. The 2500-lb (1200-kg) car, 82 in. wide by 51 in. deep (2100 by 1300 mm) inside is suggested and 42-in. (1100-mm) two-speed doors should be used. Light-ray devices are required for door protection in addition to the safety edge to avoid closing before the entrance way is clear. This size elevator can easily handle wheelchairs and mobile stretchers. A separate service elevator, if provided, should be of hospital shape.

Additional time should be allowed at each elevator stop for the expected slow transfer of passengers, resulting in a standing time inefficiency of about 25%. A prominent signal at each floor to indicate whether the elevator is going up or down is required. Raised floor numbers on the doorjambs or sight guards are necessary in addition to the normal in-car position indicator.

Elevator calculations can proceed as for normal apartment houses with the necessary extra time values. The elevator should never be considered for more than about six passengers per trip.

Housing for the elderly and nursing homes are similar, the essential difference being the increase in personal service and the operating staff requiring more elevator capacity. A nursing home should be considered as an institutional building and is discussed in Chapter Twelve.

ELEVATORS IN RESIDENTIAL BUILDINGS

Operation and Control

In any modern apartment house it has become essential that the elevator control system provide leveling. This is both because of good practice and legislation for the handicapped making leveling mandatory. All hydraulic elevators are equipped with leveling and alternating current control elevators only require two-speed motors to provide leveling. The hydraulic elevator is perfectly acceptable for rises of up to 50 ft (15 m) and speeds of 150 fpm (0.75 mps).

In buildings of six or more floors, elevators with generator field control or solid-state motor drives should be used to provide the higher speeds expected and necessary for that height. Even in the lower building, generator field control or solid-state motor drive for a traction elevator should be considered for the long-term economical elevator installation.

For apartments, dormitories, and housing for the elderly, normal operation during the day will include long periods of minimum elevator traffic. For this reason an on-call type of operation should be used. Additional features for heavier traffic periods should be provided if heavy elevator use is expected. For example, if three elevators are required to provide an adequate interval of operation in a tall building and the average passenger load per trip is expected to be no more than four or six people, long periods of intensive use are unlikely and a minimum operating system should suffice. On the other hand, if the elevators are expected to carry 10 or 12 passengers per trip and the expected interval will be near the maximum, special attention to incoming and outgoing peaks is recommended.

In hotels and motels whose occupancy must be 80% to provide a good investment return, intensive elevator traffic frequently occurs and more refinement in group operation is necessary. Programmable features such as incoming and outgoing traffic operations plus two-way traffic operations are required.

Special floors may require special operations. For example, if the ballroom floor is served by the main elevators and is on other than the main floor, it will be necessary to change the dispatching floor to the ballroom floor when meetings are breaking up even when escalators are provided to a main lobby.

Traffic flowcharts for the various types of buildings (Figure 11.2) are typical and will vary for particular buildings. Necessary group control can be determined in a manner similar to that described for commercial buildings.

For example, typical traffic in a hotel–motel follows an in-and-out pattern of movement as shown, depending on the activities of the guests during the day. If a convention is in progress guests generally go to and from their rooms between meetings and business sessions and frequently after meals. In the late afternoon and evening there is considerable traffic between floors and to and from the lobby, consisting of guests visiting hospitality rooms or local friends and people leaving the building for evening entertainment and other activities.

If no convention is in progress and the motel or hotel is patronized by business people, substantial two-way peaks occur in the morning and outgoing peaks in the evening. In addition there is a check-in peak in the late afternoon whereas checkout activity takes place as early as 7 a.m. in some areas or as late as 3 or 4 p.m. in others.

Traffic curves shown for apartments and dormitories are also altered by the location and use of the facility. If the dormitory is convenient to the classrooms it is frequently used by the students during the day. If it is remote, morning outgoing and late afternoon incoming traffic peaks may be similar to those in an in-town apartment occupied by business people. Each building has its unique traffic.

Traffic in housing for the elderly approximates that found in an apartment house but with greatly reduced intensity. This is partially offset by the relatively long time spent at each stop by the elevators designed to accommodate the elderly.

Features of Residential Elevators

Because elevators are used by many people not familiar with them, installations in motels and hotels should have the maximum features that make them easy to use. Such features include prominent landing buttons that illuminate when operated, readily seen directional indicators (landing lanterns) at each floor for each elevator, and well-lighted lobbies and elevator cars. Car operating panels should clearly present the floors the elevator serves and both a position indicator in the car showing where the car is stopping and a floor numeral on the sight guard of the hoistway door are recommended. Requirements for the handicapped include use of raised designations adjacent to the car operating buttons, and use of a single-stroke gong in the up landing lantern and a double-stroke gong in the down landing lantern to audibly indicate the direction of the elevator.

In other residential buildings, because turnover is minimized, only the minimum informational features are necessary. These are a landing button that lights to acknowledge the registration of a call, position indicators in the car, and prominent directional lanterns in the car or at the landing to show the entering passenger which way the elevator is headed.

Service elevators should be large enough to carry their intended loads: carts, tables, or baggage. Additional considerations must be given to moving furniture in all residential buildings, including cars with ceilings 10 or more ft high as well as higher and wider hoistway and car doors.

If the service elevator is normally a passenger elevator, protective pads should be provided. An independent service switch to remove the elevator from group operation with other passenger elevators should be provided and the elevator should operate with an attendant in response to car calls only. If frequent service use is expected, a separate landing button system and an annunciator in the car is an economical means to provide the special operation.

Other chapters in this book also include suggestions for special operations or features that may be necessary for a particular building. Many of them can be used in any building to solve a present or expected problem. Competent elevator engineers should be consulted to help solve operational or other problems. In a later chapter on special elevators, a discussion of platform lifts for wheelchairs and stair climbers for the handicapped will be presented which will be required in some buildings to meet requirements for the handicapped.

ELEVATORS IN PRIVATE RESIDENCES

One area of residential elevators which is growing in popularity with the introduction of economical equipment is the elevator in a private residence. These have long been prohibitively expensive and were found only in the most palatial homes. Increased use as well as the popularity of multifloor houses has created the demand and many elevator manufacturers offer specialized equipment.

The average home elevator is a small car of no more than 700-lb (320-kg) capacity and operates at a speed of 35 fpm (0.18 mps) maximum. It is designed with a collapsing car gate and swing-type hoistway doors and can either be supported by the building structure or be installed in an independent tower. The car interior is designed to accommodate a standard-size wheelchair.

The installation and design varies with each manufacturer. The A17.1 Elevator Code, recognizing that the elevator will be infrequently used, has allowed certain easing of rules. One important rule is imposed, however: the elevator must contain a telephone capable of reaching a central switchboard should the need ever arise. This is a life safety rule for any elevator in any unattended building.

Elevatoring Institutional Buildings

DEFINITION

Institutional buildings are defined as those in which people obtain a particular care or service. They are generally public and are designed to perform specific functions. Those functions influence vertical traffic, which consists of a combination of staff, visitor, and vehicular traffic, as in a hospital, or primarily visitor traffic, as in a museum or exhibition building.

The major types of institutional building are hospitals, both general and specialized, nursing homes, schools, courthouses, museums, sports arenas, exhibition halls, observation towers, and jails. Combinations are occasionally found, for example, a governmental office building and courthouse. In such a case vertical transportation must serve both the office and courthouse traffic.

Today's institutional buildings usually have extensive parking facilities that may be served. Parking integrated in the building is best served by shuttle elevators. Parking located adjacent to the building requires vertical transportation similar to commercial parking discussed in a previous chapter. Residential areas, such as living quarters on upper floors in a hospital, may be integrated into the building and the impact of these areas on vertical transportation must be determined and served.

Existing facilities, especially in hospitals, may often be expanded to serve growing needs. A basic rule in planning the expansion of any facility is to determine the adequacy of its existing vertical transportation. Too often the vertical transportation load of the entire complex is placed on the new facility, which overwhelms otherwise adequate service. One of the best ways of evaluating the impact of the new facility is to assume that the entire facility is being built today and, perhaps by expansion of existing equipment, to develop a total solution. A traffic study of the existing facility is required.

POPULATION

The amount and type of service to be rendered often determines the population of a facility. This is best described by examples of each type of institutional building and is summed up in Table 12.1.

Hospitals have more critical vertical transportation demands than most other types of institutional building. Hospital staffs have been expanding even more rapidly than patient loads. Statistics show that the average staff per bed increased from 1.98 in 1954 to approximately 2.4 in 1973 and is reported as 3.7 in 1981.* Concurrently, hospital visiting hours have been liberalized until many hospitals now allow visiting at any time between 9 a.m. and 9 p.m.; restrictions are placed on critical areas only. Staff per bed has proved to be a good indication of the elevator traffic in hospitals and an ideal criterion for population.

In nursing homes and mental institutions the staff per bed is somewhat lower than hospitals and the elevator requirements less severe. The staff per bed is a good population criterion unless other information is available.

Population ratios in schools vary with the type of facility. In primarily classroom buildings, population densities of 15 ft² (1.5 m²) of net area per student are common. The total density in the building varies with the classroom utilization factor, which school authorities are striving to maximize and which may be as high as each seat being used about 80% of the time. At present, use factors of 50 to 75% are attained.

Laboratory buildings provide more space per student than do classrooms. The average is about 100 ft² (10 m²). In advanced laboratories and graduate schools; averages of 200 ft² (20 m²) of net area per person are often allowed.

Population in library buildings varies with use. A general reading room may have densities as high as 15 ft² (1.5 m²) per person or, in some specialized areas where many book racks or files of reference material are located, density may be as slight as one person per 100 ft² (10 m²) or more.

Courtrooms are designed for a specific primary function—a jury trial. There will be spectators (varying in number with the importance of the trial); the judiciary staff (about 12 to 16 people); and the jury (12 to 16 people). By using a value of about 30 people per courtroom plus expected spectators, a reasonable population factor may be established. A probable use factor of 75 to 80% may be assumed with the prospect of improved judiciary scheduling.

Museums, exhibits, sports arenas, observation towers, and so on, are designed for a stated spectator capacity. In addition an anticipated "turnover" or entering and leaving of these spectators in a given time is considered in the overall design. Vertical transportation must be directly related to expected population and turnover.

Jails, especially holding facilities, are often located in high-rise buildings or on the top of governmental buildings. Population is easy to determine by the number of cells, but the major vertical transportation needs are related to the

---

Journal of the American Hospital Association, Annual Statistical Report.

Table 12.1. Institutional Buildings

Type of Building	Population Criteria	Recommended 5-min Capacity (%)	Interval Range (sec)	Expected[c] Inefficiency (%)
Hospitals	3 to 5 per bed (per facility type)	12	30 to 50	5
Long-term care facilities	1.5 to 2 per bed	8	40 to 70	10
School				
Classroom	15 ft^2 (1.5 m^2) per student	25 to 40	See text	0
Laboratory	100 ft^2 (10 m^2) per student	20	40 to 50	0
Library	Depends on seating	15	40 to 50	0
Courthouse	30 people per courtroom plus spectators	15	40 to 50	5
Jail	Varies	See text	See text	
Museum, sport arenas, exhibition, observation tower	Varies	See text	See text	

Type of Building	Passenger Elevators Size;[a] Doors, Type, and Size	Service Elevator Size;[a] Doors Type, and Size
		Recommended Elevator Sizes
Hospitals	5000 lb, 54-in. center-opening	6500 lb, 60-in. center-opening
Long-term care facilities	3500 lb, 48-in. center-opening	4500 lb, 48-in. center-opening
School	6000 lb, 60-in. center-opening	6500 lb, 54-in. two-speed
Courthouse	4000 lb, 48-in. center-opening	6500 lb, 54-in. two-speed
Museum, sport arenas, exhibition, observation tower	6000 lb, 60-in. center-opening	8000 lb, 60-in two-speed center-opening
Jail	3500 lb, 48-in. two-speed	6500 lb, 54-in. two-speed

284

Recommended Speeds (fpm)

Building Height	Hospital	Nursing Home	Courthouse	Museum	Jail
2 to 6	200 to 400	200	400	400	200
6 to 12	400 to 500	400	500	500	400
12 to 20	700	500	700	700	500
20 to 25	800[b]	500 to 700[b]	800[b]	—	—
25 to 30	1000[b]	—	1000[b]	—	—

Observation towers (two-stop elevators)

Exterior Elevators	Inside Elevators
100 ft (30 m)—200 to 400 fpm	500 fpm
200 ft (60 m)—400 to 500 fpm	700 fpm
300 ft (90 m)—500 to 700 fpm	1000 fpm
400 ft (120 m)—500 to 700 fpm	1200 fpm (6 mps)
500 ft (150 m)—700 to 800 fpm	1400 fpm (7 mps)
600 ft (200 m)—800 to 1000 fpm	up to 2000 fpm (10 mps)

[a] Metric equivalent: 3000 lb (1400 kg), 3500 lb (1600 kg), 4000 lb (1800 kg), 6000 lb (2700 kg), 6500 lb (3000 kg), 8000 lb (3600 kg), 200 fpm (1 mps), 400 fpm (2 mps), 500 fpm (2.5 mps), 700 fpm (3.5 mps), 800 fpm (4 mps), 1000 fpm (5 mps).

[b] For buildings of this height, local–express elevators should be considered.

[c] Inefficiency percentage is in addition to door type and platform size inefficiency.

staff, number of visitors, type of visiting, security, service needs of the prisoners, and the type of prison. The best determination of vertical transportation need will be based on the prison staff provided plus allowances for visitors, transfer, trials, and so on.

ELEVATOR TRAFFIC IN INSTITUTIONAL BUILDINGS

Considerations

Elevator traffic characteristics vary with the type of institution. In hospitals emphasis is on interfloor traffic with considerable periods of two-way traffic. In a courthouse two-way traffic is the major form.

In classroom buildings traffic is primarily interfloor, especially during class change periods. Laboratories and libraries have substantial incoming and outgoing traffic as well as two-way traffic similar to that found in office buildings.

Traffic in arenas, theaters, and other entertainment facilities is entirely in at the start of a show and out at the end. In exhibits and observation towers flow is primarily two-way and vertical transportation systems must be designed to turn over the expected number of visitors in the time allotted to view the exhibit or panorama. If a restaurant is located in the facility the average time to be seated and served is an important consideration.

Because traffic in each type of institutional building varies with the facility, few general rules can be stated, but each building type is individually evaluated. Table 12.1 gives the expected intensity of the critical traffic and is followed in the text by examples of particular elevator problems.

As mentioned previously, parking or horizontal transportation facilities affect the vertical traffic in institutional buildings as in other types. Often it is necessary to locate such parking on various levels, which creates multiple entrance floors to the building. In this case it is best to provide shuttle transportation to the single lower terminal of the main elevators, or a necessary time penalty must be included in elevatoring calculations. Further consideration must be given to how quickly autos can fill or empty a parking facility. In sports arenas especially, little is gained by excellent elevator or escalator service if bottlenecks cause slow departure from the parking exits.

All elements of the circulation system must be planned as an entity. Each form of transportation: auto, escalator, elevator, bus, or subway, has a definite utilization factor that must be considered when each of these elements is integrated into the system. The vertical transportation system should be optimized in relation to the effective utilization of the facility.

Layout

Special traffic-handling requirements in institutional buildings, especially in suburban areas, are posed by visitors who may never have ridden an elevator or escalator before and may never do so again. These are the people who visit a hospital once, occasionally go to a tourist attraction, have never been in a courtroom, and seldom go to a large city. They are often confused by complex buildings, may be at a peak of emotion, and need the maximum in guidance.

More so than in an office building, the elevators must be a center of attraction. The directions and the signage associated with elevators and escalators must be clear and well defined. Doors should be wide and car interiors should be inviting. Ample signals should be provided and the direction of elevator travel and elevator destinations plainly indicated. Desirable features include large floor markings at eye height on the jamb or sight guard edges of each landing door, large numerals to designate floors or function adjacent to each car button, and directional lanterns with arrows and lantern gongs which sound once for up and twice for down. Codes enacted in many areas for the handicapped require visual and audible features plus the provision that floor markings be tactile, i.e., raised or depressed 0.030 in. (0.8 mm) so the visually impaired can feel them.

For the staff of institutional buildings less elaborate signaling and information devices are necessary but vertical transportation must be located to minimize time consumed in horizontal movement. The transportation core should be central, located near the facilities it is expected to serve, and designed to save employee time.

When heavy public use is expected and elevators serve few floors, large cars of 5000- to 6000-lb (2300- to 2800-kg) capacity are often desirable. This procedure maximizes handling capacity because the average tourist or spectator does not expect high-frequency service or, in other words, accepts the quality of bus rather than taxi service.

Large elevators necessitate ample queuing space instead of alcove arrangements, and deep lobbies are required in front of the elevators where people may congregate.

In hospitals at least some of the elevators will have to carry beds and stretchers, a common requirement for which a special hospital-shaped elevator has been designed. Because the hospital car is narrow and deep it is unfortunately not the best shape for passenger traffic. In a large hospital the vehicular or bed traffic should be separated from the pedestrian traffic. The passenger elevators can also be designed to serve stretchers during a major emergency. In a smaller hospital both functions are usually performed by the same elevators.

Observation towers, long popular, are being built in more and more localities to provide the sensation of viewing the area from a point hundreds of feet above the ground. Vertical transportation is essential to the success of such a tower. One of the most dramatic uses of an elevator is its operation on the outside of the tower with a glass-enclosed car so that people may experience visual ascent and descent. Outside elevators require specialized design to avoid the hazards of wind and weather and to preserve architectural simplicity. These goals have been achieved by the Seattle Space Needle, the Skylon Tower at Niagara Falls,

Figure 12.1.   Old and new towers depend upon elevators to function: (*a*) Eiffel Tower, Paris, 1899; (*b*) Canadian National Tower, Toronto, 1975; (*c*) Seattle Space Needle.

the Hemisfair Tower in San Antonio, and the Canadian National Tower in Toronto (Figure 12.1).

Suggestions for elevator layout are offered in the following pages. As with building designs, variations are infinite and competent elevator engineers should be consulted whenever a problem is perceived.

HOSPITALS

Vertical traffic in hospitals may be separated into two distinct parts: (a) pedestrian traffic, consisting of staff, doctors, technicians, volunteers, visitors, and ambulatory patients; and (b) vehicular traffic, made up of patients on stretchers or in wheelchairs, linen carts, dietary carts, supply carts, portable equipment, and so on. Separating these two types of traffic is by far the best approach to hospital elevatoring.

If pedestrians and stretcher patients use the same elevators, patients suffer delay and discomfort. People must squeeze past a stretcher patient to get in and out of the elevator, thus slowing the entire trip. If all the elevators in a hospital must be designed to carry patients on stretchers they must be narrow and deep, a shape that reduces the efficiency of passenger transfer and slows the elevator trip.

Although most of the elevator traffic in a hospital may be of a pedestrian nature, i.e. staff and visitors going to and from patients, the most important traffic is the movement of patients to and from treatment or to evacuate them in the event of a building emergency. A compromise elevator design is required which dictates a wide elevator to expedite passenger transfer but with sufficiently wide doors so that a stretcher may be moved and with a car depth to accommodate that stretcher. This size forms the basis of passenger elevator service in a hospital and leads to a unique "hospital pedestrian elevator" which will be described as opposed to the conventional, narrow and deep "hospital elevator."

For these reasons the first approach to planning proper hospital elevatoring should begin with two separate groups of elevators: pedestrian-shaped cars for pedestrian traffic and hospital-shaped cars for vehicles. Because the first rule is that more than one elevator in a hospital must be capable of handling patients, the minimum plant for any multistory hospital is established as two hospital-shaped cars.

The second rule is that elevators be sufficient to provide an operating interval of from 30 to 50 sec. This interval is attained by the design of the elevator plant, as is shown by Example 12.1.

A third consideration is to have a sufficient number of elevators so that at least one is accessible within a minimum time. If food service requires exclusive use of an elevator during a particular time a separate food service elevator should be provided. This elevator can be used at off times for linen delivery and other supplies that can be scheduled at convenient times. A major priority is that any elevator may be commandeered for use to transport a critical patient.

Automated material and food-handling systems are available and, because they can save many employee man-hours, such systems are frequently applied to hospital requirements. Such equipment does not materially decrease the need for stretcher- or vehicle-type elevators in hospitals, but allows controlled scheduling of necessary deliveries. These systems will be suggested in this chapter and discussed fully in the chapter on material handling.

Lifts with automated loading and unloading ability are recommended for use between surgical supply and operating suites and between various nursing floors and pharmaceutical supply. A simple, manual dumbwaiter connecting the kitchen and cafeteria or coffee shop is also useful. If the loading dock is at a different level from the kitchen, a kitchen service elevator should be provided, of the same type as described for hotel kitchen service.

Vehicular Traffic

Vehicular traffic in hospitals consists of patients on stretchers or in wheelchairs, dietary carts, equipment such as portable x-ray machines, pharmaceutical, linen and exchange carts, library and sundry carts, plus maintenance carts. Surveys have shown that vehicular traffic can be as high as 4 carts up and 4 carts down for each 100 beds in a hospital in a 5-min period. Unless the cart delivery system is automated, each cart or vehicle will require an attendant, who, for elevator purposes, is considered as a partner to the vehicle and not part of the pedestrian or staff traffic. Vehicular traffic usually requires a longer transfer time on and off an elevator than pedestrian and often causes elevator delays especially when elevators filled with pedestrians stop and there is no room for the vehicle.

The reverse is the more frequent occurrence. An elevator can be filled in bulk with an empty stretcher which limits the space to pick up additional pedestrians; hence, a useless stop is created.

Elevator manufacturers should be challenged to develop a means to reliably recognize when an elevator is filled by area rather than by weight as is the current practice. In this way many useless stops could be eliminated and hospital elevator efficiency improved.

Layout

Various elevator layouts are effective in separating vehicular and passenger traffic. In small hospitals, which may require only three or four elevators, the efficiency of the passenger-shaped elevator is sacrificed to obtain the necessary flexibility of operation. In any hospital requiring five or more elevators, the two separate types of elevators are generally feasible and should be investigated.

Hospitals are often planned for future expansion and the layout should always include additional elevator hoistways or framing so that the elevator plant may be expanded. The ultimate expansion should be determined and the elevator plant projected to that end. If vertical expansion is contemplated, elevator machinery of the basement type can be installed initially. The overhead equipment can then be installed with minimum interference to the operating plant and the transfer made in minimum time. Alternatively, the elevator machinery can be installed in its permanent location atop a shaftway tower for the building's full ultimate height and adjusted when the upper floors are activated.

A sample layout in Figure 12.2 shows a suggested 6000-lb hospital-shaped elevator. The center-opening sliding doors provide the optimum door opening and both efficient vehicle and passenger transfer.

The 6000-lb car is the minimum that accommodates a motorized hospital bed with a patient in a body cast with traction devices and leaves room for people riding alongside the bed. A 6500-lb car should be considered if architecturally possible.

Passenger elevators can be of the conventional size and shape as for commercial buildings. For minimum efficiency in passenger handling, the 3500-lb elevator, shallow and wide, is recommended. As an additional feature, this size elevator accommodates a mobile stretcher (22 × 76 in.) in an emergency. A 5000-lb elevator with a platform 102 in wide by 82 in deep with 54 in wide center opening doors is an ideal size for a larger hospital.

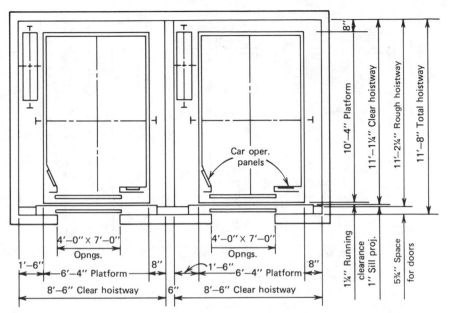

Figure 12.2. Hospital vehicular-type elevator designed to accomodate pedestrians: capacity, 6000 lb, speed, 400 fpm.

The U.S. Department of Health, Education and Welfare requires a combination passenger and stretcher elevator. This is a compromise 3500-lb (1600-kg) size with a platform as shown in Figure 12.3.

Groups of four or more cars should be installed in alcoves as recommended in Chapter Two. Service elevators require additional lobby space if elevators are placed facing each other. A minimum lobby of 14 ft (4.2 m) is recommended for sufficient room to swing two stretchers unloading simultaneously.

In hospitals over 16-18 floors, local/express elevators may be suggested, especially for passenger-type elevators that would have an excessively long trip time if called on to serve more than 16 floors. Service elevators may be in a single group if sufficient capacity is attained with six or eight cars. If more service cars are needed consideration should be given to separating them according to their functons, possibly by restricting all food service and supply to a single group and reserving another group for stretcher and staff traffic. Automated cart handling supply systems become a definite requirement for hospitals of more than about 12 floors.

Outpatients

Facilities for outpatients are receiving increased emphasis. To reduce the time spent in hospitals and to serve community needs, separate outpatient

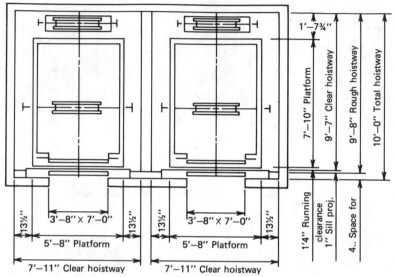

Figure 12.3. Hospital passenger-type elevator designed to accomodate a stretcher: capacity, 3500 lb, speed, 500 fpm.

facilities are usually established and should be served by separate vertical transportation.

If the facility is large, escalators may be considered to enhance the use of floors above the lobby for this function. A smaller facility may be served by an elevator or elevators with limited stops. The cars should be of ample size, 3500 lb (1600 kg), with 48-in. center-opening doors to facilitate the loading and unloading of wheelchairs and people on crutches. Any upper floor outpatient facility requires service by an elevator in view of the limited mobility of most of the patients.

Calculating Elevator Requirements

As noted in Table 12.1, the expected passenger traffic demand will amount to 12% of the hospital population in a 5-min period. This population, for elevatoring purposes, should be based on a minimum of 3 people per bed and adjusted to reflect expected staffing in the facility under study whether it be community hospital, special treatment, teaching or a central regional hospital.

Example 12.1A shows how the service demand is translated into elevator facilities. The characteristic hospital traffic is a combination of both two-way and interfloor traffic. Calculations are based on a two-way traffic approach.

Critical traffic periods occur at a number of times during the day. One of the most important periods occurs at about 3:00 p.m. when hospital nursing staffs are changing shifts and a coincidental visitor peak may occur.

Critical vehicular traffic reaches a peak at about 8:30 to 9:00 a.m. when patients are transferred for operations and therapy, the cleanup from the morning meal is in progress, supplies are transferred, and maintenance activity is in full swing.

Example 12.1. Institutional Buildings—General Hospital

A. Given: 350 beds, 10 floors, 12-ft floor heights. Passenger traffic: 350 × 3 per bed = 1050 × 12% = 126 people per 5 min; vehicular traffic: 400 × 4 per 100 beds = 16 vehicles per 5 min.

1. *Passenger traffic*

Assume: 10 passengers up, 10 passengers down, 500-fpm elevators, 48-in. center-opening doors

Probable stops: up, 10 passengers, 9 floors = 6.2 stops × 0.7 = 4.3 down

Time to run up, per stop: $\dfrac{9 \times 12}{6.2}$ = 17.4-ft rise per stop

$$17.4 \text{ ft} = 5 \text{ sec}$$

Time to run down: $\dfrac{9 \times 12}{4.3}$ = 25 ft = 5.8 sec

Elevator performance calculations:

Standing time

Lobby time 10 in + 10 out	= 18.0 sec
Transfer time, up stops 6.2 × 2	= 12.4
Door time, up stops (6.2 + 1) × 5.3	= 38.2
Transfer time, down stops 4.3 × 4	= 17.2
Door time, down stops 4.3 × 5.3	= 22.8
Total standing time	108.6 sec
Inefficiency, 15%	= 16.3
Total	124.9 sec

Running time

Run up 6.2 × 5	= 31.0
Run down 4.3 × 5.8	= 24.9
Total round-trip time	180.8 sec

$$\text{HC} = \frac{(10 + 10) \times 300}{180.8} = 33 \text{ people}, \quad \frac{126}{33} = 3.8, \ 4 \text{ elevators}$$

Interval: 180.8/4 = 45.2 sec

2. *Vehicular traffic*:

Assume 400-fpm vehicular-service elevators.

Time to load vehicle	15 sec
Run one-half building height (average trip) 5 × 12 = 60 ft	11.5 sec
Unload vehicle	15 sec
Run	11.5
Door time, two stops 2 × 5.3	10.6 sec
Average trip	63.6 sec

$$HC = \frac{300}{63.6} = 5 \text{ vehicles per elevator per 5 min}$$

$$\frac{16}{5} = 3 \text{ elevators required}$$

Minimum Elevators:

Four 3500-lb passenger elevators @ 500 fpm, 48-in. center-opening doors.

Three 5000-lb service elevators @ 400 fpm, 48-in. center-opening doors.

Recommendations:

Four 5000-lb passenger elevators @ 300 fpm, 54-in. center-opening doors

Three 6500-lb service elevators @ 400 fpm, 60-in. center-opening doors.

B. Given: 200 beds, 6 floors, 12-ft floor heights. Passenger traffic: 200×3 per bed $= 600 \times 12\% = 72$ people per 5 min; vehicle traffic: 200×4 per 100 beds $= 8$ vehicles per 5 min; combine passengers and vehicles. Assume: 6 passengers up, 6 passengers down, 1 vehicle per trip, 300-fpm elevators

Probable stops: up 6 passengers, 5 floors, 3.7

$$\text{Time to run up, per stop: } \frac{5 \times 12}{3.7} = 16.2\text{-ft rise per stop}$$

$$16.2 \text{ ft} = 6.6 \text{ sec}$$

$$\text{Time to run down: } \frac{5 \times 12}{2.6} = 23 \text{ ft} = 8 \text{ sec}$$

Elevator performance calculations:

Standing time

Lobby time 6 in + 6 out	= 11.0 sec
Transfer time, up stops (3.7 × 2) + vehicle 15	= 22.4
Door time, up stops 3.7 × 5.3	= 19.6
Transfer time, down stops (2.6 × 4) + 15	= 25.4
Door time, down stops 2.6 × 5.3	= 13.4
Total standing time	91.8 sec
Inefficiency, 15%	= 13.8
Total	105.6 sec

Running time

Run up 3.7 × 6.6	= 24.4
Run down 2.6 × 8	= 20.8
Total round-trip time	150.8 sec

$$HC = \frac{(6 + 6) \times 300}{150.8} = 24 \text{ people}, \quad \frac{72}{24} = 3 \text{ elevators for passengers}$$

$$HC = \frac{1 \times 300}{150.8} = 2, \quad \frac{8}{2} = 4 \text{ elevators required for vehicles}$$

Interval: $150.8/3 = 50.3$ sec

The conservative recommendation will be to provide four elevators. The interval is at the upper limit of recommended elevator service and, with three elevators, there will be times when vehicles will have to be delayed so that passengers can be accommodated or vice versa. Additional investigation should be made to determine if the nonpatient vehicle transfer can be accomplished with an automated cart lift system. These vehicles would include dietary needs, supplies, and linens. With only three elevators the judgment of the elevator engineer must be shared with the hospital administration and the limitations recognized. One important consideration will be the necessity of operating the hospital if one of the three elevators is out of service for maintenance or repair.

The vehicular peak may amount between three to four vehicles per 100 beds in a 5-min period depending on the scheduling of treatment and delivery of supplies. For utmost efficiency, vehicle traffic should be scheduled so that the fullest possible elevator utilization is attained.

Example 12.1A shows the most efficient way to serve vehicles, that is, separate vehicular and pedestrian travel. With a mixture of personnel and vehicles, considerable elevator inefficiency and delay will occur. Example 12.1B attempts to account for most of the expected delay by allowing 15 sec for one-way vehicle transfer. With exclusive use of the elevator, this transfer time can be substantially reduced.

Examples 12.1A and 12.1B are generalized studies of new hospitals. Many other vertical transportation considerations need to be made and will involve the overall and material management in the proposed hospital. Material management is of great concern since transporting necessary supplies and food from the stock room and kitchens is labor intensive, i.e., an unsupervised person is often used to move carts from place to place. In a moderate size hospital, it is not unusual to have 10 to 20 people involved in the material movement activity and with 24 hr a day, seven days a week operating requirements and with liberal employee benefits, a staff of 30 or more people is required to effectively accomplish this activity. Automated material-handling systems properly applied and managed can effectively reduce the number of personnel. This aspect will be discussed fully in a later chapter.

Another aspect of hospital design which is becoming prominent is the construction of a new hospital adjacent to and integrated with an existing facility. The existing facility may remain or may be phased out when the new facility is complete. A typical program is to partially build the new facility and operate both the new and existing facility together and then complete the new facility once new patterns of operation are established. For example, the new facility may contain all the operating rooms, supply rooms, and laboratories, while the old facility retains the patient rooms and beds.

Considerable study will be required to ensure the continuing operation of the old facility, recognizing that some of the vertical transportation load is transferred to the new facility. In the design of the new facility, the eventual vertical transportation system must be planned even though it may not be finally installed until a future date.

INSTITUTIONS FOR LONG-TERM CARE

Institutions for long-term care include hospitals for the chronically ill, mental institutions, postoperative care centers, nursing homes, homes for the aged, and other facilities in which patients or residents remain for an extended period and receive less intensive care than in a hospital. Such facilities can be all considered as nursing homes for elevator requirements.

Elevatoring such buildings is approached much the same as for general hospitals. If the facility is large, with 400 to 500 beds, complete separation of vehicular and pedestrian traffic should be considered. If many patients are expected to be ambulatory, elevator traffic will be heavy especially if recreation and dining facilities are concentrated on the lower floors.

If definite staffing plans are not established for the facility, a population factor of from 1.5 to 2 people per bed may be used, the higher figure related to the higher degree of patient care. For peak 5-min elevator capacity, a handling capacity of from 8 to 10% of the population, two-way traffic should be provided. A longer interval range is permissible because of little urgency in day-to-day activities.

With many ambulatory patients, the hospital pedestrian elevators should be considered as the main elevators. This will minimize the required hospital-type cars needed for the necessary transfer of supply carts and occasional stretcher movement. Elevator design should include all the features of elevators designed for the handicapped. This includes properly located elevator operating controls, markings for the visually impaired, audible and visible signaling, extra protection while doors are closing, and additional time at stops for passengers to transfer. In calculating elevator requirements a standing time inefficiency of at least 20% should be used.

In mental institutions security should be provided by locked elevator lobbies. The practice of keying all elevator call buttons has led to various abuses including forced locks, toothpicks and chewing gum in key ways, and forged keys, and is not recommended.

SECONDARY SCHOOLS, UNIVERSITIES AND COLLEGES

As land for horizontal expansion becomes unobtainable and the inefficiency of a sprawling campus becomes apparent, more and more schools are expanding vertically. Possible height requires proper attention to vertical transportation. The University of Pittsburgh and the University of Moscow are both vertical schools in single buildings of about 30 floors with the living space, classes, and libraries all in one building.

Six- or seven-story buildings with classroom space for 2500 to 3000 students are perfectly feasible when served by escalators plus two elevators for necessary freight and handicapped student needs. Science departments, with their long laboratory periods, are ideally located in high-rise buildings and require approximately the same elevator service as an office building. School libraries, by proper allocation of floor use, can be multistory buildings. General reading rooms can be located on the lower floors served by escalators, and specialized book collections can be located on upper floors with elevator service designed for the expected use which is often minimal.

Classroom Buildings

In classroom buildings the demand for vertical transportation is the most severe during the time alloted to changing classes. From 3 to 15 min may be allowed for this purpose, depending on the type of school. Secondary schools, usually in buildings of limited height, allow the least time between classes to discourage student mischief in the halls. Longer class change times are required for larger schools and extensive campuses.

During the peak 5-min of the class change period, as much as 50% of the student population may seek vertical transportation. As an average, 40% will be considered. Severe vertical transportation demands also occur when evening classes start in downtown colleges. These classes usually start about half an hour after businesses close, by which time the students must travel to the school and have something to eat. They will arrive about 5 to 10 min before classes start and create an incoming 5-min peak on the elevator system of from 25 to 40% of the student population.

Student population is determined by multiplying the total number of classroom seats by a utilization factor. For day sessions in most colleges this factor ranges from about 50 to 80%, depending on classroom scheduling. For high schools and evening colleges, a use factor of from 80 to 90% of the available seats is not uncommon. The use factor is an important determinant in elevatoring a classroom building and should be established by study of the expected or past practice.

Escalators are by far the best means of vertical transportation for a school with 1500 to 2000 students. They provide service to every floor, can be reversed to serve heavy incoming and outgoing peaks, and can be used for building heights up to about 11 floors. Typical applications appear in the chapter on escalators. In addition to the escalators at least two elevators are required for the handicapped and for the movement of furniture and supplies. The elevators should be of substantial size (3500 lb minimum) and should be readily accessible. It may be necessary to provide an attendant during class change periods to limit use of the elevators to authorized personnel.

Many state laws and local building codes are requiring all public buildings which include schools to be either built with or retrofitted with means to serve the needs of the handicapped. A grade level entrance and conventional elevator service to each floor is the necessary solution. For an existing facility this can usually be accomplished by adding a shaft adjacent to the building. In situations where this is not feasible, ramps, platform lifts, or stairway type elevators may be applied. A discussion of these specialized lifts can be found in a later chapter.

Laboratory Buildings

Science, computer, or other types of laboratory sessions are generally longer than normal classroom periods. A student with a project is required to complete it and usually may leave when finished. The net result is that the vertical transportation problem is one of an incoming peak with very little other concentrated traffic during the day.

If the science building is elevatored as an office building, based on sufficient elevator capacity to fill the building in a 30 to 40 min period, service should be adequate for all other traffic periods. Elevators should be designed for a 5-min incoming peak of from 10 to 15% of population, based on expected occupancy ranging from 50 ft^2 (5 m^2) to 200 ft^2 (20 m^2) of net area per student, depending on the type of facility.

Laboratory sessions are generally well attended so that the utilization factor is high. In many laboratories, large and unwieldy apparatus is often moved in and out, requiring an extra large passenger elevator. Rooftop observatories and greenhouses may also require elevator service, which can be provided by one of the main elevators with the necessary penthouse structure or by a two-stop underslung or hydraulic type of elevator of limited speed and capacity, operating between the top main elevator stop and the roof.

Library Buildings

A high-rise building is ideal for a library. The lower floors, connected by escalators, can provide ready access to the common reference works and general reading. The upper floors can house specialized references and rare book collections and could require minimum elevator service. The very nature of a library, the long-term storage of information, permits this segregation of function and traffic.

The elevatoring of any library depends on the expected use of the various floors and will require extensive preliminary planning.

In an undergraduate library, considerable student turnover is expected in the general areas. Observations have shown that a full turnover of the available seats or spaces for students occurs about every hour. Therefore, from 12 to 15% of the population of the library will seek vertical transportation in a 5-min period, two-way traffic.

Population is based on the number of reading spaces provided and the nature of the carrels, either locked or open. Population density also depends on the provisions for and use of seminar rooms, special collections, special libraries, either visual or audio, and other factors peculiar to modern libraries.

Elevator calculations should proceed as for office type of two-way traffic. Handling of book carts and delivery of books to the checkout desk may warrant consideration of an automated material handling system or automated cart lift.

COURTHOUSES

Elevatoring a courthouse is directly related to the activities in the courtrooms as well as the various functions connected with a courtroom trial. The court clerk's office, where necessary pretrial papers are filed, is a critical area with a constant flow of lawyers and messengers between it and the entrance to the building. Circulation is expedited by locating this facility on the main floor.

Another critical area is the reporting room for jurors. All the prospective jurors gather in this area and are moved in groups of 20 or so to the various courts for examination.

A usual requirement in courthouses is the positive separation of the judicial from the general public traffic, which calls for separate vertical transportation systems.

Transportation of prisoners in a criminal court requires security arrangements and may have to be restricted to a special elevator or elevators. Alternatively, this function may be combined with the service needs of the courts, which include normal maintenance and, possibly, the transportation of displays and records.

Finally, the general public and participants in civil suits as well as the courthouse staff require vertical transportation. If the courthouse also houses extensive office facilities such as those for the district attorney and investigative staffs, vertical transportation requirements in part resemble those of other office buildings.

These requirements suggest that elevatoring a courthouse begins with study of the general plan of the building and extent of its various facilities. A major determinant is the number of courtrooms and the percentage likely to be in full use at a given time. Full usage is improbable but a factor of 75% is reasonable.

For each court in session about 30 people will take part in a criminal trial or in a civil trial with a jury. In addition, spectators must be considered. In civil trials without juries only a few people are involved; hence, if jury trials are always considered as population criteria, conservative elevatoring will result.

Observation has shown that the starting and ending or recessing of trials overlap by about 30 min, so that about half the population enters the courtrooms while another half leaves. For example, if there are 20 courtrooms in a building about 15 will be in use at one time. At 30 people per court a population of 30 times 15 or 450 people plus spectators is possible. The demand for vertical transportation is 450 divided by 6 or 75 people every 5 min, two-way traffic. If elevators are filled to only half their capacity and operated at the acceptable intervals of 40 to 60 sec, capacity should be ample for spectators, who can be controlled.

Although judges may require separate transportation, their elevators can also be used to carry juries, either prospective or charged. At least two separate elevators in a private alcove may accomplish these tasks. Judges, because of their status and limited number, should receive priority service. Because jurors are usually escorted and each group may receive exclusive service, elevators large enough to transport the expected number of jurors in each group should be provided. This may be 16 or 20 people so a 4000-lb (1800-kg) elevator or larger is recommended.

If the building includes office functions as well as courthouses and their related functions, additional elevator service may be needed. Elevators established for the courtroom traffic should be checked to determine if they can also serve the incoming requirements of the office staff in a manner similar to that for a single-purpose building. A further check for possible conflict during lunchtime should be made and the number of elevators and intervals adjusted accordingly.

JAILS

Short-term jails for holding people awaiting trial, indicted offenders, or short-term prisoners are possibly located above the courthouse. Transfer problems are minimized and, in a high-rise structure, all necessary facilities can be provided with maximum security. Consideration must be given to evacuation during a building emergency.

Short-term prisons located on the upper floors of courthouses or municipal buildings require a separate group of jail shuttle elevators for access to and from the street. Security is strengthened by separate local elevators from the jail skylobby to the cell levels (Figure 12.4). During a building emergency, secured

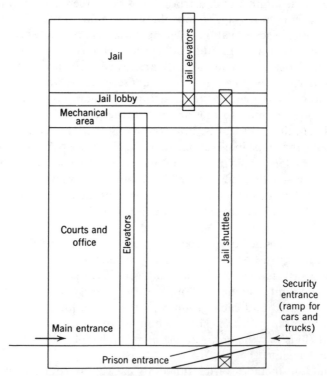

Figure 12.4. High-rise civic center complex.

stairway escape routes to refuge areas is an important consideration as in any high rise structure.

Determining the elevators necessary for a prison depends on staff, operation, and number of transfers in and out during the course of a day. Transfers consist of people being moved to long-term prisons, prisoners reporting for trial, and new arrivals. Lawyers are allowed to visit prisoners in either the cell area or a visiting room. This entails additional transfer.

Recreation and sick call, if these facilities are located away from the cell floors, require vertical transportation. Meals, however, usually take place in the cell area.

Security is accomplished by locked elevator lobbies and by attendants on the elevators. Prisoners are usually escorted, with the escort calling the elevator, which should be equipped with a vision panel so the attendant can see what is taking place before opening the doors. A form of riot control, such as a foot-operated or elbow-operated switch to call for aid or send the elevator to a secure lobby, is advisable.

Closed-circuit television can be provided for in-car surveillance. In the event of difficulty, special controls to call the elevator to a protected floor should be provided.

The extent of vertical transportation requirements must be established and sufficient elevators furnished. Because of security requirements operation efficiency is poor, so ample extra transfer time should be allowed. Larger elevators, 3500-lb minimum, should be used to facilitate cart handling and the transportation of groups. The elevators should be used for all purposes, that is, by both passenger and vehicle, to minimize attendant staffing.

MUSEUMS, EXHIBITS, SPORTS ARENAS, AND OBSERVATION TOWERS

The value of such buildings depends, in part, on how quickly people can be moved into, through, or out of the facility. The vertical transportation demand in a sports arena is obvious: to put people in their seats in the shortest possible time for prompt presentation of the event. "The shortest possible time" varies with each facility, depending on such factors as parking, ticket selling, and other attractions, such as betting at a horse race. As a guide, transportation should be sufficient to serve an average crowd in 30 min both in and out.

In a museum or exhibition hall a more leisurely pace is expected. Aside from performances at particular times, an hour or two may be allowed for visitors to fill the building to capacity. Vertical transportation needs are for turnover rather than rapid filling or emptying as in a sports arena.

The key to the success of an observation tower is the vertical transportation and the area provided for viewing or dining. A visible indication of appeal is a line of people waiting at the ground level for elevator service up into the tower and crowd control may require that people must come down before additional people are allowed up.

Observation areas often placed atop taller buildings and served by additional capacity in the high-rise group so that an elevator may be used for the visitors during peak hours without affecting tenants in the building. A separate entrance lobby is recommended for the observation-deck elevators.

Escalators

For low-rise, heavy-traffic applications, escalators are the recommended means of transportation. They can serve the most people, can be reversed for incoming or outgoing traffic, and require the least space in relation to their traffic-handling capacity. Local and express escalators have been furnished for some sports arenas to provide the necessary service to a succession of upper levels.

When escalators are applied it is essential that adequate queuing area be provided at the entrance and exit to the escalator. An unobstructed area at least as wide as the escalator and from 10 to 15 ft (3 to 5 m) should be provided at the entering and exiting areas. Escalators that feed into another rise of escalators or into an area that may be locked at times should be electrically interlocked so that the exiting area is open before the escalator can be operated.

Elevators

Two-stop elevators are the other preferred means of high-capacity vertical transportation. Because everyone boards at one level and exits at the other, they can serve structures of any height and minimize passenger confusion. Introducing a third stop exacts a substantial time penalty besides creating confusion for passengers.

If an observation tower facility has more than two main upper levels, serious consideration should be given to shuttle elevators between the various levels. All people would then go to one of the upper main levels by the major elevators and transfer to the shuttle elevator for the other upper levels.

Refreshment Service

Service needs, especially refreshments, are of prime importance in any facility. Refreshment facilities should be stocked between events, using service or freight elevators provided for that purpose. These elevators should be located near loading docks or storage areas on the lower level and close to the refreshment stands on the upper levels to minimize truck transfer time. Elevators should be large and heavy enough to accommodate either industrial truck or handcart loading. Freight-type, vertical biparting doors can be used to gain the maximum width in the elevator interior.

Elevator Capacity

Passenger capacity rather than frequency of service is the prime objective in elevatoring facilities of this nature. Elevator speed should be proportional to the height of the structure as in Table 12.1. Because elevators are the most critical and expensive equipment for observation towers, consideration must be given to using the same elevators for both service and passenger use. Deliveries should be scheduled at off-peak visitor periods and ample cab interior protection should be provided. Outside elevators are popular for observation towers and can be used as an attraction in other spectator-type facilities. A full discussion of outside elevators will be found in a later chapter.

GENERAL

All the general rules for equipment location and passenger information and guidance apply to the elevatoring of institutional buildings. The large number of visitors, many unfamiliar with elevators or escalators, requires that particular care be taken to avoid confusion and operating inefficiency and that ample passenger information be provided.

Elevator operation must be simple and the signals explicit. An occasional visitor to a tall building is more inclined than the regular occupant to take any elevator that stops at a floor, no matter which way it is going. For this reason signs and directional arrows must be prominent and unequivocal in indication. All buildings should have elevators equipped to serve the handicapped.

OPERATION

As described in Chapter Ten on commercial buildings, an operational study must be combined with equipment analysis to gain full utilization of the vertical transportation plant. It is essential to provide traffic-responsive operating features and the programmable type of elevator control is most promising for the unique types of traffic expected in institutional buildings.

In Figures 12.5, 12.6, and 12.7 a series of traffic flowcharts is presented for principal types of institutional buildings. The charts also indicate some of the expected traffic conditions and the operations necessary to meet those conditions. People responsible for design as well as the operation of the building should see that these requirements are fulfilled.

Each chart shows the percentage of building population expected to be traveling in each direction at a particular time. This percentage will vary for the individual institution and should be studied relative to the capabilities of its vertical transportation system. If the system has substantial capability in relation to the expected traffic, minimum operating features for each traffic period are required. If traffic is heavy compared to system capabilities, additional operating features are necessary.

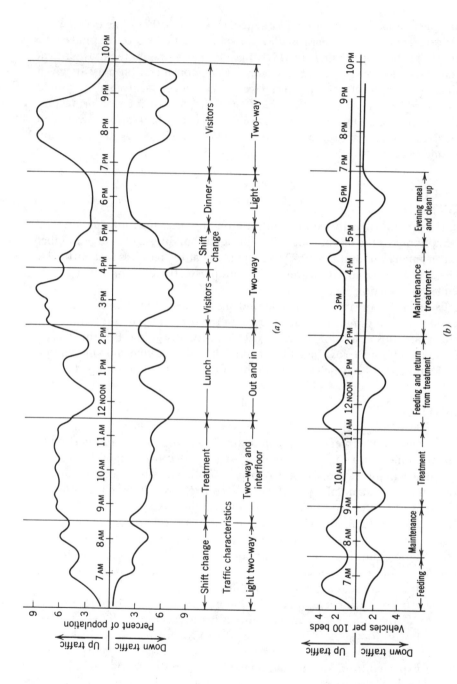

Figure 12.5. Hospital elevator traffic—in and out of the elevators at a main floor: (*a*) pedestrian traffic; (*b*) vehicular traffic.

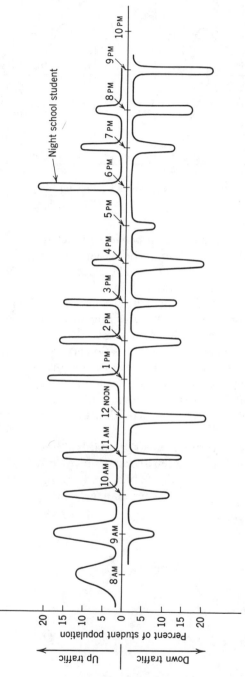

Figure 12.6. Elevator traffic—school, classroom building, day and night classes. Based on hourly class changes, 10-min class change time allowance. Traffic in and out of the elevators at the main floor.

305

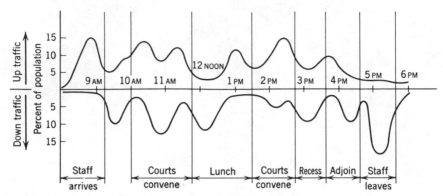

Figure 12.7.   Combination courthouse and civic office building elevator traffic in and out at the main lobby.

Note in Figure 12.5, the relation between expected activity in a hospital and its need for vertical transportation. The activities shown on the chart are typical for most hospitals but the times they take place may be different. Observations of local practice should be made to construct a chart for the new or existing facility. Constructing a chart will help relate the building's activities to its vertical transportation system. This is clearly borne out in the school situation, Figure 12.6, where the class change periods create the critical traffic demands.

A relation between planning and operation is very important in institutional buildings. Customary practices are fewer than in office buildings, in that management can largely establish its rules as long as the desired function is performed. For example, if a sports arena is not opened until 15 min before a performance, vertical transportation needs are entirely different than in a facility that is open 1 hr before.

The successful elevatoring of institutional buildings is predicated upon understanding relationships between all factors involved and early involvement of the people who will operate the facility. Any compromises need to be well documented for the guidance of later management.

CHAPTER THIRTEEN

Service and Freight Elevators

SERVICE AND FREIGHT ELEVATORS

There are only two types of elevators which are recognized by elevator codes. One is a passenger elevator which has well-defined requirements regarding the usable area of the car platform and the load that must be carried. The other type is a freight elevator which, depending upon its classification, can have various platform loadings versus area depending upon the intended use of the elevator. Freight elevators are prohibited by elevator codes from carrying any passengers other than those required to handle freight. They can carry passengers if the platform area related to load meets the requirements for passenger service.

A service elevator is a passenger elevator usually designed with a rugged interior and intended to carry both passengers and freight. All the rules for passenger elevators must be met and the equipment can be designed to meet certain freight loading requirements, such as the use of industrial trucks or to serve one-piece loads. These will be discussed later in this chapter.

The classification of service elevator is generally used to designate an elevator in an office building primarily intended to move material rather than people, a stretcher or vehicular elevator in a hospital, or the "back of the house" elevator in hotels.

FREIGHT ELEVATORS

Size and Capacity

If a building requirement calls for the ability to lift large, light weight loads such as furniture or clothing, a freight elevator may be an ideal application. The various loading classifications are explicit in the A17.1 Elevator Code and are amply illustrated in Figure 13.1.

GENERAL INFORMATION FOR
FREIGHT ELEVATORS WITH
CLASS "A","B" & "C" LOADING
(REFER TO ANSI-A17.1, RULE 207.2b)

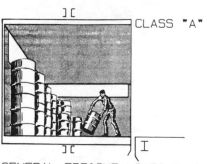

CLASS "A"

GENERAL FREIGHT LOADING
WHERE NO ITEM (INCLUDING LOADED TRUCK)
WEIGHS MORE THAN 1/4 RATED CAPACITY

RATING NOT LESS THAN
50 LBS/SQ-FT - 244.10 kgs/SQ-M

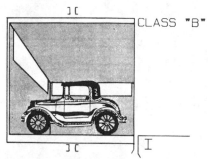

CLASS "B"

MOTOR VEHICLE LOADING
(AUTOMOBILES, TRUCKS, BUSES)

RATING NOT LESS THAN
30 LBS/SQ-FT - 146.46 kgs/SQ-M

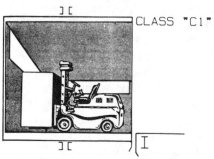

CLASS "C1"

INDUSTRIAL TRUCK LOADING
WHERE TRUCK IS CARRIED

RATING NOT LESS THAN
50 LBS/SQ-FT - 244.10 kgs/SQ-M
MUST HAVE AUTOMATIC LEVELING

THIS LOADING APPLIES WHERE CONCENTRATED
LOAD INCLUDING TRUCK IS MORE THAN 1/4
RATED CAPACITY BUT CARRIED LOAD DOES NOT
EXCEED RATED CAPACITY.

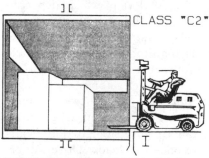

CLASS "C2"

INDUSTRIAL TRUCK LOADING
WHERE TRUCK IS NOT CARRIED, BUT IS
USED FOR LOADING AND UNLOADING

RATING NOT LESS THAN
50 LBS/SQ-FT - 244.10 kgs/SQ-M
MUST HAVE AUTOMATIC LEVELING

THIS LOADING APPLIES WHERE CONCENTRATED
LOAD INCLUDING TRUCK IS MORE THAN 1/4
RATED CAPACITY BUT CARRIED LOAD DOES NOT
EXCEED RATED CAPACITY.
THIS LOADING ALSO APPLIES WHERE INCREMENT
LOADING IS USED, BUT MAXIMUM LOAD ON CAR
PLATFORM DURING LOADING OR UNLOADING DOES
NOT EXCEED 150% OF RATED LOAD.

Figure 13.1. Classes of freight loading on elevators. Note: Both freight and passenger eleva-

NOTE

THE PICTORIAL FREIGHT LOADING
SHOWN FOR THE DIFFERENT CLASSES
OF LOADING IS INTENDED FOR BOTH
ELECTRIC AND HYDRAULIC ELEVATORS.
SOME DIFFERENCES OCCUR IN RAIL
REACTIONS.

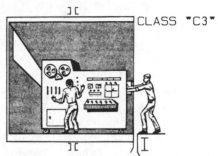

CLASS "C3"

CONCENTRATED LOADING
(NO TRUCK USED) BUT LOAD
INCREMENTS ARE MORE THAN 1/4
RATED CAPACITY. CARRIED LOAD
MUST NOT EXCEED RATED CAPACITY

RATING NOT LESS THAN
50 LBS/SQ-FT - 244.10 kgs/SQ-M
MUST HAVE AUTOMATIC LEVELING

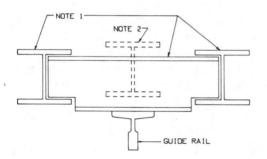

NOTE 1

NOTE 2

GUIDE RAIL

NOTE 1   VERTICAL RAIL COLUMN SUPPORTS AND CROSS TIE MEMBERS ARE REQUIRED AND PROVIDED BY OTHER
THAN THE ELEVATOR SUPPLIER WHEN RATED LOAD EXCEEDS 8000 LBS./3628.7 kgs. THE SIZE OF THE
RAIL COLUMNS ARE DETERMINED BY OTHERS FROM RAIL REACTIONS FURNISHED BY THE ELEVATOR
SUPPLIER.

NOTE 2   ALTERNATE METHOD OF RAIL COLUMN SUPPORT, WHEN RATED LOAD IS 8000 LBS./3628.7 kgs. OR LESS.
THE SIZE OF THE RAIL COLUMNS ARE DETERMINED BY OTHERS FROM RAIL REACTIONS FURNISHED BY THE
ELEVATOR SUPPLIER.

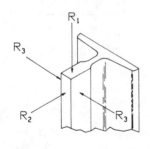

R_1

R_3

R_2

R_3

GUIDE RAIL REACTIONS (FOR A SINGLE RAIL)	ELECTRIC ELEVATOR		HYDRAULIC ELEVATOR	
	U.S. LBS.	S.I. kgs.	U.S. LBS.	S.I. kgs.
R_1	FOR THESE REACTIONS,			
R_2	CONSULT ELEVATOR SUPPLIER			
R_3				

tors can be designed for these various loadings (Courtesy National Elevator Industries, Inc.).

As may be noted from Figure 13.1 most freight elevators have a minimum platform loading of 50/lb ft² (245 kg/m²). Where the elevator is intended to carry motor vehicles such as the elevators in a garage or a truck elevator serving a basement loading dock area from the street, a loading of 30 lb/ft² (145 kg/m²) may be used. With a truck elevator it will be important to determine the loaded weight of the truck intended to be carried, which may require loadings as high as 150 lb/ft² (750 kg/m²) as may be experienced with armored trucks in a bank or fully loaded trash trucks in an office building.

In contrast, passenger elevator platforms are rated from about 70 to 130 lb/ft² (105 kg to 200 kg/m²), depending on total platform area. If a freight elevator is intended to be used for passenger service it must be rated as a passenger elevator and special door operation provided, as will be described later. In European countries both passenger and freight elevators have the same area load requirements.

Freight elevators are available in sizes from 2500 to 25,000 lb (1200 to 12000 kg) or more (Tables 13.1 and 13.2). Some can carry loads of 100,000 lb and others can carry fully loaded trailer trucks. It is quite common to provide 60,000- or 70,000-lb (27500- or 32000-kg) capacity hydraulic elevators with platforms 12 ft by 40 to 45 ft (3 m by 12 to 13.6 m) long to take trailer trucks from street level to a basement loading dock in congested downtown areas.

Because of the relatively long time required to load or unload a freight elevator, speed is generally of secondary importance. Elevator travel at high speed is not cost effective if 4 or 5 min are necessary to transfer its load. Practical speeds are proportional to building height and required loading time, with quicker transfers warranting higher travel speed. Speed should be sufficient so that traveling will consume only about 25% of total round-trip time.

Hydraulic elevators are especially attractive for freight elevators. Lower speeds are the most cost effective since motor horsepower is directly propor-

Table 13.1. Loading by Hand or by Hand Truck (Class A Loading)

Capacity, lb (kg)	Platform Size		Maximum Speed (Standard Equipment)
	Width, in. (mm)	Depth, in. (mm)	
2,500 (1,200)	64 (1,650)	84 (2,250)	No limit
3,000 (1,400)	76 (1,930)	96 (2,450)	No limit
3,500 (1,600)	76 (1,930)	96 (2,450)	No limit
4,000 (1,800)	76 (1,930)	96 (2,450)	No limit
5,000 (2,250)	100 (2,540)	120 (3,050)	Usually 700 fpm (3.5 mps)
6,000 (2,700)	100 (2,540)	120 (3,050)	Usually 700 fpm (3.5 mps)
8,000 (3,650)	100 (2,540)	144 (3,660)	500 fpm (2.5 mps)
10,000[a] (4,550)	124 (3,150)	144 (3,660)	300 fpm (1.5 mps)
12,000[a] (5,500)	124 (3,150)	168 (4,270)	300 fpm (1.5 mps)

[a]Elevators of this size should always be considered for industrial truck loading.

Table 13.2. Loading by Industrial Trucks (Class C[b] Loading)

| Capacity, lb (kg) | Platform Size | | Maximum Speed (Standard Equipment), fpm (mps) |
	Width, in. (mm)	Depth, in. (mm)	
10,000 (4,550)	100 (2,540)[a]	144 (3,660)	300 (1.5)
12,000 (5,500)	164 (3,150)[a]	168 (4,270)	300 (1.5)
16,000 (7,300)	124 (3,150)[a]	168 (4,270)	300 (1.5)
18,000 (8,200)	124 (3,150)[a]	192 (4,880)	300 (1.5)
20,000 (9,100)	148 (3,160)[a]	240 (6,100)	200 (1.0)

[a]Vertical biparting doors of this width less 4 in. (100 mm) can be accommodated.
[b]Class C1 or C2 depending on static load requirements.

tioned to speed. The load imposed on a hydraulic elevator is directly transferred to the pit through the plunger, hence less structural cost is incurred versus a overhead electric elevator (Figure 13.2). In 1980, progress is being made using telescoping hydraulic jacks. With advancement in this technology it is predictable that hydraulic elevators will be applied in excess of 70 ft (21 m) or so of rise which is an approximate limit of the single-length hydraulic jack at present. Telescoping jacks also mean much less involvement in drilling problems and costs. In European countries the use of a combination of roller chains and hydraulic jacks is being used to develop a "holeless type" of hydraulic elevator.

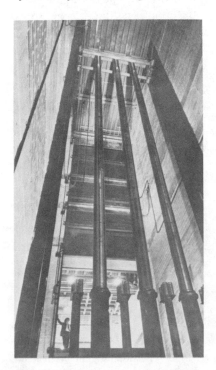

Figure 13.2. Hydraulic freight elevator capable of handling trucks (Courtesy Dover Elevator Company).

A car safety device is required with this approach whereas a direct plunger does not presently require such a safety.

Loading and Unloading

Loading and unloading of a freight elevator usually take the greatest percentage of the freight elevator round-trip time depending on the loading method used. Loading pallets by hand truck may require from 15 to 20 sec to transfer each pallet load on or off the elevator. If the car is to be loaded with four pallets, over a minute will be spent in loading or unloading. This can be improved if power truck loading is employed. Time for door operation must be added to the loading time.

Freight elevators usually have vertical biparting doors, consisting of upper and lower panels that counterweight each other (Figure 13.3). With manual operation, a door of this type can be opened or closed in 6 to 8 sec plus about 6 sec to open or close the vertical lifting gate on the elevator. With power operation the door and gate can be simultaneously opened or closed at a speed of about 1 fps. The average 8-ft-high opening can thus be closed or opened in about 4 sec, as each panel only has to travel half the height.

Vertical biparting doors allow an opening almost the full width of the platform and the structural steel frames of the opening are less susceptible to damage than the formed metal frames of passenger-type entrances. The door panels are completely hidden before loading can begin and thus are less likely to be struck

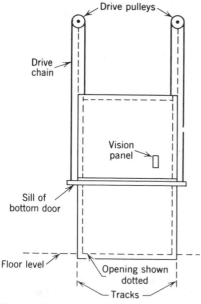

Figure 13.3. Vertical biparting doors.

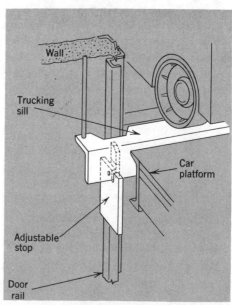

Figure 13.4. Truckable sill for vertical biparting door.

by a moving truck as can happen with horizontally sliding doors. Sills can be reinforced to withstand truck wheel impact (Figure 13.4).

The time required to transfer a single pallet load with an industrial truck obviously varies with the transfer distance. If we assume transfer directly off or on the elevator, each pallet can be moved in 10 sec and total loading or unloading of four pallets will require no more than about 40 sec. The typical cycle would be as follows:

Door opening time	4 sec
Unloading (or loading) time	40 sec
Door closing	4 sec
Time per stop	48 sec
Inefficiency, 15%	7 sec
Total time per stop	55 sec

If loading at one stop and unloading at another takes place, running time is simply added to twice the time per stop indicated above to determine the time required to handle four pallet loads. If a running time of 1 min is assumed (which will vary with speed and rise), the total round-trip time is 3 min, or 20 trips per hour or 80 pallets per hour. If the elevator is to be unloaded and then reloaded at the same stop, the additional loading time should be allowed and the longer round-trip time calculated.

Personnel on Freight Elevators

If people are to ride on any freight elevator, certain provisions are required to be made by elevator codes. If only personnel necessary for freight handling are allowed on the car and prominent signs to this effect are displayed, the elevator is considered strictly freight. If other employees use the elevator for transportation to upper floors, the A17.1 Elevator Code requires that the elevator platform area be rated the same or greater than a passenger platform area.

Biparting doors must have sequence operation so that the car gate closes before the hoistway door can close and the hoistway doors must open before the car gate opens. A protective safety edge must be added to the car gate so that the closing gate will stop and reopen if it encounters an obstruction in its descent. Additional protection can be a light ray across the entrance that will stop and reopen the gate when interrupted. When sequence operation is provided, an elevator with biparting doors may have an automatic closing operation. This means that, as with a passenger elevator, the doors may begin to close a time interval after they are opened so that the elevator can be called from floor to floor without an attendant. Without these provisions the doors must be closed by a constant pressure button after each stop. Because people tend to be careless about reclosing doors, a door-open bell should be provided which automatically rings when someone calls for an elevator at another floor and the elevator doors are not closed.

The door-close button is often hung on a pendant in front of the elevator so that the driver of an industrial truck does not have to dismount to close the door.

Structural Requirements

Freight elevators with class A loading are loaded and unloaded manually and require structure in the hoistway equal to that of a passenger elevator. If the elevator is to be rated for class B or C loading with vehicular loading, tractor and trailers, or industrial truck loading and unloading, extra rail support must be provided (Figure 13.5).

The entire impact of the truck stopping, running, and reversing is transmitted to the building structure through the elevator rails. The diagram in Figure 13.5 shows how these forces react. An example of an extra rail structure to absorb these forces is shown in Figure 13.6. The necessary structure for each installation must be calculated and engineered for the particular elevator size and capacity.

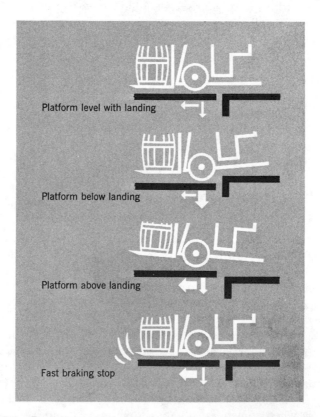

Platform level with landing

Platform below landing

Platform above landing

Fast braking stop

Figure 13.5. Reactions on the edge of a freight elevator platform during loading.

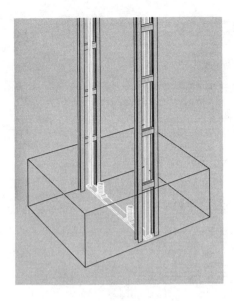

Figure 13.6. Rail column support.

The elevator engineer normally calculates the forces while the architect or structural engineer is responsible for providing the building supports. The memorandum below and Figures 13.7 and 13.8 show one form for transmitting the necessary information.

All other parts of the elevator must be designed to withstand the extra loads of industrial truck loading. The platform, for example, requires extra-heavy flooring either of steel plate or industrial-type wooden blocks on end; the platform structure requires a multiplicity of stringers to distribute the load and prevent distortion; and the car frame and hoisting machine or hydraulic pistons must be of heavy duty design. With extremely large loads, multiple guide shoes and, possibly, two sets of rails may be required.

Manufacturing Company
Anywhere

Subject: Industrial Truck Loaded Elevator
           Building Number
           Anywhere

Freight elevators carrying industrial trucks, automobile trucks, or passenger automobiles exert forces of large magnitude on the building structure. In order to assure a safe and satisfactory installation, it is essential that the supports you provide for the guide rails will be of adequate strengths and stiffness. We are therefore furnishing you the rail forces and stiffness requirements for your elevator. A brief explanation of the effects of these forces is outlined below.

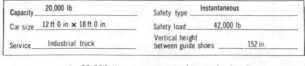

Capacity	20,000 lb	Safety type	Instantaneous
Car size	12 ft 0 in × 18 ft 0 in.	Safety load	42,000 lb
Service	Industrial truck	Vertical height between guide shoes	152 in.

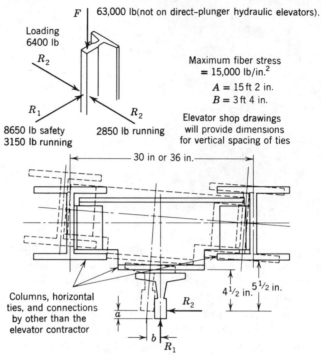

F | 63,000 lb(not on direct–plunger hydraulic elevators).

Loading
6400 lb

R_2

Maximum fiber stress
= 15,000 lb/in.^{2}

A = 15 ft 2 in.
B = 3 ft 4 in.

R_1 R_2

8650 lb safety   2850 lb running
3150 lb running

Elevator shop drawings
will provide dimensions
for vertical spacing of ties

— 30 in or 36 in. —

Columns, horizontal
ties, and connections
by other than the
elevator contractor

$4\frac{1}{2}$ in. $5\frac{1}{2}$ in.

R_2

a

b

R_1

Section X–X
See table below for values

Condition	Net forces	Maximum allowable deflection
Loading [a]	R_1 = 3150 lb	$a = \frac{1}{8}$ in.
(as indicated)	R_2 = 6400 lb	$b = \frac{3}{16}$ in.
Running[a]	R_1 = 3150 lb	$a = \frac{1}{8}$ in.
(at S)	R_2 = 2850 lb	$b = \frac{3}{16}$ in.
Safety[a] *	R_1 = 8650 lb	$a = \frac{5}{16}$ in.
(at S)	F = 63000 lb [b]	–

[a] These conditions need not be considered as acting simultaneously.
[b] For design of pit supports, this force must be doubled for impact.

*These forces are not present with
direct–plunger hydraulic elevators

Figure 13.7.   Freight elevator rail reactions—plan.

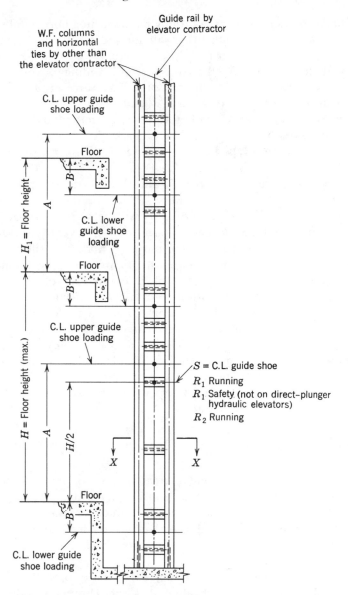

Figure 13.8. Freight elevator rail reactions—elevation.

Forces at Loading. When the loaded truck enters the elevator, it exerts heavy loads at the front edge of the platform. These loads produce a force couple—the top guide shoe exerts a force on the guide rail toward the loaded edge of the platform and the lower guide shoe exerts a force away from it. With entrances at the rear of the elevator, these forces will be reversed. These rail forces (designated R_2) produce bending moments in the rail and its support, the double H columns, act as a beam placed vertically.

The R_2 forces also produce turning moments in a horizontal plane, bending one column toward the hoistway wall and the other away from the hoistway wall.

In addition to force R_2, there is a force R_1 tending to spread the rails apart due to eccentric loading of the platform. This force produces direct bending moments in the columns (Figures 13.7 and 13.8).

Since these forces are at loading and are transmitted to the columns through the guide shoes and the rails, they occur at determinable points above and below the loading levels. Refer to Figure 13.7 for dimensions and table for values of forces.

Forces at Running. Forces R_1 and R_2 also occur during running of the car and have the same effects. Force R_1 can be the same magnitude as for loading or somewhat larger. Force R_2 will be somewhat smaller. However, since the running forces R_1 and R_2 occur during the entire travel of the elevator, it must be assumed for design purposes that they will be applied to the center of each column span and each support for the column (floor beam, spreader beam, etc.). Refer to the table in Figure 13.7 for magnitude of forces.

Forces at Safety Application. At safety application when the car is provided with a safety device, the stopping action of the safety exerts a large force F vertically on the rails in a downward direction. These forces are transmitted through the rail to the pit floor. There is also a large force R_1 due to eccentric loading, which tends to spread the rails (similar to loading R_1 and running R_1). The loads due to safety application can occur anywhere in the entire car travel. See the table in Figure 13.7 for size of these forces. These forces do not apply on direct-plunger hydraulic elevators.

ELEVATOR DESIGN

Car interiors deserve particular attention. Steel or oak rubbing strips should be mounted along the sides and adequately supported to withstand the blows from carelessly driven trucks and designed to be renewable. If trucks are to be carried, the location of the landing button fixtures demand special consideration such as pendant mountings. Car operating panels should always be recessed into the side of the elevator.

Car lighting should be bright to make the interior of the elevator fully visible which will tend to reduce the likelihood of trucks hitting the side or rear. A curb can be effectively used to limit the distance a truck can be driven into the elevator. An effective protection for the sides of the car is to mount one or two 12-in (300 mm) channels on end on the lower sides. In that way, the forks of a fork lift truck cannot damage the sides.

The car platform should be provided with durable or easily renewable flooring. Checkered steel plate or aluminum plates with abrasive surfaces have been used. These should be designed in sections of about 200 lb (100 kg) which can be removed and replaced. Many safety codes require the use of high-visibility markings such as yellow and black striping on movable parts such as doors and gates. These markings act as a warning and the contrast in color may act as a caution and minimize impact.

Materials-handling elevators also have to be installed in "problem" environments ranging from an explosive atmosphere to an area in which normal cleaning with hot water and steam requires a completely waterproof installation. Corrosion control and other elements of maintenance of structural steel in the elevator installation must also be considered. All safety devices as well as ropes and electrical equipment must operate properly in these difficult environments, and special provisions must be made for each hazard. Hazards must be identified and precautions taken in the specifications for elevator equipment to guard against premature wear or danger.

The contractor or structural engineer must provide the necessary structural framing at each level which will become the frame and sill for the vertical biparting door. Because these doors as well as the lifting car gate may require extra space above the top landing or in the pit, elevator overhead and pit must be structured to suit.

Structural requirements for freight elevators are considerably more critical than those for passenger elevators. They are numerous and variable and require early consultation with elevator engineers when large freight elevators are planned.

SERVICE ELEVATORS

Service elevators in commercial, residential, or institutional buildings are generally passenger elevators with special provisions to handle oversize loads and both hand- and motor-assisted trucks. They should have abuse-resistant interiors, must comply with code requirements for carrying passengers, and should be equipped with horizontal sliding doors which allow full automatic operation with maximum efficiency. The service elevators should be classified with the types of loading expected: class A freight loading if hand- or motor-assisted trucks will be used; class C 1 or 2 if powered industrial trucks are used (the elevator should be large enough to accommodate such a truck); or class C3 if an occasional heavy single-piece load, such as a transformer in a commercial building, is to be carried. Such classification ensures that the structure and the platform of the elevator will be designed to accommodate such loading.

As general rules of application, at least one service elevator should be provided for each 300,000 ft^2 gross (3000 m^2) of office space, each 100 beds in a hospital, and each 200 rooms in a hotel or any apartment house that deserves the "luxury" classification. These are general rules and should be modified by the specific requirements of the building under study and the needs of the occupants. For example, a headquarters building of a large corporation of 300,000 ft^2 (3000 m^2) may require two service elevators due to the expected activity, or a small single-purpose office building of 100,000 ft^2 (1000 m^2) may have a definite need for a service elevator. An automated materials handling system may modify this requirement and is discussed in Chapter 15.

Figure 13.9.   Service elevator with high cab and hoist eye.

If enough information is available, a time and motion study can be made of the expected number of trips required to be made on the service elevators. For example, if it is known that mail will be delivered twice daily, coffee service will take place to each floor, supply requisitions will be delivered on schedule, and a certain amount of maintenance movement and a predictable number of moves and renovations are possible, a time allowance can be assigned to each movement which can be totaled and compared against available service elevator time. A judgment must then be made as to the consequence of not having the service elevator available versus an alternate means of moving essential material.

Service elevators can be any practical platform size or ceiling height provided the platform area conforms to elevator code requirements. Elevators can have an interior height of up to 15 ft (4.5 m) if sufficient elevator hoistway overhead exists. Such a height can accommodate long carpet rolls in an office building and oversize furniture. To aid in loading and unloading, a hoist beam or lifting eye can be mounted in the ceiling so a chain fall may be used to haul the load in the car (Figure 13.9). Elevators of this height have contributed to the early completion of buildings by allowing the prefabrication off site of large building components and office interiors. For the larger office building, an elevator of 6500-lb (3000-kg) capacity with 54-in. (1400-mm) wide center-opening doors is recommended. A minimum recommended size service elevator is 4000 lb (1800 kg) with 48-in. (1200-mm) wide doors.

The size of a building service elevator should be established by determining the most frequently expected loading. This may be movable partitions for office renovations or large-size supply carts. The area of the car inside will establish the rated load carrying capacity of the elevator. As a second consideration, if this size area capacity can be moderately increased to carry the heaviest load such as a transformer or compressor, it should be done.

OPERATION OF SERVICE AND FREIGHT ELEVATORS

The grouping and operation of service and freight elevators depend on their role in the overall building industrial or storage system. Service or freight elevators are seldom used in large groups, the usual installations being individual units. Two major forms of automatic operation are available as well as a number of manual operations.

Automatic operations are either collective or single automatic. With collective operation each call is remembered and the elevator answers all the up landing calls on the up trip, reverses at the highest call, and answers all the down calls on the down trip. The operation is useful if one loading is not expected to fill the elevator and additional loads can be taken on at other stops. If full loads are expected, a bypass feature is used, allowing an attendant to use a special switch within the car to bypass landing calls.

Attendant operation is almost a necessity on a busy freight elevator in an industrial setting and can often be used to an advantage on a service elevator in a large commercial building or hospital. The attendant can move material on and off the elevator into an adjacent lobby and can act as a supervisor if a great deal of material delivery is being made to avoid individual messengers tying up the elevator. Where a move in or move out is in progress in a building, the attendant can supervise and provide service for the movers plus allow other material movements to take place when the movers are away from the elevator.

If "with attendant" is to be used on a building service elevator, an annunciator is recommended to indicate the floors at which loads are waiting. In addition, an illuminated sign in the landing button or landing lantern fixture should inform prospective passengers that the car is "In Freight Service" and not available for general use. A "house" phone should be provided in the service elevator so that the attendant can be called if a special material movement is necessary.

If more than one service elevator is installed, the elevators should be next to each other and provided with a duplex collective or group operation. If security is a consideration, the loading dock attendant's office can be provided with elevator position indicators and means to recall and monitor the elevators, as well as a dedicated intercommunication system. All service elevators should be provided "with attendant" operation.

Single automatic operation allows the person who has control of the elevator exclusive use of the car for that trip. The landing call buttons incorporate illuminated signs to indicate if the elevator is in use. When the light goes out the elevator may be called to a particular floor and be used by the next person. This operation is preferred whenever the load is such that it would fill the elevator and additional stops would be unproductive. A variation of this operation is to have a central dispatching station through which a dispatcher would control the elevator use and destinations. The dispatcher would be called by telephone or intercom to send a car to a given floor and to dispatch it to the unloading floor. This operation is often used in garages when a number of elevators in a group are required to serve the garage capacity. For any multiple automatic elevator group, a landing lantern or indicator light associated with each entrance should be provided to inform the truck operator which elevator will be available.

A semi-automatic operation called "double button" is often used with freight elevators as a cost savings approach. The operation is such that a car can be called to a floor by constant pressure of an up or down button at the landing. The elevator can also be operated from within the car by similar constant pressure on either an up or down car operating button. Leveling may be automatic or accomplished by means of "inching" buttons used to jog the car to floor level within a restricted zone at each floor with the doors open, and only to operate the car toward floor level, never away from it.

An optional operation for any freight elevator which opens on a street level is a system known as "tail-board inching," which can be used to stop the elevator platform above the actual floor level, at the same height as a truck backed up to the elevator. An extra long guard is installed below the elevator platform to avoid the danger of a person or object falling into the hoistway.

If the industrial plant or warehouse requires a group of elevators, many of the operations outlined for passenger service may be adapted for freight service. The final choice would depend on the door-operating system employed. Because it is difficult to ensure that people will close biparting doors after they use an elevator, the accepted system is to employ a door-open bell that will sound if someone is calling the car while the door is open. Use of sequence operation with biparting doors and a safety edge on the car gate is one means of providing automatic operation.

Large openings with horizontal sliding doors will ensure prompt automatic operations of the elevators. If loads are light and bulky, such as racks of garments in a department store or clothing factory, the horizontal sliding entrances are desirable.

ELEVATORS IN INDUSTRIAL PLANTS

Years ago, before the electronic revolution in musical entertainment, a typical piano factory was a multistory building. Light parts of the piano were manufactured on the upper floors, all raw material being taken to the top. The foundry and finishing shops were on the lowest floors. The myriad small wooden pieces were assembled as they traveled down, becoming larger and larger sections until they became the completed piano on the shipping floor.

Many modern industrial processes now follow a similar sequence. Because land costs were moderate when many manufacturers started, their plants were usually horizontally oriented. Transportation of raw materials and finished products was also economical. Transportation and land costs have been rising, so that the vertical factory is becoming more and more economical and, with dependable vertical transportation, quite feasible.

By engineering vertical transportation on the basis of the production process, a completely integrated factory can have compact vertical design. Large elevators can carry raw materials to upper floors, and conveyors, dumbwaiters, or small elevators can bring the finished parts to the lower floors for assembly. Personnel can be moved swiftly to and from their jobs by elevators or escalators.

The role of gravity cannot be ignored in a vertical factory. In a cannery, for example, raw tinplate sheets are moved to upper floors for shaping into cans and rolled down to the food-processing floors for filling, sealing, labeling, and shipping.

The number, size (area), and capacity, as well as required speed of manufacturing plant elevators can be determined by simple calculation based on indicated need. Each area requiring transportation must be studied and the expected volume estimated for an applicable time period. Elevators to handle this volume can be based on the following factors:

Time to Load

This time varies with the type of loading: industrial truck, hand truck, cart, or hand. Industrial truck loading is the fastest but also creates the greatest stress on the elevator equipment.

Door-close Time

This time depends on the type of door, the height or width of the opening, and the use of power or manual operation. Vertical biparting power-operated doors can be operated at 1 fps (0.3 mps) per panel average speed which allows an 8-ft (2400-mm) opening to be opened or closed in 4 sec. Horizontal sliding doors must comply with the kinetic energy limitations of the A17.1 Elevator Code and door-open and door-close times are listed in Table 4.3.

Time to Start and Run the Elevator

This will be the same as calculated for passenger elevators of the same given speed and is shown in Chart 4.2. Speed is not a critical factor for freight elevators, 50 fpm (0.25 mps) being acceptable for a two-story building and 100 fpm (0.5 mps) for four stories. Most freight elevators are of the low-rise hydraulic type due to the usual large loads they carry. The disadvantage of the hydraulic elevator is the relatively large electrical demand imposed on the power supply during the starting in the up direction. If frequent use of the elevator is expected, i.e., many trips per hour, the more cost effective elevator may be the counterweighted electric-type elevator.

Unloading Time

This time depends on the factors considered in loading. Front and rear entrances will expedite loading and unloading of the elevator, especially if industrial-type small tractors and trailers are used to pull loads around the plant.

Typical industrial elevator situations are analyzed in Examples 13.1 and 13.2. An infinite number of variations are possible to satisfy the requirements of various plants, but using equipment in line with manufacturer's standard equipment is recommended as being the most cost effective.

Example 13.1. Elevators in Industrial Plants—Pallet Loads

Required to move: pallets 4 × 4 ft × 6 ft high, each weighing 2500 lb, approximately 200 per day. Loading by industrial truck, truck weight 8000 lb, 80% of weight on front wheels.

Distribution: 50% load to fourth level from dock, 36-ft rise; 50% load to second level from dock, 12-ft rise

Determine: elevator size and speed

Size: assume 4 pallets per trip; minimum platform size 8 × 8 ft interior, 10,000 lb; recommended standard size 8 ft 4 in. × 12 ft 0 in.

Rating:

1 pallet and truck	2500 + 8000	= 10,500 lb
2 pallets and truck	5000 + 8000	= 13,900 lb
3 pallets and 0.8 truck	7500 + 6400	= 13,900 lb
4 pallets and 0.8 truck	10,000 + 6400	= 16,400 lb

16,400 static load required. 12,000-lb car with static loading (50% over capacity). Number of elevators required (assume 100 fpm).

To fourth level:

Load at dock 4 pallets at 15 sec per pallet	60 sec
Door close	4 sec
Run to fourth level, 36 ft @ 100 fpm = 23.8 sec	24 sec
Door open	4 sec
Unload	60 sec
Door close	4 sec
Return	24 sec
Door open	4 sec
Total time, 4 pallets	184 sec, say 3 min

Total time, 100 pallets, 25 trips × 3 min = 75 min

To second level:

Same as above except running time changes:

12 ft @ 100 fpm = 9.4 sec, 24 − 9. = 15 × 2 = less 30 sec

Total time, 4 pallets 184 − 30 = 154 sec = say 2.5 min

Total time, 100 pallets 25 × 2.5 = 62.5 min

Total time, 200 pallets 75 + 62.5 = 2 hr 17.5 min

One elevator, 12,000 lb @ 100 fpm with static loading will provide ample service.

Example 13.2.   Elevators in Industrial Plants—Carts

Required: Determine the number of industrial carts that can be moved an average 6 floors (72-ft rise) by one elevator per hour. Maximum load each cart, 6000 lb, tractor load 8000 lb. Front and rear openings on elevator.

Assume: tractor can pull and deposit 2 carts on elevator; alternate tractor and 1 cart each trip (sufficient tractors)

A.   2 carts per trip.

Tractor and 1 cart 6000 + 8000 = 14000 lb

½ cart, cart and ½ tractor 3000 + 6000 + 4000 = 13000 lb

2 carts 2 × 6000 lb = 12000 lb

Time to load 2 carts	15 sec
Uncouple	15 sec
Drive off, close doors	10 sec
Run 72 ft (assume 100 fpm)	45 sec
Couple tractor and drive off	25 sec
Close doors	15 sec
Run back to lower level, doors open	45 sec
Total per 2 carts:	170 sec

Number of carts per hour $\dfrac{3600 \times 2}{170} = 42$ carts

B.   1 cart per trip and tractor:

Drive on	15 sec
Close doors	4 sec
Run	45 sec
Open doors	4 sec
Unload tractor and cart	15 sec
Second tractor drives on while first is unloading	
Close doors	4 sec
Run	45 sec
Open doors	4 sec
	136 sec per cart

Number of carts per hour $\dfrac{3600 \times 1}{136} = 26$

Result: 2 carts per elevator best way. Elevator should be rated at 14,000 lb class C1 loading for occasional cart and tractor trip. Static loading not required. Platform size depends on size of cart and tractor.

ELEVATORS IN WAREHOUSES

The nature of the warehouse operation determines the type of freight elevator required. If it is for a long-term, low-turnover storage with bulky but light loads a large, slow-speed elevator with manual operation and manual doors may be sufficient.

If the warehouse is a fast turnover facility making extensive use of industrial forklift trucks, its elevators must be so designed. Their number and size will be a direct function of the expected turnover.

As indicated earlier in this chapter, a time cycle of elevator loading and operation must be established to determine the average time per load. This is projected for the number of loads that must be transported and the elevator requirements determined accordingly.

A number of automated elevator systems are applicable to material handling and may be considered for a particular warehouse problem. The systems may be divided into two categories: those in which the load is moved by means of a horizontal conveyor to the elevator and those in which the elevator moves, in essence, to the load.

Conveyor systems to feed a stationary elevator are generally supplied by conveyor specialists (Figure 13.10). Initiating and limit switches on the conveyor control elevator response. The elevator arrives and conveyor equipment on the elevator platform transfers the load. A limit switch signifies the completion of transfer. Disposition of the load is controlled either manually or by program, with the elevator moving the load to another floor and automatically unloading. A programmable controller may be employed to keep track of and direct multiple loads to various floors and to return waiting loads at upper floors to a main floor or another destination.

One type of automatic transfer includes the use of an elevator with a tilting platform. Cylindrical loads, like rolls of newsprint, are rolled onto the elevator at the loading floor by either a pusher or a floor tilter. When the elevator arrives at its destination the elevator platform is titled and the load rolls off.

An elevator in a tower moving on tracks has been applied to a number of automated material-handling systems. The elevator contains a platform with a pallet transfer device and by a combination of vertical elevator motion and horizontal tower movement, the platform is indexed to various stalls and the transfer device either deposits or extracts the load from the stall.

Figure 13.10.   Automatic pallet loading on an elevator by means of pallet conveyors.

Such systems have been utilized in warehouses handling pallet loads and one such unit is operating in a frozen food storage warehouse at $0°F$ ($-15°C$). A programmable computer system is employed to direct the tower and elevator to various stalls for loading and unloading and to keep inventory current. Inputs to the computer are used to assemble truck loads consisting of different pallet loads at an unloading station as orders for the material are received. A similar system has been applied in an air freight terminal to receive and stage loads to minimize the turnaround time of large cargo carrying aircraft.

ELEVATORS IN GARAGES AND FOR OFF-STREET LOADING

The history of garage elevators can be traced to the early 1900s when both horses and carriages were carried on elevators to upper floors for stabling or storage. Early horse elevators had a gate between the horses and the elevator operator and channel troughs in the platform so it could be washed down. Modern garage elevators do not need the trough, but protection against the vehicle being driven into the extreme end of the elevator is advisable.

A garage elevator should be designed for operation by an attendant who parks or retrieves automobiles at an upper or lower floor. Landing buttons should be pendant mounted and the operating panels in the elevator located so that the attendant can operate them by leaning out of an automobile window.

Time can be saved by a garage elevator with both front and rear openings. The attendant can drive in the front and out the back, thus saving 10 to 20 sec per elevator trip.

A typical sequence of operation of a garage elevator with a single entrance, drive in, back out is as follows:

Door open	4 sec
Drive in	10 sec
Door close	4 sec
Run to upper stop depends on rise and speed of elevator	
Door open	4 sec
Back out	20 sec
Door close	4 sec
Total standing time	46 sec to park one automobile

The cycle is then repeated by adding the running time to return for the next automobile.

With the drive-through type of elevator, back-out time is replaced by drive-off time, which can be as little as 10 sec.

The number of elevators required for a particular garage will be a function of the number of floors, the number of automobiles to be turned over (parked and unparked in a given time), the number of attendants available, storage space at the entrance floor, and the speed with which customers can pay their bills and drive out to the street.

All the factors are important and are reflected in the economics of the elevator installations. Little is gained by an elevator plant that can deliver a car a minute if the local streets are so congested that it takes two minutes to leave the garage.

If the garage depends on elevators for its operation, a minimum of two is recommended to maintain continuity of service. A garage elevator usually has a 7000-lb capacity to handle large limousines. The dimensions of the 7000-lb elevator are a platform 10 ft (3050 mm) wide by 24 ft (7300 mm) long. If it is intended to operate the elevator without leaving the automobile, the width should be 8 ft (2500 mm); 21 ft (6400 mm) is the minimum length for a garage elevator. If larger cars or trucks are expected to be handled, capacity and dimensions should be increased.

An important outgrowth from garage-type elevators is the use of off-street loading facilities in office buildings to minimize the area needed for street-side loading docks. Because street space is valuable, trucks can be taken by an elevator to a basement area, turned around by turntable if necessary, and unloaded at a basement loading dock. The size of the elevator is exceedingly important and most of the truck elevators are hydraulic because of the loads involved, their location, and the short travel distance involved.

Truck type elevators can range in size from a platform 10 ft (3050 mm) to 12 ft (3700 mm) wide by 24 ft (7300 mm) long which will handle small delivery trucks to a platform 12 ft (3700 mm) to 14 ft (4300 mm) wide by about 40 ft (12,200 mm) long to handle a large trash truck. If tractors and trailers are to be served, lengths up to 60 ft (18,300 mm) long are necessary. Capacities range from 20,000 lb (9100 kg) for the smaller trucks to 70,000 lb (32,000 kg) for a fully loaded trash truck with wet trash. An important consideration in the design of a larger building is the way they intend to handle their trash. With extensive food service facilities, wet garbage can present a problem and the considerations may include the use of compactors and associated containers which must be loaded on and off special trucks.

If the building is dependent upon the basement-level off-street loading dock facility, two elevators are necessary. Traffic control must be provided so that a truck leaving the loading dock area will not be in the way of a truck entering either at the street level or at the elevator exit level below.

Consideration should also be given to emergency operation in the event of a power failure. The building's standby generator can be designed to have sufficient capacity to run one of the truck elevators at a time or a low-horsepower auxiliary pump can be designed into the elevator system so that one of the hydraulic-type truck elevators can be raised to street level at a much lower speed than normal operation to get a truck out of the building.

OTHER FREIGHT AND SERVICE ELEVATOR CONSIDERATIONS

In addition to the total load and methods of loading freight and service elevators, the environment in which loading takes place must be considered.

Unless special hazards in a plant dictate otherwise, durability is a prime requirement. If hazards such as chemical atmospheres, abrasive or explosive dusts, or moisture exist, the elevator equipment is governed by the same considerations as any other electrical and mechanical installations in the hazardous area. This is more fully discussed in the chapter on special installations.

To gain the greatest efficiency from an elevator plant, sufficient space for access to the elevator and unloading must be allowed. This is one advantage of an elevator with both front and rear openings, provided one side is used for unloading and the opposite side for loading. A minimum consideration must be sufficient room to maneuver either hand or industrial lift trucks. With a loading dock, sufficient space to hold material to be moved must be provided and related to the capability of the elevator.

Elevators in industrial plants are production equipment and must be treated as such. They are subject to downtime for maintenance and repair just as a lathe or a press would be. The fact that elevators are unusually vital, not only to the productive function but to the movement of people, must be recognized. If the entire operation of a plant depended on one machine, a standby would probably be considered, and the same logic should be applied to plant elevators. The alternative would be to have available spares for any parts that might fail to permit replacement in minimum time.

A systematic consideration of all the factors that the use of a service or freight elevator must fulfill should be made before the actual equipment is specified. The choice of equipment and subsequent structural and building requirements will follow once the quantity and quality of the loads are known. The next consideration is a failure analysis—what would be the consequence or alternative if the elevator were not available.

As with any vertical transportation analysis, sufficient documentation should be provided so that the people responsible for building operation are aware of the basis of original design.

As may be seen in this chapter, a variety of designs are available to the architect for making a proper application of vertical transportation to a new facility. In chapter 15 automated means of materials handling will be fully discussed.

CHAPTER FOURTEEN

Nonconventional Elevators and Elevator Applications

NONCONVENTIONAL ELEVATOR APPROACHES

The previous chapters described the bulk of the conventional elevator applications wherein the elevators were within a building and the building was used for a single function, i.e., commercial, residential, or institutional use.

A growing trend toward multipurpose buildings is apparent. A hotel or apartment can be built on top of an office building and the same office building built on top of a store. A school can be located on the lower floors of an office building or an apartment house. By multiple use of the building the efficiency of a 24-hr-a-day operation can be gained from a single capital investment. This chapter will discuss the vertical transportation aspects of such multipurpose buildings.

Other nonconventional elevator applications include the trend toward observation-type elevators in either the atrium of a building or on the outside of a building, either exposed to the weather or in a glass enclosed hoistway. Such elevators not only provide the necessary vertical transportation but serve to enhance the uniqueness of the building they serve.

In addition to vertical elevators, inclined elevators are being favored for a number of applications both commercial and scenic. This type of elevator will be described later in this chapter.

The needs of the handicapped are receiving increasing attention and many special-purpose types of lifts are being developed to fulfill this need. Such areas covered are the use of platform lifts to supplement stairways to replace ramps and the application of stairway lifts.

Many nonelevator related vertical transportation systems are being increasingly used. Such applications include stage lifts and platform lifts. The latter is often needed to develop multipurpose room areas such as raising the end of a gymnasium to create a stage or providing a platform for a TV camera.

Special industrial applications are frequently found and include the use of rack and pinion elevators alongside smokestacks for service personnel or inside the caissons of offshore oil production platforms to serve maintenance needs.

MULTIPURPOSE BUILDINGS

With the increasing emphasis on land usage and energy conservation, the concept of the multipurpose building is gaining favor. A building may well combine lower floor commercial space and residential space above. Hotel or apartments on top of office buildings and/or stores may permit 24-hr use of the structure.

The secret to successful multiuse of a building is separation of its multiple functions by separate vertical transportation and lobbies. People entering the residential area, for example, need not interfere with people using other areas of the building.

Elevatoring of such multipurpose buildings should be based on separation of the several functions and elimination of interfering traffic. If the expected elevator traffic patterns for the functions do not coincide, elevatoring may be established for the major function and sufficient service made available for the minor function. If the two functions are expected to coincide, as in combination office and apartment buildings, separate groups of elevators should be provided (Figure 14.1).

Figure 14.1 shows two alternatives for the residential section. The apartment elevators can start from the ground floor and operate express to the first apartment floor with local stops thereafter. Alternatively, shuttle elevators can be provided stopping at the ground floor and operating nonstop to the sky lobby. Apartment amenities such as a package room, a concierge, health clubs, or swimming pool can be conveniently located at the sky lobby.

The service elevator functions for the office building and the apartment should be provided on two separate elevators. The nature of apartment service needs which consist of maids, deliveries of clothing, furniture and food, catering services, and movers are usually in conflict with office needs such as supplies, mail, and renovations. In addition, apartment tenants expect a greater degree of anytime security whereas office tenants expect greater nighttime security. This can be effected by locking out the separate entrances to the elevators except for emergency purposes.

The sky lobby concept has been a successful approach to the office–apartment combination. A separate lobby is located on the lowest apartment floor and connected to the street by shuttle elevators. Apartment tenants ride these elevators to the sky lobby and change to the local elevator, which takes them to their floor. The sky lobby is enhanced by swimming pools, shops, or a restaurant. This is the elevatoring arrangement of the John Hancock Center in Chicago, where 44 floors of apartments are located above a 40-story office, store, and garage building (Figure 14.2).

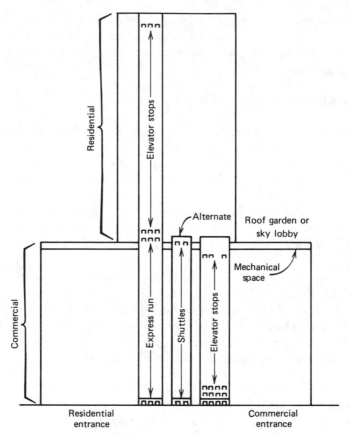

Figure 14.1. **Multipurpose building—office and residential.**

Water Tower Place, also in Chicago, combines stores on the lower levels surrounding an atrium which includes three observation-type elevators in glass hoistways, office space above the stores with separate elevator service, and a luxury hotel with a sky lobby above the office space. A third similar-type building is planned in the same area.

A hotel with extensive convention facilities, such as ballrooms and meeting rooms often used by others than the hotel guests, can be considered as a multipurpose building. Separate vertical transportation for outside guests increases the value of the meeting facilities and minimizes their interference with hotel guests. Guests, even when they are attending the meeting room functions, appreciate the reduced congestion on the main passenger elevators.

In estimating the vertical transportation requirements of the separate functions in a building, the expected maximum usage of each facility and the possible time of use of each must be considered. If periods of maximum use do not overlap it may be possible to combine functions and the vertical transportation is calculated for the maximum use. In addition to the quantity of service re-

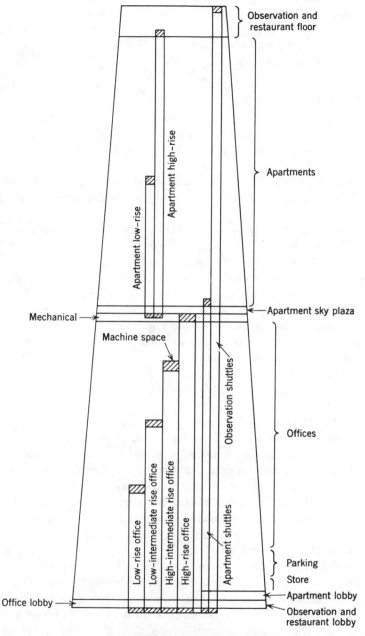

Figure 14.2. John Hancock Center, Chicago.

quired, the nature of the traffic must be determined. Its direction is either in unison or opposed and proper elevator provisions must be made. An important example is the inclusion of restaurant facilities on the top of an office building. The normal lunchtime traffic of the tenants is usually in direct opposition to that of the restaurant patrons.

SKY LOBBY ELEVATORING

In addition to multipurpose buildings, sky lobbies have been used in many large office buildings. The space starting at the sky lobby is elevatored as a conventional building, as if the sky lobby were the first floor. The elevators from the ground to the sky lobby are calculated as two-stop elevators for a single sky lobby. This has been done in a 17 story building as well as 100 story buildings.

It is possible to have the same shuttle elevators serve more than one sky lobby. A 28 story building has been built with two sky lobbies and a single group of shuttle elevators serving the ground and the two lobbies.

In calculating shuttle elevators the total handling capacity of the elevators must match the handling capacity of the local elevators and the required handling capacity for the population of the sky lobby floor. Elevator calculations should be based on the nominal full load requirements and then the elevators increased in size to reflect what is required if one of the shuttle elevators is out of service. The high-rise elevators of the section below can be arranged to serve the sky lobby floor to provide an alternate route to the street in the event of an emergency, or alternatively, suitable stairways can be provided to connect the sky lobby with the lower floor.

The high-rise elevators of the lower section which are arranged to serve the sky lobby should not be arranged to serve the sky lobby during peak traffic periods. If they do so, people will find it more convenient to ride to the sky lobby and take the lower section elevators down to their floor, creating a two-way traffic on the elevators during an up-peak period.

Service elevator needs for a building with sky lobbies can be served either by conventional all-stop service elevators from the ground or by creating "sky-docks" with shuttle service elevators from the street and conventional service elevators starting at the sky lobby level.

Each of the World Trade Center Towers in New York City consists of approximately three 30-story "buildings" on top of each other. The lobby of each of these buildings is connected to the street by single deck shuttle elevators. A person wishing to go, say, to the upper third of the building, will ride a shuttle elevator nonstop to the seventy-seventh floor sky lobby and there change to a local or express elevator to the desired floor. Without this sky lobby arrangement the necessary shafts for conventional local and express elevators serving all floors would consume almost the total area of the lower floors (Figure 14.3).

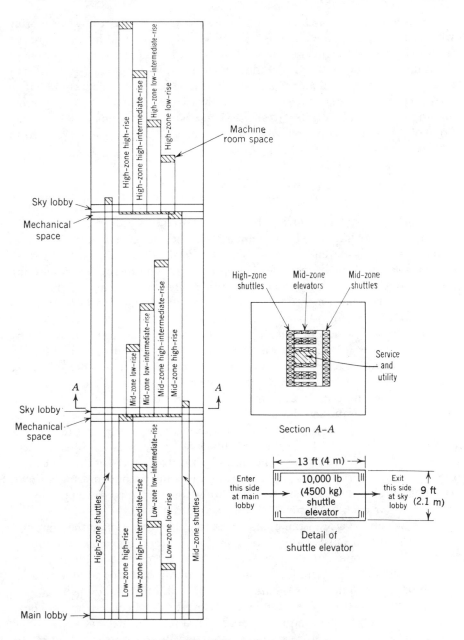

Figure 14.3.   World Trade Center, New York.

335

The Sears Tower in Chicago is another example of sky lobby elevatoring with double-deck shuttle elevators. People entering the building at the ground floor and wishing to travel to floor 34 or floor 67 will enter the upper deck of the shuttle elevators at the ground floor while those wishing to go to floor 33 or floor 66 will ride an escalator to a lower shuttle elevator lobby corresponding to the lower deck of the shuttle elevators. At the two-level sky lobby, the terminal landing of the local elevators to the desired floor corresponds to the existing floor of the shuttle. If people wish to interchange, escalators are provided to connect the two levels of each sky lobby.

DOUBLE-DECK ELEVATORS

Another approach to reducing the space required by elevators in taller buildings is the use of multideck or compartment elevators. Here the upper and lower deck of each elevator is loaded simultaneously (during the incoming rush, for example), with passengers destined for the odd-numbered floors entering the bottom deck and those for the even-numbered floors entering the upper deck. When the elevator stops, passengers are discharged from both decks simultaneously (Figure 14.4).

The restricted operation is maintained for passengers entering at the lobby floor at all times. If the car stops for a landing call during its up trip, the operation of both decks is made unrestricted so that a person entering either deck at an upper floor is able to exit at any floor served by the elevator.

After operation becomes unrestricted, all the car buttons in the car on both decks are operative and the elevator is arranged so that the trailing deck, i.e., the lower deck on an up trip and the upper deck on a down trip, is the one that responds to a landing call. In this way advantage may be taken of coincident stops, i.e., stops wherein both a landing and car call are served at the same time. The operation on the down trip is unrestricted and people entering a down traveling, double-deck elevator do not know whether they are entering the upper or lower deck until they arrive at the lobby.

Double-deck elevator installations can be found in operation in the Time and Life and Standard Oil Buildings in Chicago, the John Hancock Building in Boston, International Building in Dallas, Citicorp Center and Philip Morris Headquarters in New York, First Canadian Place in Toronto, and the Nationwide Insurance Building in Columbus, as well as in other buildings.

The round-trip time of a double-deck elevator is calculated in a manner similar to a single-deck elevator. The probable stops will usually be equal to the number of stops the elevator can make. Double-deck elevators are rated by the load in each deck, for example, 3500 lb/3500 lb (1600 kg/1600 kg) is indicative that each deck is the size of a 3500-lb conventional elevator. Each deck will have a nominal capacity of 16 people. Quite often, one of the decks may be stopped for a passenger while no one transfers from the other deck. If the elevator stops for a landing call, the person entering may only wish to travel one floor, causing an additional stop. If it is assumed that the probable stops are equal to the

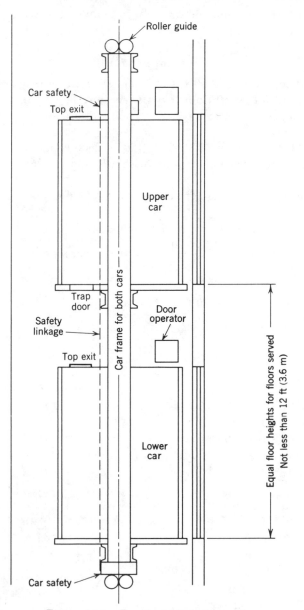

Figure 14.4.   Double-decked elevator car.

minimum possible stops, these contingencies are recognized. Transfer time should be calculated as if an equal number of people leave each deck at a single stop. Door time is the same as for a single-deck elevator. Since both doors must be closed and locked before the elevator can move, some delay in coordination may take place. For this reason, an extra 10% standing time inefficiency should be used when calculating the standing time of a double-deck elevator system.

Interfloor traffic can be quite detrimental to double deck operation especially during peak incoming traffic.

Example 14.1

Given: an office building, 20,000 ft^2 (2000 m^2) net usable area per floor on floors 1 to 20, 12-ft (3.6-m) floor heights. Assume single-purpose occupancy, 150 ft^2 (15 m^2) per person average. How many double-deck elevators are required?

Population: 20,000 × 20 = 400,000 ÷ 150 = 2666 people

Required handling capacity: up peak with 10% down, 2666 × 15% = 400 people.

Assume: 3500/3500 lb double-deck elevators, @ 500 fpm, 48-in. center-opening doors

Probable stops: 20 upper floors, 10 probable stops

Time to run up, per stop: $\dfrac{12 \times 20}{10}$ = 24-ft rise per stop

$\qquad\qquad$ 24 ft $\qquad$ = 5.8 sec

Time to run down: $\dfrac{240}{2}$ = 120 ft = 17.2 sec

Elevator performance calculations:

Standing time			Transit Time
Lobby time 16 + 16	=	⌈4 sec	14
Transfer time, up stops 10 × 2	=	20	20
Door time, up stops 10 × 5.3	=	58.3	58.3
Transfer time, down stops 2 × 4	=	8	—
Door time, up stops (10 + 1) × 5.3	=	10.6	—
Total standing time		110.9 sec	92.3
Inefficiency 15%	=	16.6	13.8
Total		127.5 sec	106.1
Running time			
Run up 10 × 5.8	=	58.0	58.0
Run down 2 × 17.2	=	34.3	—
Total round-trip time		219.9 sec	164.1 sec

$HC = \dfrac{(16 + 16 + 2 + 2) \times 300}{220} = 49 \qquad \dfrac{400}{49} = 8.1,\ 8$ elevators required

Interval: 220/8 = 27.5 sec

A suggested elevator scheme would be eight, 4000/4000 lb (1800/1800 kg) double-deck elevators @ 500 fpm (2.5 mps) serving lower lobby, upper lobby, and floors 1 to 20. The slight excess transit time over the desired 150-sec maximum would preclude locating executive offices at the top unless arrangements are made for special service or additional executive elevators. The total building requires escalators and a shuttle elevator to connect the two lobbies and sufficient service elevator capacity. Before a final recommendation is made, conventional and sky lobby elevators should be calculated and an analysis of the difference in cost and space required for the conventional, sky lobby, and double-deck elevator arrangements developed. The final decision should be a result of a resolution of all the members of the building team.

When double-deck elevators are considered for a building, a number of building restraints must be effected. All floor heights must be equal within a tolerance that can be adjusted by the setting of the landing door sills.

Lobbies for double-deck elevators require special considerations. They can be designed with either the upper or lower deck at ground level and must have both escalator and shuttle elevator service between both lobbies. Both lobbies should be equally attractive and clearly visible signs should be provided to guide visitors to the proper elevator entrance level for the destination floors sought.

The space requirements for double-deck elevators must consider the increased structure and larger elevator machinery that is required to handle the heavier loads. The space in the hoistway allowed for rails and car frame structure is increased to 12 in. (300 mm) rather than the 8 in. (200 mm) allowed for a conventional elevator. Pit and overhead spaces are slightly more than conventional elevators for the speed specified.

Operating systems are the same as conventional elevators with additional features to accommodate the double-deck operation. When an elevator stops at a floor it can be arranged so that the doors on both decks open even though a passenger may transfer at only one deck. The people on the other deck, if they are not aware it is a double-deck elevator, will think a false stop has occurred. Some buildings prefer to only open the required doors and light a sign in the other deck with a legend "other deck loading."

In chapter 16, the economic differences between double deck and sky lobby elevatoring are discussed. In general, a full comparison between three schemes, conventional, sky lobby, and double deck elevatoring must be made for any major project before a final elevatoring recommendation is made.

OBSERVATION AND OUTSIDE ELEVATORS

An elevator traveling up the outside or in the atrium of a tall building or hotel is a dramatic sight. For the rider the thrill of seeing the scenery in motion is unsurpassed; for the observer on the ground or in the lobby the smoothness and majesty of the moving mass is incomparable. From the engineer's point of view observation and outside elevators pose many challenging problems. Notable applications of observation and outside elevators have been made at the St. Francis Hotel in San Francisco, the observation towers mentioned in Chapter Twelve, hotels in Atlanta, Chicago, Nashville, San Francisco, and Cambridge (Figure 14.5), and a host of other installations throughout the world.

Observation elevators in a building atrium are less critical in design than an outside observation elevator exposed to the weather. The important consideration in design is making the parts of the elevator other than the cab unobtrusive.

Figure 14.5.   Outside elevators, St. Francis Hotel, San Francisco.

Counterweights are usually hidden at the side, the door mechanism and the backs of the hoistway doors are painted a dark color, and the top and bottom of the car are shrouded to hide operating mechanisms. The elevator should be arranged with a minimum, preferably one, traveling cable and compensating chain or rope eliminated by the use of higher horsepower hoisting motors or traveling cable compensation to the counterweight to match the fixed traveling cable to the car.

From the design aspect the outside elevator must be weather and windproof. Windproofing is helped by eliminating all possible ropes and troughing traveling cables and governor ropes. The compensating ropes are eliminated and extra motor horsepower provided for the uncompensated load. The top and bottom of the elevator cab are enclosed and streamlined, and if the elevator car doors do not face the inside, they are positively locked while the car is away from the landings.

All hoistway switches and electrical installations are completely moisture-proofed to withstand driving rain. When wintertime operation is expected, rails are electrically heated and ice scrapers are mounted on the guide shoes. The glass cab needs to be air conditioned for passenger comfort on hot days and heated on cold days. For windy days automatic-bridge-type rope guards may be provided to hold the hoist ropes to keep them from swaying and entangling the elevator structure.

Observation elevators are usually of the wall-climber design, in which the car frame and guide rails are set as close to the back of the car as possible. The final effect provides maximum viewing area for the passengers (Figure 14.6).

Some observation-type elevators are designed to operate within a glass enclosed hoistway. Any glass that is used must be of the laminated safety type and should be tempered so that, even if broken, openings do not appear which may endanger passengers. Consideration needs to be given to cleaning the outside glass and both the inside and outside of the elevator enclosure.

Variations of observation elevators can be found in many applications (Figure 14.7). Some are found in shopping malls to serve upper and lower levels and to provide an attraction to patrons. Others are found in racetracks to serve the visitors to the upper clubhouse. Observation-type elevators can be found serving the floors facing an atrium and continuing into a glass enclosed hoistway to serve a rooftop observation lounge and restaurant. One notable installation is in the vertical Assembly Building at Cape Kennedy where employees and guests can observe the assembly activity from the elevator which rises over 500 ft to the roof of the building.

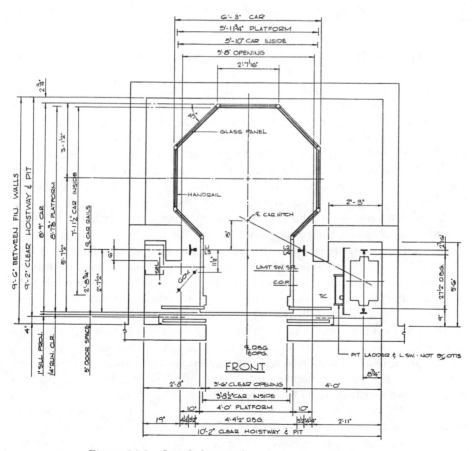

Figure 14.6. Sample layout of an observation-type elevator.

Figure 14.7.   Observation-type elevator, Tropicana Hotel, Atlantic City (Courtesy Globe Van Doorn).

INCLINED ELEVATORS

Elevators that travel in other than a vertical path are necessary for a wide variety of applications. Supplementing escalator service, especially for rises of 50 ft or more (15.2 m or more), to serve the people unable to use the escalator is a necessary application for inclined elevators. Many such inclined elevators have been provided in the Stockholm, Sweden, subway system for that purpose.

The development of a hillside for residential property indicates a need for inclined elevators. The alternative is to cut the side of the hill for roadways which wastes space and creates the possibility of earth slides.

A resort located on a bluff above a beach is an application for inclined elevators which has occurred in many places in the world.

Inclined elevators are an outgrowth of the funicular railways that were quite prominent in the early 1900s and many have survived. An inclined railway is built on railway tracks with the car moved by a driving machine and ropes. The car traveling up is counterbalanced by a car traveling down with a simple passing system at the middle of travel. The outside wheels have flanges which guide the car on the outside track. At the turnout the up car moves to one side and the down car to the opposite side as the inside wheels without flanges slide over on the rails and the outside wheels guide the car. The trip continues on the single track. The turnout had the formal name of "A Brown Turnout," named after the inventor (Figure 14.8).

Inclined railways had safety devices designed to dig into the rail ties or wooden strips along the rails if the tension on the ropes was released for any reason or if the attendant released an emergency brake. Surviving examples of inclined railways can be found in Chattanooga, Tennessee; Hong Kong; Bergen, Norway, and many European and South American countries.

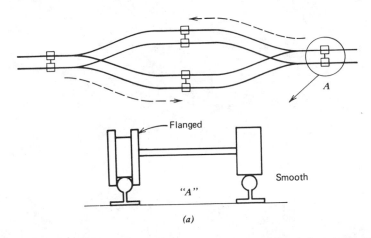

Figure 14.8. (*a*) Inclined railway "Brown Turnout;" (*b*) inclined railway, Pittsburgh, Pa.

An inclined elevator is a descendant of the inclined railway. It is presently designed to be automatic in operation, and the counterbalancing car and turn-out are gone, replaced by a counterweight with its own set of rails. The car doors and landing doors are power operated and equipped with interlocks, and the attendant is gone (Figure 14.9).

An inclined elevator can be considered for an application depending upon the angle of incline if certain limitations are recognized. Up to about 10° from the vertical, conventional elevator equipment may be adapted to the incline;

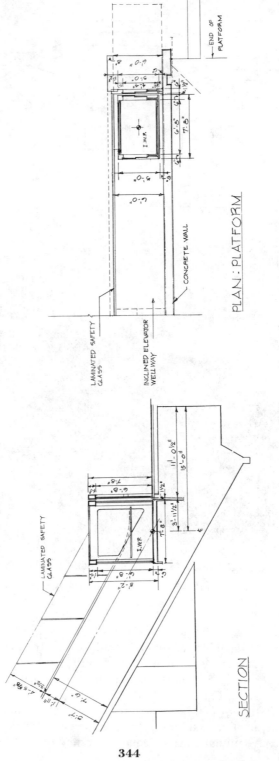

LAMINATED SAFETY GLASS

I.W.P.

PLAN : PLATFORM

CONCRETE WALL

LAMINATED SAFETY GLASS

INCLINED ELEVATOR WELLWAY

END OF PLATFORM

(a)

LAMINATED SAFETY GLASS

I.W.P.

SECTION

344

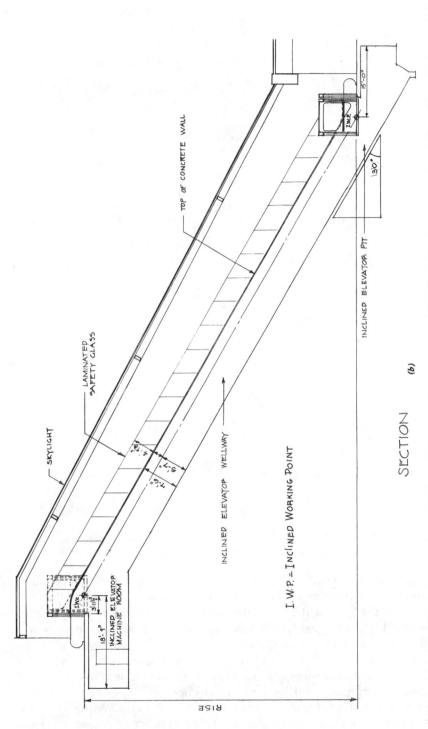

SECTION

(b)

I. W. P. = INCLINED WORKING POINT

Figure 14.9. Inclined elevator to be installed parallel to an escalator to serve mobility-limited in the same travel path. Note: The entrances are in line with the elevator travel as opposed to side entrances as found on most inclined elevators.

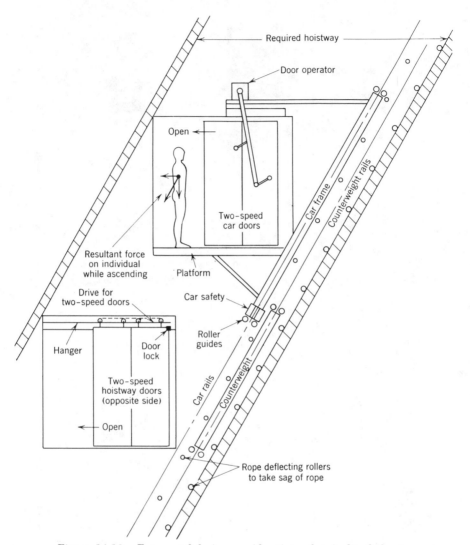

Figure 14.10.   Forces and design considerations of an inclined elevator.

with more than 20°, considerations must include doors at the side, guide rollers for ropes to keep them from rubbing the back of the hoistway, and guide rollers reinforced for the load of the car and counterweight leaning in one direction (Figure 14.10). Traveling cables will have to be provided with a guiding system to keep them from dragging on the structure.

The motion control of an inclined elevator will have to be designed for limited acceleration and deceleration which will provide gradual starts and stops, and the stopping distance lengthened upon the application of brakes during emergency stops depending upon the horizontal accelerating forces expected to be exerted on standing people. These forces are recommended to be

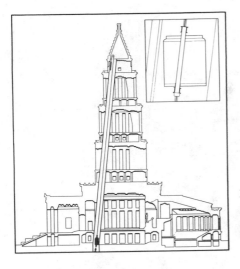

Figure 14.11. Inclined elevator at the Washington Masonic Monument in Alexandria, Virginia. Note: This elevator travels at an inclined angle in two planes similar to the edge of a pyramid.

about 0.3 fps² (0.l mps²) or as required by elevator safety codes which are being revised to reflect inclined elevator applications.

The term "inclined elevator" assumes travel in a single inclined direction in a single plane. Once a compound angle is introduced, such as the corner of a pyramid, multiple complications occur and an extremely difficult equipment layout is necessary. Such was the situation that was encountered in the Washing-

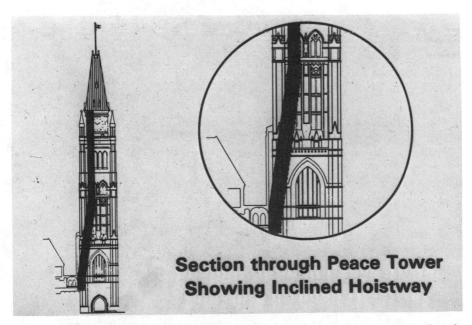

Section through Peace Tower Showing Inclined Hoistway

Figure 14.12. Combination inclined and vertical elevator, Peace Tower, Ottawa, Canada (Courtesy Otis Elevator Company).

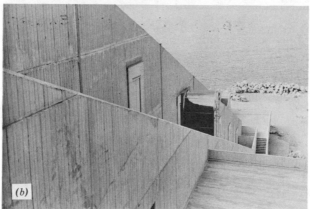

Figure 14.13. (*a*) Inclined elevators—Post Office and Telephone Exchange, Genoa, Italy (Courtesy Telefuni S.A. Zurich). (*b*) Inclined elevator, President Hotel, Dubrovnik, Yugoslavia (Courtesy Elevator World Magazine). (*c*) Inclined elevator, California (Courtesy Dwan Elevator Company).

ton Masonic Monument in Alexandria, Virginia (Figure 14.11).

Inclined elevators can also transverse multiple angles in the same plane. The elevators on the legs of the Eiffel Tower in Paris travel up the approximate parabolic path of the legs of the tower. An installation that travels part of the way at an angle and then proceeds vertically is located in the tower of the Parliament Building in Ottawa, Canada. The compound angle was necessary to avoid a memorial in the base of the tower and to fit the elevator within the confines of the tower (Figure 14.12). Many commercial applications of inclined elevators may be found in Europe and residential applications in California (Figure 14.13).

PLATFORM LIFTS AND STAIR ELEVATORS

Increased concern and legislation requiring means to serve the needs of the "mobility impaired" segment of the population have added emphasis to the need for vertical transportation for short rises and the need to supplement stairways in buildings without elevators.

Many buildings do not have level access from the street to the main floor which requires the use of a ramp or short-distance lift if a person in a wheelchair wishes to gain access. In many buildings, even with a street level entrance, stairs are often the only means to gain access to a main lobby, and considera-

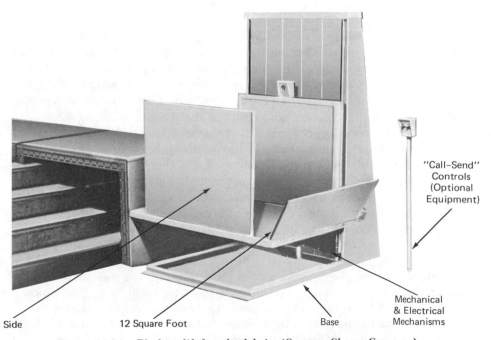

"Call–Send" Controls (Optional Equipment)

Mechanical & Electrical Mechanisms

Side   12 Square Foot   Base

Figure 14.14. Platform lift for wheelchairs (Courtesy Cheney Company).

**Figure 14.15.   Stairway inclined platform lift for wheelchairs
(Courtesy Cheney Company).**

tion of a stairway lifting means is one way to accommodate a wheelchair. In residences, an accident or illness may necessitate the addition of a stair climbing lift or residence elevator so the upper level becomes accessible to the disabled or handicapped person.

Recognition of these considerations have led to the development of a variety of equipment which may be employed to solve the particular needs. Many types have certain user hazards and should only be used by people familiar with their use in a restricted area such as a private home. Others are suitable for public places, but require supervision and restriction in their use by the general public. The only totally public conveyance remains the elevator or escalator.

Platform lifts are designed primarily for wheelchairs and serve very limited rises up to about 6 ft. Gates should be provided top and bottom which must be closed and should lock after the lift moves from its landed position. The operation is by user key and the speed is exceedingly slow at about 30 fpm (0.15 mps). The shaft area must be smooth and shear hazards minimized. If the application of such a lift is contemplated, manufacturers' details and local building and code requirements must be investigated. Figure 14.14 shows one model of a platform lift.

Stairway inclined lifts have been developed to supplement stairs so that a wheelchair can be occasionally moved up and down. The path must be clear and pressure sensitive switches should be used to stop the lift if an obstruction is encountered. The lift should be designed to fold out of the way when not in use. A key-operated switch needs to be employed to restrict its use. Figure 14.15 shows one type of a stairway lift presently available.

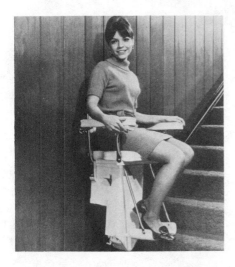

Figure 14.16. Stairway chair lift (Courtesy Cheney Company).

Stair climbing lifts are similar to stair inclined lifts except that a seat is provided for the mobility impaired person to sit while they are lifted. They have been designed for both temporary and permanent installations and models are available that curve to follow stairways around landings. Various manufacturers offer different features and should be consulted. Figure 14.16 is an example of one type.

RESIDENCE ELEVATORS

Residence elevators can be considered as being available in two levels of sophistication. One type is installed in a fully enclosed permanent hoistway with landing doors and a car gate with most of the attributes of a conventional elevator. A second type is installed in an open space and a pressure sensitive pad is provided under the car to stop the car if an obstacle is under the car as it descends. Figure 14.17 shows an enclosed residence elevator. Such elevators are often provided in luxury duplex (two-floor) and triplex (three-floor) apartments in high-rise buildings. If there is occupied space below the elevator, both car and counterweight safety devices need to be provided.

New sections in the A17.1 Elevator Code are available or in preparation for the various types of lifts. The code outlines various minimum safety considerations and should be consulted when an installation is planned. With an enclosed residence elevator, an outside telephone connection to a central exchange is an absolute necessity.

ELEVATORS IN PROBLEM LOCATIONS

Safe vertical transportation must often be provided in areas where hazards exist detrimental to either the equipment or its operation. Such locations may

Figure 14.17. Residence-type elevator. Note: small size and folding gate.

Figure 14.18. Columbia launch, Cape Kennedy. Elevators in the launch service tower (left) are often damaged and must be rebuilt after each launch.

be found in storage facilities for flour or other dusty, often abrasive materials, or in petroleum refineries or chemical processing plants handling corrosive or explosive substances. Wet locations include elevators in mines, those exposed to the weather, as in the observation towers previously mentioned or located at waterside, or, perhaps, used to handle wet ashes in a power plant. Elevators operating in the vicinity of rockets fueled with liquid hydrogen and liquid oxygen must endure special environmental hazards (Figure 14.18).

Whatever the hazard or location, conventional elevator equipment would be subject to deterioration from the elements or which present eventual dangers to personnel.

Classes of hazard are recognized by the National Electrical Code, which sets forth rules for the treatment of electrical equipment in such locations. Common sense and engineering considerations must be applied to the other parts of the elevator exposed to hazards. For example, if all the electrical equipment is required to be watertight, all the structural parts of the elevator including the ropes, machine, and rails should also be protected against deterioration from moisture. Similarly, if the elevator equipment must operate in a corrosive atmosphere, as in a fertilizer or chemical plant, all metal parts as well as the electrical equipment should be treated to withstand corrosion. These considerations may often lead to installing wooden rather than steel guide rails and providing a safety mechanism that operates on wood. A damaged section of rail can be replaced, as in the early days of "safety" elevators. Safety devices with nonsparking metals such as beryllium or aluminum have been developed and successfully applied to steel rails.

Specifying elevator equipment for installations require that the type and extent of the hazard be established. As an example, *Class I*, *Division 2*, Group B would indicate a possible explosive hazard of hydrogen gas as outlined in the National Electric Code ANSI/C1. Equipment manufacturers should be consulted to determine the necessary precautions and elevator specifications prepared accordingly.

SHIPBOARD ELEVATORS

When elevators are installed aboard ships many special factors must be considered. The equipment must be moisture-proof and designed to resist corrosion, as these conditions are to be expected on a ship. Traveling cables must operate in troughs to avoid tangling when the ship rolls, pitches or yaws. The elevator control equipment as well as the elevator car must be able to operate during these movements in moderate seas. All the equipment must be able to withstand the extra forces imposed by the motion of the ship.

Counterweighted elevators are usually used aboard ships for passenger cars. The counterweight must be equipped with a safety for the obvious reason that if it ever broke loose it may go right through the bottom of the ship. Layout conditions on shipboard are more exacting than on land. Space is at a considerable premium, and fitting a reasonable-size elevator in a hoistway trunk is quite a challenge. As with any special installation, competent manufacturers' representatives or consulting engineers should be called on for aid.

Cargo can be handled by elevator and, as in a warehouse on land, by industrial power truck (Figure 14.19). Because the elevator may be loaded by a truck weighing from 8000 to 10,000 lb, the elevator must be rated for its duty load plus the load imposed by the truck. This may mean elevators capable of handling 15,000 lb or more. If a counterweighted elevator were used, the deadweight of the counterweight would decrease the loading capacity of the ship. For this reason cargo elevators on ships are usually drum machines, rated to lift the entire deadweight of the platform plus the capacity load as well as impact loads of the truck during loading and unloading operations.

ELEVATORS IN MINES

Increasing recognition is being given by mine operators of the value of providing high-speed passenger-type elevator service for the miners. Time spent traveling is nonproductive and the well-being of the individuals contributes to

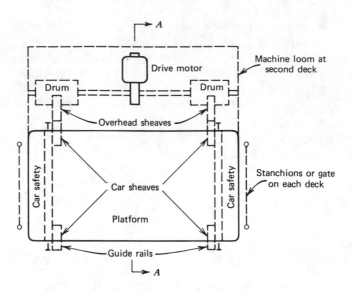

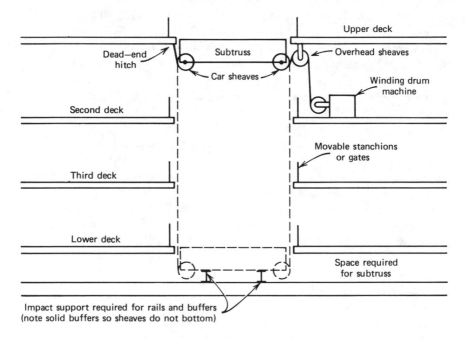

Section $A-A$

Figure 14.19. Shipboard cargo elevator. Subtruss and underslung design allow platform to be flush with the top deck. This design is ideal to lift helicopters to the landing pad.

Figure 14.20.   **Mine cars loaded on a double-decked mine elevator; upper deck serves personnel and lower deck serves mine cars.**

their productivity. More stringent safety requirements recognize that the old mine "cage" hoist could be considerably improved (Figure 14.20).

Elevators in mines require additional operational and environmental considerations. Since temperature variation may be extreme, heaters are often required in door sills to prevent icing and the car equipment must be both temperature and moisture resistant. In a deep mine, considerable air pressure differential may be present, creating a stack effect in the elevator shaft. These air currents can cause traveling cable and ropes to sway and precautions must be taken to prevent snagging. Very often the elevator shaft will also double as the ventilation shaft and very high air velocities are intentionally employed.

Standby electrical power and means to move the elevator in case of unexpected shutdown or during an emergency are important considerations. In many mines, a single elevator installation in a single shaft requires an adjacent emergency ladder or stairs.

INDUSTRIAL-TYPE ELEVATORS

Elevators are often required to transport personnel on a periodic basis to the height (or depth) of such installations as smokestacks, cable shafts at the bottom of a dam, the operating cab of a high gantry crane, grain elevators, the legs of a deep-sea oil production or drilling platforms, and in many other extraordinary locations.

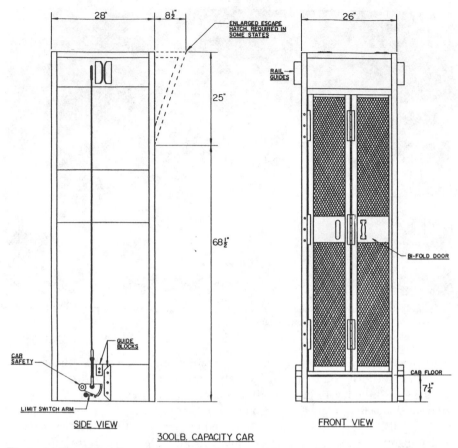

SIDE VIEW

FRONT VIEW

300LB. CAPACITY CAR

Figure 14.21. Special purpose personnel elevator (Courtesy Sidney Grain Machine Company).

A variety of small personnel elevators designed for these special purposes is available. This specialized field includes elevators of 600-lb (270-kg) capacity with a platform about 3 ft × 3 ft (900 mm × 900 mm) to carry maintenance personnel to the top of grain elevators to service conveyor equipment, as well as other applications. These elevators are of an overhead-traction-machine type with counterweight installed with simple rails and a "telephone booth" type cab integrated with the car guides and safety devices (Figure 14.21).

Another type is a rack and pinion elevator. A steel tower with guide rails and a continuous gear rack is erected adjacent to a smokestack, a crane, in a TV tower, or in a caisson, and the elevator is driven by a pinion gear and machine on the platform. Such elevators have unlimited travel capabilities and are designed for a load of 600 lb (270 kg) or more at 100 fpm (0.5 mps) or greater. They can be adapted to travel vertically and then be inclined at an angle to follow the contour of the structure. Further discussion is given in Chapter 20.

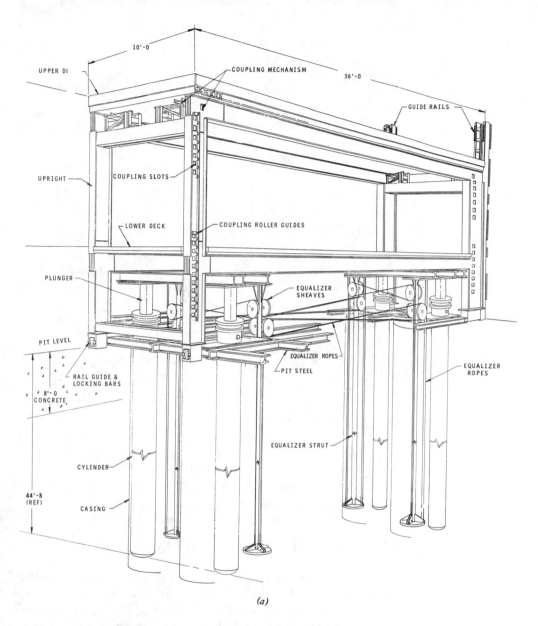

Figure 14.22. Stage lift: (*a*) a single section of a three-section stage lift showing the main components.

357

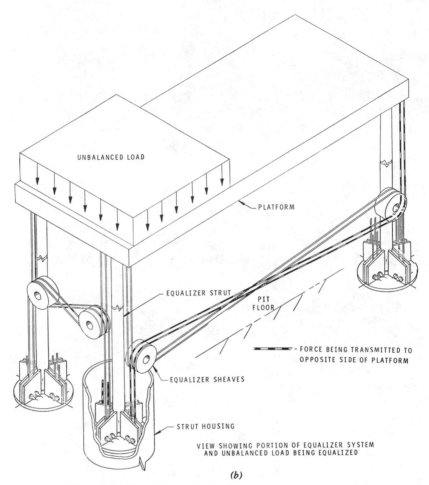

UNBALANCED LOAD

PLATFORM

EQUALIZER STRUT

PIT
FLOOR

- - - FORCE BEING TRANSMITTED TO
OPPOSITE SIDE OF PLATFORM

EQUALIZER SHEAVES

STRUT HOUSING

VIEW SHOWING PORTION OF EQUALIZER SYSTEM
AND UNBALANCED LOAD BEING EQUALIZED

(b)

Figure 14.22. **(b)** details of the rope equalization system.

NONELEVATOR VERTICAL TRANSPORTATION

Safe lifting means utilizing other than a fully enclosed elevator in a fully
enclosed elevator hoistway is often desired for many applications. Such installations can include stage lifts in theaters and auditoriums, platform lifts in industrial applications, raising and lowering the bottom of swimming pools for various activities, lifting aircraft on an airplane carrier, plus lifts for extremely
special applications.

Conventional elevator lifting approaches have been used including: elevator-
type traction machines with counterweights; drum-type machines both with
and without counterweight; hydraulic plungers, both direct and indirect drive
using chains or ropes, screw jacks, and hydraulic or electric rack and pinions.

Detail descriptions are beyond the scope of this volume, but a few examples
will be described.

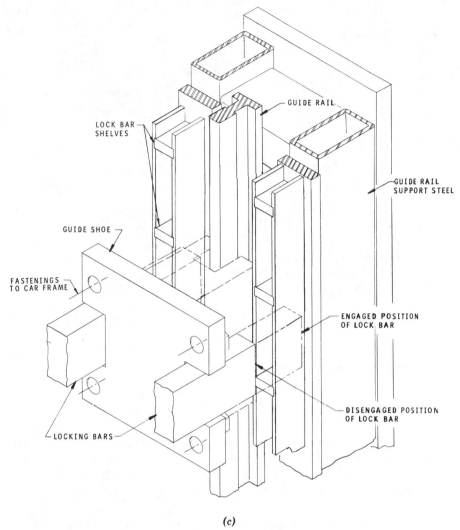

(c)

Figure 14.22. (*c*) **locking bar arrangement to lock platform so heavy loads can be transferred.**

Stage Lifts

Stagelifts are generally platforms ranging from 10 to 20 ft (3 to 6 m) wide by 30 to 60 ft (9 to 12 m) long with a vertical rise from a few feet to two or three floors. Multiple direct-hydraulic plungers are the preferred means of lifting and two, four, or more such plungers may be employed depending upon depth and width. When more than one plunger is provided, displacement of each plunger must be equalized and this is done either by rope or rack and pinion equalization.

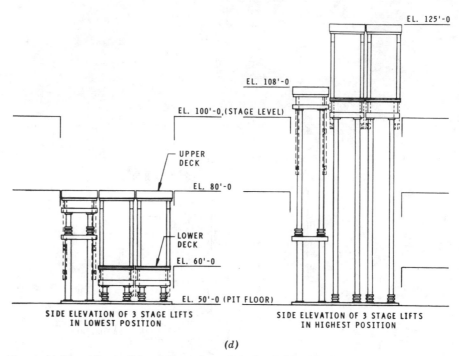

Figure 14.22. (*d*) **possible configurations of single- and double-decked stage lifts. (Photos courtesy of the Otis Elevator Company.)**

Figure 14.22*a* and *b* shows the rope equalization system between four plungers lifting a 10 × 36 ft (3 × 11 m) platform. Note the "four-way" considerations. The same equalization can be accomplished by rack and pinions. Two racks would be located next to each plunger and pinions and shafts would connect the racks in the same plane.

If two such platforms are to be located next to each other as shown in Figure 14.22*c* and *d*, the platforms can be pinned together by toggles if rope equalization is used or the locking can be accomplished by clutching the pinion shafts on each lift together if rack and pinion equalization is used. In one such installation, three familiar platforms are clutched together and caused to move as a single unit. The configuration of the platforms could be flat or steplike, depending upon the location of the platforms when they are clutched together.

Stage lifts can be quite elaborate. In Vienna, a stage lift is mounted on a three-story-deep turntable and the entire stage can be revolved as the two-section lift raises and lowers. The effect is spectacular in that the four different scenes can be shown in a matter of minutes by the combinations of lifting, lowering and revolving the stage.

Moving Floors

Gymnasiums can be transformed to auditoriums with stages by lifting one end of the room and creating a stage. Screw jacks are recommended since they are self-locking, can be simply equalized, and require very little horsepower since screw jack speed is very low.

Aircraft Carrier Elevators

Aircraft carrier elevators employ roped hydraulics to lift the heavy load of an aircraft, which can be as high as 300,000 lb (140,000 kg), at a high speed of 300 fpm (1.5 mps). Figure 14.23 shows an arrangement of such a lift.

Platform Lifts

Platform lifts are often needed to provide a means to unload a truck or to position heavy equipment from one level to another which is a short distance within a floor height. Such lifts are usually hydraulic, equalized as described for stage lifts with full skirting to minimize shear hazards. They are frequently found in industrial plants and often in the loading dock space of commercial buildings.

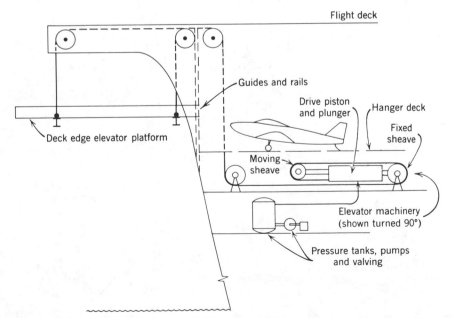

Figure 14.23. **Aircraft carrier deck—edge elevator. High-pressure hydraulic fluid moves piston to actuate the roped hydraulic arrangement.**

Other Lifts

The equipment available to provide vertical transportation can generally be adapted to any lifting need. Familiarity will suggest means of employing such equipment and the knowledge of the elevator consultants and elevator engineers can be called upon to add to the development of the solution to fulfill the need. The foregoing is but a brief description of some of the applications.

CHAPTER FIFTEEN

Automated Material Handling

AUTOMATED MATERIAL-HANDLING SYSTEMS: DEVELOPING TECHNOLOGY

The first dumbwaiters have unknown origins but it is known that Thomas Jefferson had one while he lived in Monticello in the early 1800s. Dumbwaiters graduated from a hand-powered affair to an electric powered lift in the early 1920s and were often applied in hospitals and hotels to move food from a central kitchen to an "on floor" pantry with a reasonable assurance that an attendant was present to unload the food (Figure 15.1). As the name implied, the device was not only silent but needed to be waited upon.

About 1960 an innovation was made which led to second and later generations that can hardly be called "dumb." This innovation was a means to automatically eject a tote box load when the car reached the desired floor. That coupled with the addition of powered doors which automatically opened and then closed after the load was ejected has led to a complete line of automated material-handling systems. The lowly dumbwaiter, both hand powered and electrically powered, is still in demand and readily available for less stringent applications.

Later generations include both automatic loading and unloading of wheeled carts as well as tote boxes so that systems that are both dumbwaiter size [maximum: 9 ft^2 (0.83 m^2) platform area, 48 in. (1200 mm) high and 500 lb (225 kg)] and elevator size are available. A separate section of the A17.1 Elevator Code has been developed for such systems and is entitled Part XIV, Material Lifts and Dumbwaiters with Automatic Transfer Devices. Systems have been developed where overhead track conveyors or in-floor conveyors provide the horizontal transportation and the material lift, the vertical transportation. Under further continued development are systems wherein a self-propelled robot vehicle follows a path along the floor and automatically calls a material lift, waits, boards and automatically exits at a destination floor, and then travels to its programmed destination.

Figure 15.1. Cutaway of a dumbwaiter showing principal parts (Courtesy Matot).

AUTOMATED MATERIAL HANDLING IN PUBLIC BUILDINGS

The scope of this chapter is limited to the types of automated material-handling systems that are found in public buildings which will require minimum training for the users who should be employees who work in the building. In industrial plants and warehouses, elaborate automated material-handling systems are found and skilled people are expected to be involved. The public building use generally employs elevator technology and is regulated by local building and elevator codes. In industrial applications, conveyor and crane technology is used and regulated by industrial safety codes which generally recognize the high degree of skill of the system operators. Industrial material handling, both manual and automated, is a field in itself and is a continuous growth technology. It is expected that many of the developments in that field will be applied to the public building area as the trend toward a more automated commercial or institutional building continues.

SELECTIVE VERTICAL CONVEYORS

One aspect of the conveyor industry can be found in many office and hospital buildings. This is the selective vertical conveyor which is designed to automatically load and unload tote boxes at various floors throughout a building. It can be interfaced with horizontal belt conveyors, gravity slides, both straight and spiral, and branch lifts so that the tote boxes can be delivered from and returned to a remote processing room by the system.

These systems have been used with considerable success in high-rise single-purpose office buildings for the delivery of mail and supplies from a central location. The operator places the filled box on a loading station for the conveyor, indicates the destination floor on a selector switch, and walks away (Figure 15.2). The next empty carriage on the conveyor will trip a loading mechanism and pick up the box on its upward trip. The box will rise "ferris wheel" fashion up and over to the down side to where the coding on the box indicates its destination. A trip mechanism will activate an unloading arm and the box will be deposited at the unload station at the destination floor. At various floors, people wishing to send boxes to other destinations or to return them to the central station will use the same loading process at their floor. At a central station, if it is remote from the conveyor, the box will be discharged on to a powered belt or gravity conveyor at a high level and carried to the final destination where it may accumulate with other boxes in a spiral gravity conveyor (Figure 15.3). The box is then removed, unloaded, and reused for another delivery.

Tote boxes come in various sizes named A, B, and C and the inside dimensions are shown in Table 15.1. The coding system consists of magnetic tapes on the side of the box or movable metallic strips. The approximate maximum capacity of a box is about 40 lb (18 kg) although loads as high as 60 lb (27 kg) are possible.

Table 15.1. Tote Box Inside Dimensions—Selective Vertical Conveyors

	Width, in. (mm)	Length, in. (mm)	Depth, in. (mm)	Volume, ft^3 (m^3)
A	11½ (240)	15⅞ (400)	10 (254)	1 (0.028)
B	15¼ (388)	18⅞ (480)	10 (254)	1.62 (0.045)
C	14⁷⁄₁₆ (365)	19⅜ (490)	8½ (216)	1.43 (0.04)

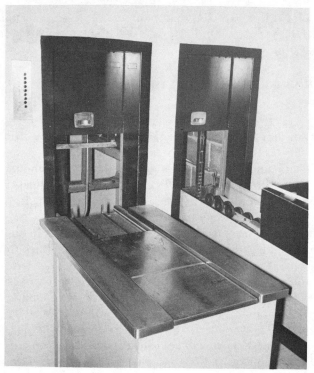

Figure 15.2. **Vertical selective conveyor showing a typical floor station. Tote-box to be sent is placed on the left side, the destination indicated on the push buttons and the tote-box is automatically accepted. Material delivered is discharged on the right side (Jaros, Baum and Bolles photo).**

The maximum speed of a vertical selective conveyor is 80 fpm (0.4 mps). The tote boxes are picked up and delivered while the chain is in motion, and at full efficiency the conveyor can pick up and deliver (throughput) 8 to 12 boxes per minute. The delivery cycle will vary with the length of the trip and can be estimated by the time required from the sending station where the box is loaded on the up traveling chain to the top of the structure and down the unloading station where it is unloaded in the down direction.

Example 15.1

Given: a 20-story building with a mail room at the basement floor. How long will the average delivery of a tote box require? Assume 12-ft floor heights. Assume sufficient empty carriers so that access to the system occurs within 10 sec after box is placed at the send station.

1. Accept box: 10 sec
2. Travel up: $\dfrac{(12 \times 19) \times 60}{80 \text{ fpm}}$ = 171 sec (to top of building)

3. Travel down: $\dfrac{(9 \times 12) \times 60}{80 \text{ fpm}} = 81\text{sec}$ (to building midpoint)

Total trip time: 262 sec

Due to the one-way nature of the chain conveyor any box that is loaded will require a full trip to the top of the building and down to the unloading floor, even though the unloading floor may only be one floor away.

Vertical selective conveyor systems presently require that the entry and exit access doors to the conveyor shaft remain open. The openings are generally equipped with a fire door that should be provided with a magnetic latch which will release if deactivated by a signal from a smoke detector system or water flow if sprinklers are activated or if the power fails. To reduce or avoid the stack effect of a conveyor system shaft with open doors in a tall building, the conveyor shaft and entry and exit stations should be located in a closed room. An arrival lantern can be located over the door to the room to indicate that a box has been delivered and is ready to be picked up.

Future generations of selective vertical conveyors may have automatically opening and closing doors as technology progresses. This will certainly enhance their acceptance in view of the serious considerations being given fire and smoke control in buildings.

An advantage of a vertical selective conveyor is its ability to interface with horizontal transportation. Conveyor systems in remote buildings can be interconnected by horizontal belt conveyors so that a tote box placed on one floor in one building can be automatically delivered to a selected floor in an adjacent building. The horizontal belt can be located in a ceiling area or a remote mail

Figure 15.3. Vertical selective conveyor spiral accumulator for delivered tote-boxes in a central receiving area (Jaros, Baum and Bolles photo).

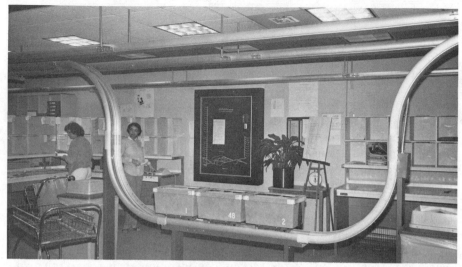

Figure 15.4. Telelift through station for loading and dispatching. Note status indicator board in background (Jaros, Baum and Bolles photo).

room can be connected to the vertical shaftway by horizontal belt conveyors. Discharge to the mail room can be accomplished by gravity roller conveyors or spiral conveyors to accumulate filled tote boxes until they can be emptied. An on-floor branch lift can raise a filled tote box to ceiling height where it can be discharged to a horizontal belt for further travel to the selective vertical conveyor serving all the floors in the building.

SELF-PROPELLED TOTE BOX SYSTEMS

Tote box transportation is a convenient means to send and receive mail, samples, small supplies, or any material of limited weight and volume from one place to another regardless of horizontal or vertical travel. It is, in essence, a brief or suitcase which can be transported by mechanical means by conveyor, as described in the previous section, or by a self-propelled carrier, as will be described in this section.

The tote box described for a vertical selective conveyor has been modified and securely attached to a motorized carrier designed to run along a dedicated track system which is arranged as intricately as a model railroad and travels both vertically and horizontally (Figure 15.4). An article is placed in the box, the destination coded on the cover, and the carrier sent from the loading station. From there it will travel horizontally with the motor driving traction wheels and, as it approaches a vertical bend, a rack section on the track is engaged by a gear on the drive wheel and the carrier will proceed vertically. At a junction, it will enter a switch, stop, and be switched right or left to a parallel track that will take it to the destination coded on the cover. Magnetic readers

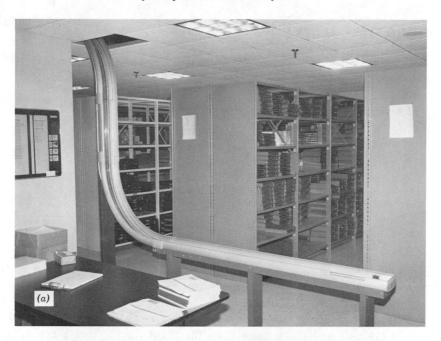

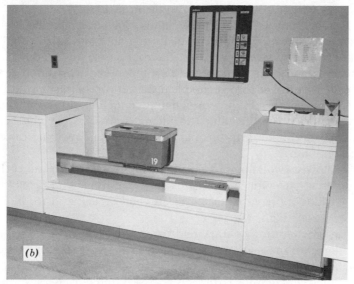

Figure 15.5. (*a*) Telelift reentry station: last container in is the first container out; (*b*) Telelift through station: first container in is first out (Jaros, Baum and Bolles photos).

along the way read the code and cause the proper switching and traveling directions to be given.

At the receiving station, which may be a dead end or a through track section (Figure 15.5), the carrier must be manually unloaded, and the next destination

RIGHT OF WAY (R.O.W.)

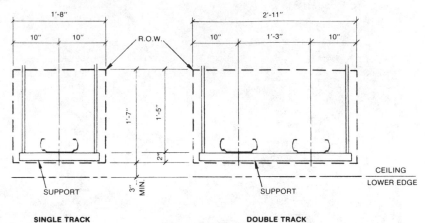

SINGLE TRACK   **DOUBLE TRACK**

Figure 15.6. Cross section of a telelift track and required right of way clearances (Courtesy Mosler Company).

coded. If the receiving station is a dead end, the carrier will reverse and travel to the switch which will place it back on the main track, reverse again, and proceed in its normal travel direction to the next destination.

Such systems may be found in hospitals delivering pharmaceuticals and blood samples or in office buildings delivering mail and small supplies. With the track section and switch machine, the routing may be readily changed to accommodate additional stations and new functions. Horizontal remote stations, even if they are at a considerable distance, can be easily configured. As with any conveying system, proper planning to include sufficient routing, stations, and equipment in the initial design and contract is the most economical approach. Alteration in the future may require system downtime and expensive building alteration.

In 1982, at least two manufacturers are supplying systems and each offers a slightly different variation in box size. Increased usage and demand will undoubtedly lead to more variations. The sizes and capacities listed in Table 15.2 are presently available.

A typical cross section of track and space requirements is shown in Figure 15.6. Various clearances are necessary both for horizontal curves and vertical bends. Many systems operate in ceiling space so early design routing to avoid structure, ducts, and lighting fixtures is necessary and must be coordinated with the other mechanical requirements.

Table 15.2. Inside Box Dimensions—Automated Tote Box Systems

	Width, in. (mm)	Length, in. (mm)	Depth, in. (mm)	Capacity, lb (kg)
A	11 (280)	18 (460)	8 (200)	20 (9)
B	6 (150)	18 (460)	12 (305)	20 (9)

Vertical shaftways must be arranged to provide space for the exiting bend. System design features and a typical layout is shown in Figure 15.7. The openings to the vertical shaft must be protected by an open fire shutter or damper which is designed to release upon a signal from a smoke detector, sprinkler water flow, or in the event of power failure. In any building, rubber baffles are employed to minimize drafts.

As with the vertical selective conveyor, development work will be necessary to design an automatic opening and closing door which will enhance the fire and smoke control life safety features of the system and minimize stack effects.

An essential system design consideration for both the vertical selective conveyor and the self-propelled tote box system is the layout of the main service area, whether it be a mail room, central supply, reproduction center, laboratory facility, or any combination. An in-depth study of functions and requirements must be made to determine the proper layout of the tracks to properly place carriers to be unloaded, the queuing of carriers waiting be unloaded, queuing of carriers waiting to be reloaded, and dispatching and reloading areas. This study should also include the number of boxes or carriers needed for processing, in transit, and stored at various receiving stations.

As a fundamental example, if a system has 10 receiving stations and it is assumed 2 boxes are needed per station, 2 or 3 boxes should be provided in the central area to be loaded for each station, so this minimum system requires at least 40 boxes. With vertical selective conveyor, the investment in tote boxes is small. With a self-propelled system, the investment is considerable since each box requires a motorized carrier. In addition, with the motorized carrier, an area for carrier maintenance should be set aside since periodic overhaul of each carrier will be necessary.

The centralized facility where the carriers are received and dispatched is also expected to be attended and provides the ideal location for a monitoring system (Figure 15.8). A panel showing the system displays indications of vacant and filled stations, occupied tracks, operating switches, and system delays so that corrective actions may be taken.

The value of any automated material-handling system is the number of employee hours that can be saved by eliminating unsupervised people carrying items vertically from place to place. A detailed study of an example of such employee involvement is given in Chapter 18.

AUTOMATED CART LIFTS

Oftentimes a tote box is not large enough to transport bulky documents required in the everyday activity of a firm. For example, in an engineering firm, large rolls of drawings are common items that are sent from place to place; in a power company headquarters, the boxes containing hand-held meter reading recorders are often transported to a central point for data processing: in other firms, bulk mail in boxes is the common requirement. All these items, being larger than the largest tote box and often too heavy to be handled in a single tote box by an individual, require cart transportation.

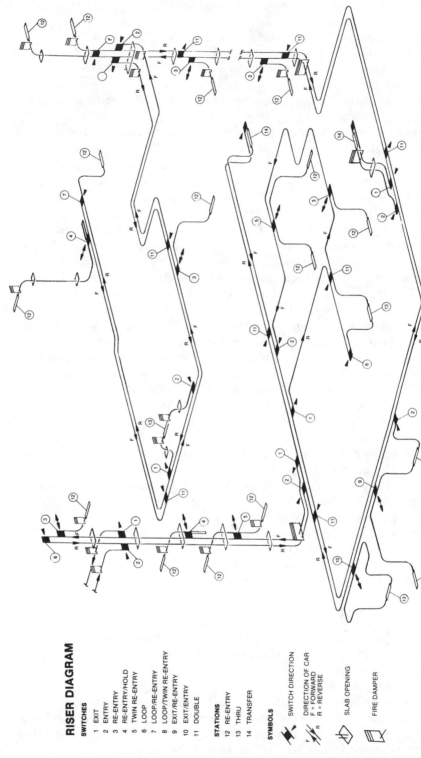

RISER DIAGRAM

SWITCHES

1 EXIT
2 ENTRY
3 RE-ENTRY
4 RE-ENTRY/HOLD
5 TWIN RE-ENTRY
6 LOOP
7 LOOP/RE-ENTRY
8 LOOP/TWIN RE-ENTRY
9 EXIT/RE-ENTRY
10 EXIT/ENTRY
11 DOUBLE

STATIONS

12 RE-ENTRY
13 THRU
14 TRANSFER

SYMBOLS

SWITCH DIRECTION

DIRECTION OF CAR
F = FORWARD
R = REVERSE

SLAB OPENING

FIRE DAMPER

Figure 15.7.   Composite layout of hypothetical telelift system showing the variety of components (Courtesy Mosler Company).

372

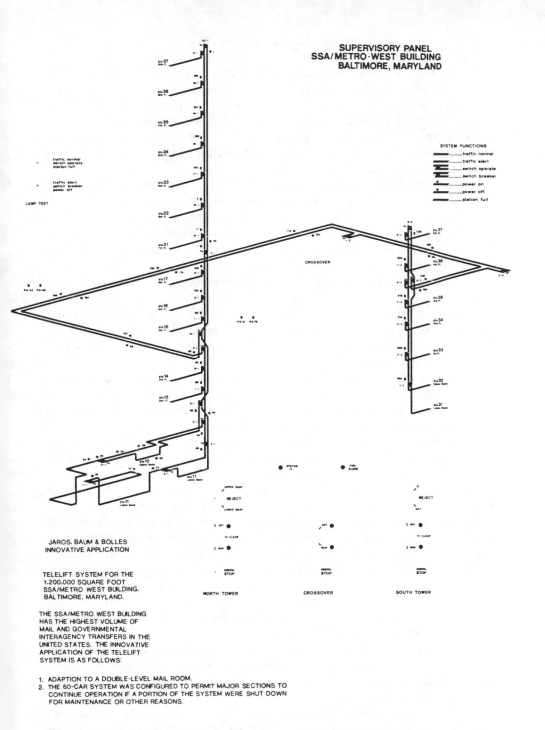

Figure 15.8. Telelift system display panel showing status of all stations, switches, occupancy of tracks, as well as alarm and trouble indications by means of colored indicator lights (Jaros, Baum and Bolles photo).

In hospitals, loads of linen, medical supplies, and the surgical case equipment often require a single cart each. It was in the hospital that the automatic unloading cart-carrying dumbwaiter was introduced and subsequently refined into an automated material-handling system which automatically loads and unloads carts on a dumbwaiter or elevator and which can be programmed to perform prearranged deliveries or pickups. This class of equipment is called lifts with automatic transfer devices.

In a delivery cycle, loaded carts are positioned in front of the entrance to the lift and a destination floor is registered on an operating panel. The lift doors are automatic power opened, and a transfer device is driven out under the cart, couples, and injects the cart into the lift. The doors automatically close, the lift travels to the destination floor, doors open, and the cart is ejected out by the transfer device, which then uncouples and retracts, doors close, and the lift is free for the next assignment. This cycle is referred to as the "dispatch" program.

In a pickup cycle, the empty carts are left in front of the entrance and a call is registered. When the lift is programmed to pick up, it will travel to each floor, pick up the waiting cart, and deliver it to the central area. This cycle is referred to as the "return" program.

If the lift is equipped with both front and rear openings, both dispatch and return programs can be performed on the same trip, the cart delivered to the front entrance and any empty car picked up at a rear entrance and returned to an entrance opposite the central loading entrance.

With a single opening, the cart lift is often arranged with a program selection switch to establish the delivery or pickup programs. By proper arrangement at the floor landings, sensing devices can be used to indicate waiting carts or areas where delivered carts have to be removed from in front of the lift. This is especially important if space limitations occur in the unloading area.

Proper layout of a cart lift system proceeds from an initial determination of the size of the cart and its maximum load. If the gross area covered by the cart is 9 ft² (0.83 m²) or less, the cart is no more than 4 ft (1200 mm) high, and the cart weighs 500 lb (225 kg) or less when loaded, a floor stopping dumbwaiter arranged as an automated cart lift can be used. Any dimension in excess of these limitations requires an elevator design including a car safety device and safe means for maintenance operation. Part XIV of the A17.1 Elevator Safety Code outlines the requirements and limitations.

Once the width and depth of the platform necessary to accommodate the cart is established, the remainder of the space required for a hoistway is similar to that for a conventional elevator. Pit and overhead vertical space is established and will be a function of cart lift speed.

The floor space, both at the central area and on each floor, is an essential consideration. The horizontal distance in front of the cart lift entrance at each floor must be at least the length of a cart, preferably two, before any obstruction is encountered. If the cart lift opens up into a public space such as a corridor, a minimum of 4 ft (1200 mm) plus the length of a cart must be provided. In many local areas, protective rails are necessary to prevent a person from walking in front of the lift entrance which may open and discharge a cart at any time.

The central loading and cart storage area must be designed to accommodate all the carts expected to be staged for a dispatch or return cycle.

All carts must be of identical chassis or base dimensions and have the coupling device located in the same position. It is recommended that the cart chassis be ordered along with the cart lift itself since the superstructure can be designed for the intended use and later placed on the standardized chassis. Caster wheels are necessarily of large diameter, 5 in. (130 mm) minimum, and one pair arranged to swivel.

As part of the cart lift installation contract, a floor plate with slightly depressed tracks should be provided and integrated with each cart lift entrance. In this way, distortions and out-of-level landing areas are avoided, the cart lift installer aligns the floor area with the transfer mechanism, and dependable cart lift transfer operation is assured. Once the floor plate is aligned it can be permanently set in concrete and the floor covering matched up to the tracks. The slightly depressed tracks in the plate ensure that the cart and casters are properly positioned, and even the most careless attendant can readily push the cart up to the lift entrance without serious misposition. Figure 15.9 shows a front view of a single cart lift of elevator size and capacity, Figure 15.10 shows a

Figure 15.9. Cart lift with cart being loaded (or discharged) at a typical floor station (Jaros, Baum and Bolles photo).

Figure 15.10. Cart lift transfer device, "ferry slips" on the entrance and guidance rollers in the car. Extended rollers actuate switches to indicate cart position (Jaros, Baum and Bolles photo).

transfer mechanism inside the lift and the landing floor tracks, while Figure 15.11 shows a cart in position ready for discharge. This also ensures in line ejection of the cart at the receiving floors.

Nominal cart sizes in the United States and Canada are 24 in. (600 mm) wide by 5 ft (1525 mm) long and the general height is 5 ft (1525 mm) so the attendant can see over the top while pushing. In hospitals, carts are of stainless-steel construction and designed so they can be washed and sterilized. For office buildings, carts built up of racks are often used and the racks allow reconfiguration for various tasks. Figure 15.12 shows the layout of cart lifts in a building.

TOTE BOX LIFTS

As mentioned in the introduction to this chapter, the first application of an automatic transfer device on a lift was an automatic unloading dumbwaiter. The device, which was mounted in the dumbwaiter, automatically ejected the tote box load onto a table after the power operated doors opened and, once the transfer was completed, the doors closed and the dumbwaiter returned for the next load (Figure 15.13).

Later developments improved the operation so the transfer device on the dumbwaiter not only automatically unloaded the tote box, but was arranged to automatically pull a waiting tote box in for a next delivery. This was done in

Figure 15.11. Cart on lift ready to be ejected. Swiveled casters will be aligned by floor tracks as cart is moved out (Jaros, Baum and Bolles photo, transfer device, tracks, and doors by Courion Industries Inc.).

two ways, the first, with a transfer device that reached out under the box, raised a pusher bar behind the box, and pulled the box into the dumbwaiter car. The second means was to provide a belt on the car and a second belt on the loading table, which was driven by the motor driving the car belt. Unloading was done by turning the belt in one direction and loading by turning in the opposite direction. These refinements led to the fully automated tote box lift.

The tote box lift is valuable in offices, libraries, hospitals, laboratories, or any place where tote box loads need to be sent from floor to floor on an intermittent basis. It is ideal in a multifloor law office where briefs and law books can be kept at a central location and sent to people who may need them. Similarly, in libraries the tote box lift is valuable where less frequently used books can be stored on upper floors and recovered when required. The tote box feature and automatic unloading provide the means to keep the lift in operation as opposed to the necessity of manually loading or unloading a dumbwaiter and losing that productive time.

Many variations can be found. A double-deck tote box lift has been built, the bottom deck to accept loads and the top to discharge using gravity to a table. Lifts with front and rear openings have been built, so loads can be picked up and/or discharged at the same floor level. As an example, a bank uses such a lift to provide secure delivery of negotiable paper from a vault area to a banking floor. A double deck lift with tote box on top and cart lift on the bottom may also be considered.

The tote box lift or a dumbwaiter require the same hoistway, pit, and overhead space. The tote box lift usually has an entrance of lesser height than a dumbwaiter since there is no need for a person to put their hands in the lift. Unlike a dumbwaiter, a tote box lift requires a table for loading and unloading in front of the lift. The table is equipped with an electrical presence detector so that a transfer will not be attempted if the table is full and a call buzzer which is intended to summon an attendant will sound.

Both automated cart lifts and tote box lifts are in fully enclosed hoistways with fire rated entrance assemblies which are kept closed and only opened when transfer operation occurs. All the rules of the A17.1 Elevator Code apply to such lifts and Part XIV of that code provides special rules with the intent of maintaining safety requirements.

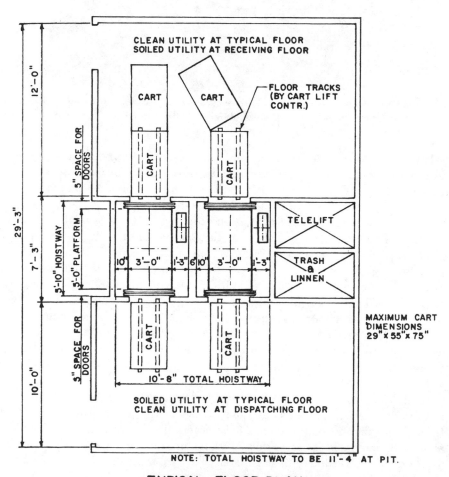

TYPICAL FLOOR PLAN
AUTOMATIC CART LIFTS 1 & 2
800 LBS. AT 500 F.P.M.

Figure 15.12. Layout of a cart lift system for a hospital showing front and rear entrances and separation of carts with clean and soiled articles.

Tote-Matic unit automatically loading container horizontally into car. After transfer is made car will travel vertically to pre-selected level, unload container onto receiving table.

Figure 15.13. Tote box lift showing transfer table and lift (Courtesy Courion Industries Inc.).

TRACK SYSTEMS—OVERHEAD

Extensive development of "powered and free" track conveyor systems has been accomplished for industrial material-handling applications and some of these systems have been applied to hospital material handling. A power and free track conveyor is an overhead track where carriers are moved horizontally by power driven chain and may be released at any point to accumulate or to travel down an incline by gravity. The innovation that allowed track conveyors to be interfaced with vertical travel was the development of a transfer device which would accept a waiting carrier at the entrance to a lift and transfer it onto a lift. The lift would travel to another level and the transfer device would automatically discharge the carrier at the new level where it would engage the powered track to horizontally transport it to a final destination (Figure 15.14).

The power chain is such that the carriers can only travel in one direction so that both front and rear entrances are required on the lift. The lift is a modified elevator reinforced to withstand the cantilevered load effects of the transfer device's reaching out to pick up the carrier and designed so that the mounting of the transfer device is at the top of the car. Car and hoistway doors are provided, the hoistway doors usually requiring a slot so the track extension can be aligned with the transfer device with a minimum horizontal gap.

The carrier attached to the track is designed to grasp the top of the cart by means of hooks. The track at the pickup and discharge points is depressed so the cart can be rolled off when the hooks are released or rolled into position for pickup. During horizontal travel as the carrier moves along the track, the cart wheels are off the floor a short distance (Figure 15.14).

Figure 15.14. Overhead powered and free conveyor track system and lift interface (Courtesy Acco Industries Inc.).

The disadvantage of the system is the fixed installation of the track system and space required, making it difficult to add to or rearrange the system. The obvious labor savings is the rapid vertical and horizontal travel without an attendant. Carts must be manually attached or released from the carrier which requires an attendant.

TRACK SYSTEMS—IN-FLOOR

Tow chain conveyors, which feature a continuously moving chain embedded in the floor wherein a cartlike carrier can be engaged by dropping a pin into a slot, have been an accepted horizontal conveying device for many years. They are extensively used in factories and mail order warehouses, are heavy duty and

Figure 15.15. Tow chain conveyor and elevator interface: (*a*) vehicles about to enter elevator; (*b*) vehicles exiting from the elevator.

rugged, and, if two- or more floor travel is required, can be arranged to be moved up a ramp. Switching means have been developed and magnetic readers on the carts can be used to route the vehicle to various loading and unloading stations.

Moving from floor to floor by ramp requires considerable horizontal distance and valuable warehouse or factory space. This led to the development of an in-floor transfer device which is mounted in the platform of a lift and is used to move carts on to or off of a lift (Figure 15.15).

The traveling cart signals the lift by means of a magnetic or photoelectric encoder mounted on the cart and a reader in the floor. The cart is brought to the lift entrance and stopped. The lift is signaled, the doors open, and a device on the lift reaches out into the floor slot in front of the lift and pulls the cart by means of its engagement pin into the lift. Doors close, the lift travels, and the cart is discharged by a discharge transfer device onto the destination floor and with the cart's pin in a back floor slot where it is picked up by the powered chain to its destination.

The system is adapted for heavy industrial use and may be found in a number of mail order warehouses in the United States. Its main function is warehouse stocking and order picking wherein loads of up to 3000 lb (1350 kg) can be moved by one cart to and from the shipping area. The lift is a freight-type elevator with vertical biparting doors and can be designed for single or parallel track operation. Front and rear entrances are required since tow chains only operate in one direction and the towed cart is engaged only by a pin at the front end.

ROBOT VEHICLES

Vehicles which are battery powered and follow a guide path on the floor can be programmed to leave from a given point, travel horizontally along corridors along with pedestrians, call an elevator, board it, travel vertically, leave the elevator at a destination floor, discharge a load which may be on a cart or pallet, and either wait or return to a central place.

Such vehicles can be found delivering mail in office buildings, supplies in hospitals, material in factories, and used for orders in warehouses. Some have automatic loading and unloading features and all may be programmed to travel to a battery charging station when their batteries are low and a recharge is needed. The vehicles are designed to follow a preestablished guide path which may be either a wire embedded in the floor or a reflective, invisible, stripe painted on the floor or on the carpet or other sensed path (Figure 15.16).

In the mail application, the vehicle is typically equipped with compartments to receive mail at a centralized mail station. The vehicle is started and travels from place to place pausing a few seconds at each programmed stop so a person can remove mail or packages and place outgoing items on board. As the vehicle travels it beeps to warn pedestrians in its path and a sensitive bumper is provided in the event an obstruction is encountered. Nonprogrammed stops and starts can be made by conspicuous switches on the sides of the vehicle.

In the hospital application, a vehicle is designed to pick up a hospital-type cart which can be loaded with supplies, linens, or medications (Figure 15.17). It is programmed for a destination and a dedicated lift system is provided to carry the vehicle and its load from floor to floor. At the destination, the cart is

Figure 15.16. Mailmobile vehicle (Courtesy Bell & Howell).

deposited and the vehicle can return to either pick up a new load or a waiting empty cart.

Where heavy traffic is anticipated, vehicles remain on each floor and a transfer vehicle may remain captive with the lift. The cart is unloaded from the vehicle in front of the lift, a signal is given, and the horizontal vehicle departs. When the lift arrives, a transfer device brings the cart on the lift and travels to the destination floor. There the cart is discharged and, either an on-floor vehicle travels to pick it up and takes the cart to a final destination or an attendant can manually move the cart.

Part XIV of the A17.1 Code allows guided vehicles to move on passenger elevators if passengers are prevented from using the elevators while they are being used for vehicles. This can be accomplished by rendering the in-car operating buttons inoperative and establishing the call for elevator service through the vehicle's electronic system. The elevator doors would not open unless a vehicle is in position to block the entrance and discourage people from entering.

Two important layout considerations must be made when self-propelled guided vehicle elevators are planned. These will be the width of corridors and space to turn the vehicle. Most vehicles are three wheeled with the single wheel

Figure 15.17.   Transcar vehicle (Courtesy Lamson Corporation).

being the steering wheel. As it turns, the trailing portion of the vehicle swings out. Where two must pass on a curve, additional space consideration must be made. The layout must include a battery charging area which must be well ventilated to dissipate the hydrogen gas generated when lead-acid batteries are charged.

The various manufacturers must advise of the necessary power supply requirements and electrical interfaces with power opening corridor doors, elevators if the elevators are a separate contract, and any life safety requirements such as stopping the vehicles in the event of a fire emergency. Manufacturers' layout information is required to properly design corridor width, turns, elevator access, storage space for vehicles, size of carts that can be carried, tote boxes on mail carriers, and other aspects. Very often, the guided vehicle system is sold as a tenant supplied equipment after a building is built and floors equipped. In that event, it will be up to the consulting engineer and the system supplier to determine if a system can be used in an existing space. Systems can usually negotiate inclines: the steeper the incline, the lower the carrying load capacity of the vehicle and shortened time between battery charges.

PALLET LIFTS

One of the most common means to package bulk material is on a pallet. The entire material-handling industry is pallet oriented: industrial forklift trucks are especially designed to handle pallets and completely automated warehouses are in operation which will automatically store and retrieve pallets with storage facilities for thousands of pallets.

Simplified pallet handling systems utilizing an elevator equipped with automatic pallet loading and unloading means can provide effective low-cost material handling in a multistory production facility. This was briefly described in Chapter 13 and shown in Figure 13.10. Pallet loads of raw material can be placed on a receiving pallet conveyor, programmed for a destination floor, and automatically unloaded on pallet conveyors at that floor. Conversely, finished products can be placed on pallets, loaded on the pallet conveyor at various floors, and automatically transferred to the loading dock floor for subsequent shipment. The forklift trucks are used at each floor but waiting time for the freight elevator to return and the cost of an attendant to unload it are eliminated. Modular-type pallet conveyors have been developed so a system can be created for a particular application. The lift is a conventional freight elevator arranged so that the pallet transfer device is placed on the platform. The pallet transfer device and the pallet conveyor consist of powered rollers with limit switches so that the pallets are transported and positioned.

Biparting freight-elevator-type doors are used to close the hoistway to provide fire integrity together with a fully enclosed hoistway. A programmable controller interfaces with both the elevator and conveyor controllers to direct their movements. As pallets are received, the attendant registers their destination and the pallet joins the queue. As the lift delivers them, the memory in the controller directs each operation. A printout can be produced which can list the various movements and can provide inventory control.

OVERVIEW

It is difficult to accept the idea that an office is becoming more and more like an industrial facility where material (paper) is received, processed, sorted, and shipped. While the increased use of computers and information processing systems may indicate the paper-free-office in the future, the present, however, requires the systematic movement of hard copy.

The foregoing describes a state of the art in automated material handling in public places, i.e., where people are not specifically trained to interface with an automated system. When such conditions exist, all the facets of life safety are a necessary consideration when the material-handling system is planned. In the industrial atmosphere, training is expected and the individuals are more aware of hazards. The line between a conveyor as regulated by the ANSI B20 Code Conveyor Code and a material-handling system as regulated by the ANSI/

ASME A17.1 Elevator Code is rather fine and hopefully will be defined with greater precision in the coming years.

The description that has been given of each system has been brief and designed to provide an overview with a minimum of specifics.

The other important aspects are the relations among the capital cost of the equipment, its annual maintenance, direct labor expense, and the reduction of man-hour equivalents the equipment affords. Too often an elaborate system may look attractive but may not return the investment fast enough to meet financial criteria as outlined in chapter 17.

CHAPTER SIXTEEN

Environmental Considerations and Special Operating Conditions

USE OF ELEVATORS

Elevators are applied in all types of buildings and structures. Elevators are found anyplace where there is a need for personnel and material to travel any distance vertically or at an incline. Applications include TV transmission towers which may be 1000 ft (300 m) high, smoke stacks to maintain emission control systems, mines thousands of feet deep, oil production platforms resting on the bottom of the sea, and nuclear plants where the elevator may be submerged under cooling water to be used for periodic plant overhaul, to cite just a few. Each represents a distinct set of environmental considerations. Even the environment in a full commercial building may represent some unique consideration including building sway, stack effect, life safety or earthquake considerations.

Not only does the environment affect the design and installation of the elevator, but also the way the elevator can be used. In a downtown building it may be desirable to use an elevator for the handling of freight and mail or to move a single large piece of building equipment from the roof to the ground. In a structure, operation during an emergency may be totally different than during normal use. For example, an emergency in an oil refinery may require access to valves located at an upper level and commandeering an elevator may be the most effective approach. If hazards are present, such as explosive gases and dust, or if the elevator is exposed to wind and weather, considerations in addition to necessary special operation must be made. In this chapter, many of these aspects will be discussed and guidelines established to suggest trouble-free elevator application and operation.

FIRE EMERGENCY

In any building or structure a fire is one of the most serious of occurrences and elevators by the very nature of being exposed on every level of the structure may be especially hazardous. As a general rule, elevators should not be used during a fire emergency unless under strict control of knowledgeable personnel. In buildings, elevator hoistways are required to be in fire rated enclosures and the elevator entrance assemblies should be designed, tested, and installed to be comparable to the fire rating of the hoistway enclosures. This was seldom a consideration when hoistway enclosures were masonry; the entrance was bricked-in solid and secure. With drywall construction using gypsum board, the interface between the entrance assembly and the adjacent wall is a critical connection which must be made in accordance with the way the entrance was fire tested. Figure 16.1 shows a typical interface with an entrance frame: the important consideration is that the connecting "J" stud be fastened to the entrance and the long face of the stud be at the shaft side.

The first step in elevator operation during a fire emergency is to recall the elevators to a main floor where they can be evacuated and arranged for super-vised operation by emergency personnel. In this way all the riders can be accounted for and elevators not returning can be located and checked by emer-gency personnel. The elevators are also at a place where emergency personnel can use them to transport fire or other emergency equipment to an upper-floor position.

Recall operation is prescribed by the A17.1 Elevator Code and essentially requires that the elevators in normal operation be stopped and immediately

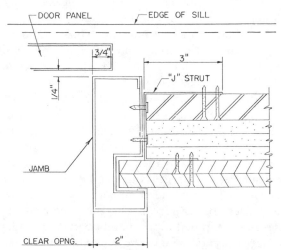

Figure 16.1. Plan of center-opening entrance as fire tested in a nonmasonry wall (dry wall)—maximum door height 8 ft. 6 in. (2600 mm) (Courtesy of Jaros, Baum and Bolles).

recalled to a main floor with minimum interference. Up traveling elevators are reversed and returned without door operation, all landing and car buttons are rendered inoperative, and emergency stop switches are disconnected. Open doors are immediately closed and door protection devices that may be affected by smoke or flame are rendered inoperative. Audible and visual signals are given to elevators manually operated, such as those on independent service, inspection, or with an attendant.

Recall can be initiated by manually using a recall switch, by the actuation of a products of combustion detector, or if water flow occurs in the sprinkler system. If a false alarm occurs, normal elevator operation can be reinstated by an override position on a supervised switch until the source of the false alarm is corrected.

Once the elevators are recalled, they will park at a main floor ready for use by emergency personnel. A keyed switch is provided in the elevator which places it in operation under the complete control of the person in the car (Figure 16.2). A desired floor may be chosen on the floor buttons and starting is initiated by pressure of a door-close button. The elevator will stop at the desired floor and to open the door requires constant pressure of a door-open button until the doors are fully opened. If smoke or heat is detected, release of the door-open button will cause the door to reclose and the car can be taken away.

The elevators can be used to transport emergency personnel and their equipment or to evacuate building personnel from designated refuge floors. These people should have been trained by means of fire drills to assemble in refuge areas since elevators would have been rendered inoperative for the floor occupants at the first signal of a fire emergency. It would be virtually impossible to evacuate from the floors served by a group of elevators within any reasonable time since the elevator-handling capacity, as described earlier in this book is, at the most, only between 10 and 20% of the population the elevators serve in a 5-min period.

Slight modification of the emergency procedure described may be provided due to requirements in local jurisdictions. For example, the A17.1 Elevator Code only allows the top emergency exit in an elevator to be opened from outside on the car top. This is based on the assumption that trained emergency personnel should be the ones to aid people exiting a stalled elevator. In some areas, a keyed lock is provided on the exit so emergency personnel can open it from the inside to inspect the hoistway for signs of smoke or flame. Opening the top exit from inside the car is also valuable after an earthquake and will be discussed later in this chapter.

Some jurisdictions advocate providing sprinklers in elevator machine rooms and hoistways. This can have a disastrous effect, since heat and smoke may not have a detrimental effect on elevator operation but soaking elevator electrical equipment may cause complete shutdown and permanent damage. With modern elevators there is a minimum of combustible material in the hoistway. Most elevators are equipped with roller guides operating on dry rails. Before such guides were extensively used, sliding guide shoes on oiled rails were common.

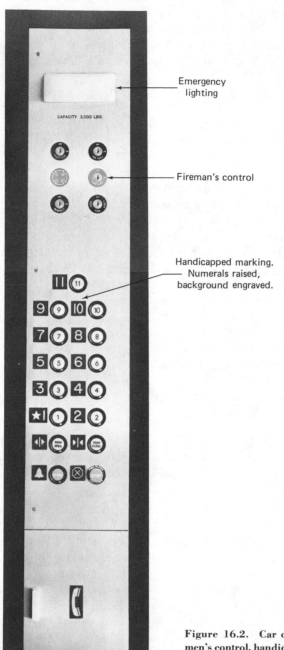

CAPACITY 2,000 LBS.

Emergency lighting

Fireman's control

Handicapped marking. Numerals raised, background engraved.

Figure 16.2. Car operating panel with firemen's control, handicapped markings, and integral emergency lighting (Courtesy G.A.L. Mfg. Co.).

Prior to 1969, elevator entrance assemblies were tested with the fire assumed on the hoistway side. At that time, sufficient application of roller guides had taken place, and for that and other reasons, the test procedure for new elevator entrances was changed to assume that fire is more likely to occur on the landing side rather than on the hoistway side.

There remains one likely area to apply sprinklers and that is in elevator pits. Unless housekeeping is stringent, paper and trash will accumulate in the pit and a cigarette dropped between the car and hoistway sill can ignite the accumulation.

In some areas, a fire-safe elevator is required. This consists of a separate elevator in a fire rated hoistway with each landing area enclosed by fire doors and a rated wall. Such a requirement is common in Europe and in some parts of the United States. An elevator of this nature is ideal as a service elevator for any building and is recommended.

EARTHQUAKE DESIGN

In a number of large cities in the world siesmic disturbances may be expected. The hazardous areas have been mapped and both probability and severity ratings are available for most areas. Any building in a siesmic risk area should be designed to withstand the earthquake shock and elevators and their hoistways require certain additional considerations over normal installation practices.

One potential danger is the possibility of a counterweight becoming disengaged from its rails and swinging into the hoistway. If an elevator is running there is the potential hazard of the car colliding with the free-swinging counterweight. A number of preventative means are available. One is to reinforce the counterweight rails by means of box brackets so the counterweight is restrained from swinging out (see Figure 16.3). Another is an electrical detector, a taut wire from top to bottom that will be electrically intercepted and cause an electrical circuit to stop the elevator if the counterweight is displaced (Figure 16.3).

In addition to the counterweight protective measures, the elevator hoisting machine and other machine room equipment is tied down with fastenings sufficient to withstand the expected shock. Rope guards are provided to prevent the ropes from jumping from the sheaves, and the car-to-counterweight compensating rope system is tied down with an arrangement to prevent the car and counterweight from bouncing upward during an earthquake shock. The elevator car is also equipped with retainer plates to maintain it within the rails. Antisnag guards are required in the hoistway to prevent swinging ropes and traveling cables from hanging up.

The rails and rail structure are required to be reinforced to withstand horizontal shocks in addition to the normal forces of loading and safety application that the rails are initially designed to withstand.

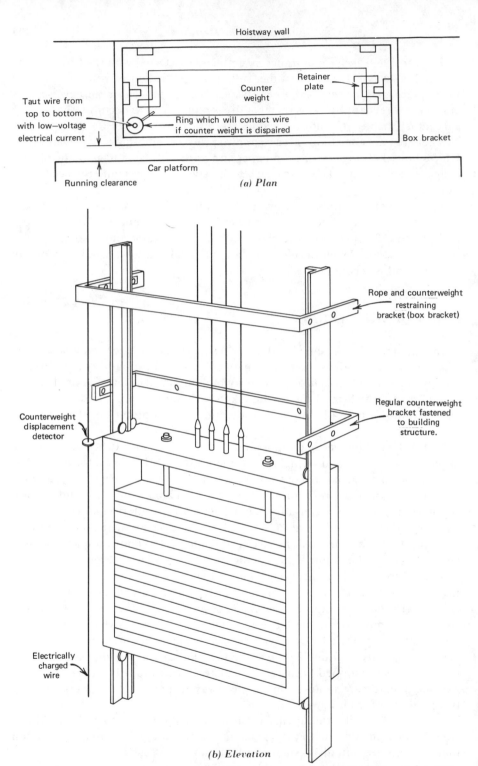

Hoistway wall

Retainer plate

Counter weight

Taut wire from top to bottom with low—voltage electrical current

Ring which will contact wire if counter weight is dispaired

Box bracket

Car platform

Running clearance

(a) Plan

Rope and counterweight restraining bracket (box bracket)

Regular counterweight bracket fastened to building structure.

Counterweight displacement detector

Electrically charged wire

(b) Elevation

Figure 16.3.   Counterweight box bracket and counterweight displacement detector.

Seismic engineers have determined that there are distinct earthquake ground waves that precede a severe earthquake shock. Detecting equipment is available to sense those waves and provide a slight early warning. This information is used to cause all elevators in the building to stop at the next available floor and remain with their doors opened. An elevator traveling in an express portion of the hoistway is stopped and designed to proceed at a slow speed toward the next landing in a direction away from the counterweight. For example, if the elevator system has an express run from floors 1 to 9 and serves floors 10 to 20, and the early warning occurred as the elevator just passed the eleventh floor in the down direction, it could possibly encounter an up traveling counterweight. It would be stopped and slowly travel up to the eleventh or twelfth floor before finally stopping. In this way the counterweight would travel down away from the elevator and the hazard of a collision would be avoided.

Once the earthquake shock waves have ceased, the elevator hoistways should be inspected for displaced equipment. Emergency personnel have the option of opening the top of car escape hatch from inside the car by key to make a visual observation before the car is moved. In the same manner, they can move the elevator upward at slow speed and observe the hoistway to determine any damage before restoring the elevator to normal high-speed operation. As a first step, a reset switch on the elevator controller is provided to restore normal operation once the early warning seismic device is reset after damage has been appraised or repairs are accomplished.

The A17.1 code has an appendix which describes earthquake considerations for elevators. The state of California has additional rules as do many other seismic risk areas. These rules must be considered as a minimum in designing elevators for those areas.

WIND EFFECTS

With tall buildings a certain amount of flexibility is present so the building will tend to sway during buffeting winds. In addition some buildings are designed to lean, ever so slightly, into the prevailing winds. The motion induced by the wind causes a certain displacement related to the fundamental building period. A well-designed elevator system is expected to operate under these conditions without hazard to the passengers.

Too often, elevator systems with their ropes at or near resonance with the building and building sway can cause excessive movement of the elevator ropes. Visualize an elevator starting up from the main floor with its ropes swaying a small amount (Figure 16.4). As it travels up, the amplitude and frequency will increase as the rope is shortened and the increased amplitude may be severe enough to cause the ropes to strike the hoistway sides.

The effect of building sway is more pronounced on the less tense compensating ropes and traveling cables than on the hoist ropes. Unchecked, these ropes can snag hoistway structures and tangle with each other. Means to protect

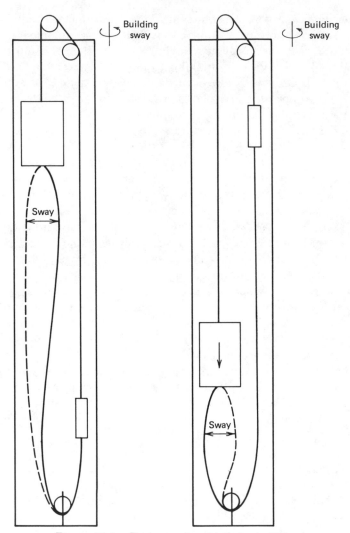

Figure 16.4.   Rope sway in a high-rise building.

against damage have been employed, such as providing a traveling rope guide carriage to contain the compensating ropes to minimize their movement. Studies are being made to "detune" the elevator system to minimize the possible adverse actions during building sway. Specifications for the elevator equipment for a particular building should include a statement of the expected period, magnitude, and other aspects of the building under wind conditions for the guidance of the elevator design engineer. A later section of this chapter on outdoor elevators will discuss the measures that have been taken where wind is an expected basic design condition.

STACK EFFECT

An elevator hoistway can be a free path for air currents throughout the building unless some special construction considerations are made. As an example, an elevator standing with its doors open and a window or door is open to the atmosphere in the elevator machine room at top. A door is open to the street at the lobby and air will flow from the street to the roof through the lobby space around the elevator entrance, through the necessary holes in the elevator machine room floor for the ropes, and out the machine room (Figure 16.5). The

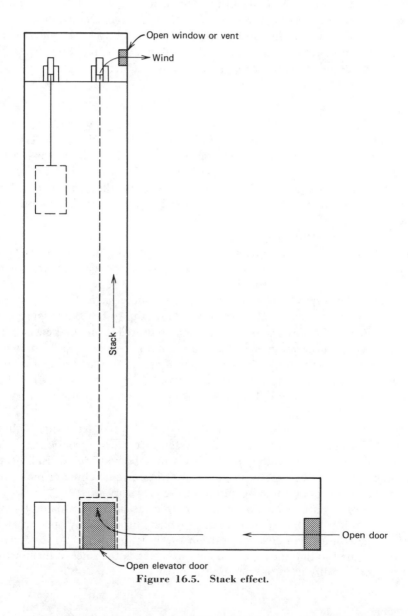

Figure 16.5. Stack effect.

effect may be a sufficient air flow so that the elevator doors cannot be closed by normal power means.

This stack effect is also a severe hazard if a fire occurs on a lower floor. Smoke or gases will be accelerated up the hoistway and, unless sufficient venting is provided at the top, will engulf upper floors.

The stack effect is particularly troublesome in winter weather when buildings are heated and service elevators may serve open loading dock areas. The use of double doors or revolving doors on the lobby floor may help to minimize such effect. Special closing devices to assist closing of lower-floor elevator doors, roller guide shoes in place of gibs in the door track, or special door operations to provide additional closing force may be employed to overcome the problem. The building design, venting, use of indirect air handling in the elevator machine room, and proper balancing of the building's air-handling systems all must be considered to reduce or eliminate the stack effect.

OUTDOOR ELEVATORS

Elevators installed outdoors are subject to all the varieties of wind and weather. Installations serving the general public may be found as far north as Niagara Falls, as well as in less severe climates. Industrial installations are found in almost any climate including Greenland above the Arctic Circle. Wind effect is a serious consideration and, in northern climates, icing may present a particular hazard.

The elevator hoist ropes, governor rope, and traveling cables are especially affected by wind. To overcome some of the effect, the hoist rope tension is increased by using smaller or fewer ropes, the governor rope is either provided with guides or located in a trough, and compensating ropes are eliminated by increasing the horsepower of the hoisting motor and electrically compensating for the changing load as the elevator travels from top to bottom (Figure 16.6).

An elevator cab exposed to the weather may require heating in the winter and air conditioning in the summer. A heat pump mounted on the car is often used and a receiver for air conditioning condensate water is provided so that it can be emptied as the car travels to the bottom to avoid dripping on pedestrians from above. For winter operation, heaters are used in the door operator motor and on the car and hoistway sills.

For extreme weather where operation may be essential, ice scrapers should be provided on the rails, and operating systems designed to periodically and automatically run the elevator up and down may be considered. Rail heaters have been employed but the wattage per foot in extreme climates may prove uneconomical versus periodic operation of the elevators.

Wind problems are always present and attempts have been made to overcome the problem by the use of wind gates which automatically open when the elevator passes and then close to restrain the hoisting ropes. An underslung elevator designed so that the ropes are semiprotected at the sides of the hoistway has been successfully used in one application.

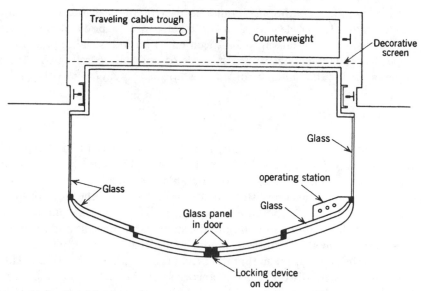

Figure 16.6. Outside observation-type elevator which is exposed to the weather. Note the traveling cable troughing.

When an outside elevator opens into an enclosed space, severe windage will occur in the gap between the elevator entrance and the car. In a number of applications this has been overcome by the use of a diaphragm around the perimeter of the door and the use of weatherstripping on the hoistway door. In one installation at a rocket launching facility, the elevator opened into an airtight clean room. There, both the car and hoistway doors were installed in pockets and an inflatable diaphragm was used to seal the entrance before the car and hoistway doors were opened. Such diaphragms impose a time penalty to inflate and require stringent maintenance.

Hoistway equipment must be weathertight and the machine room protected from the infiltration of rain water being carried by the hoist ropes into the machine room area. Gaskets surrounding the ropes have been provided and equipped with wipers to clear off water from galvanized hoist ropes. Machine rooms can be designed to operate under a positive air pressure to minimize moisture. This latter practice is also valuable in locations where a hazardous or dusty atmosphere is present such as oil refineries or cement mills.

Glass used in outside elevators must be laminated and, preferably, tempered safety glass so that the hazard of broken glass falling or the elevator enclosure being opened is eliminated. Since the glass window wall in the elevator is critical to safe elevator operation, it is always recommended that spare glass of the correct size and shape be specified and packed for storage as part of the initial elevator contract so that prompt replacement can be made.

The nature of the atmosphere must also be considered when specifying elevators for an outdoor location. Temperature, expected winds and their direction, and extreme weather conditions should be part of the specification for any outdoor elevator. If the elevator is to be located near a body of salt water, corrosion-resistant materials must be specified. If an industrial application is being specified, statements as to any hazards or unusual operating conditions should be made for the guidance of the elevator design engineer.

BUILDING COMPRESSION

Any new building as it is being built, and when completed, filled with people and furniture, slowly compresses (Figure 16.7). The difference between initial and fully compressed height may be considerable and usually requires a year of heating and cooling to accomplish final compression. Since the elevator guide rails are erected before the building is heated, provision needs to be made to compensate for this compression in both the stacking of the rails and the use of rail clips which allow the compressive movement. In addition, provisions need to be made in the elevator pit so that the excess rail length may be trimmed and the rails remain essentially bottomed to absorb safety impact if it occurs. This is accomplished by providing adjustable jack bolts under the rails (Figure 16.8).

Rail joints are usually erected with a slight gap to compensate for initial compression and rail clips must be designed to allow rail slide as compression proceeds. Failure to design for this action will cause the rails to distort and lose alignment, resulting in an elevator ride that produces excessive and noticeable horizontal accelerations. These accelerations are measured to judge the quality

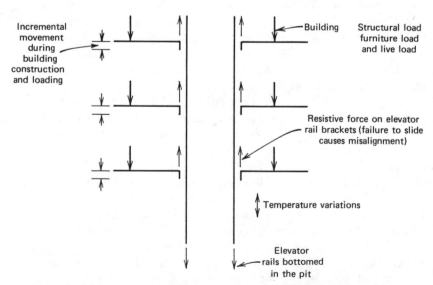

Figure 16.7. Building compression. Elevator rails erected during initial construction must adjust to the building compression during construction and furniture loading and the initial heating and cooling cycles.

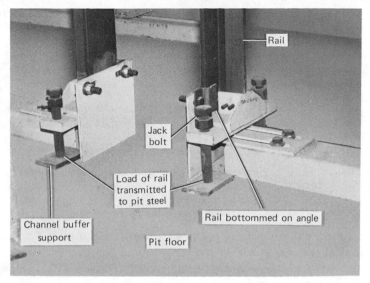

Figure 16.8. Jack bolts used under elevator rails to adjust for building compression so that rails may be maintained bottomed in the pit to transmit the force of a safety application.

of the elevator installation by which instrumentation produces a chart as shown in Figure 16.9.

The horizontal acceleration which can be felt is from about 10 milli-*g* at 500 fpm (2.5 mps) to 25 milli-*g* at 1000 fpm (5 mps). Measurements should be made both postwise, i.e., side to side between rails, and from front to back as the elevator travels up and down the hoistway at contract speed. The instrument should be mounted on the platform of the elevator at the approximate center. Readings that show excessive horizontal accelerations indicate that corrective rail alignment is required, and the length of the recording chart in relation to the chart speed and rise of the elevator will show the approximate location of

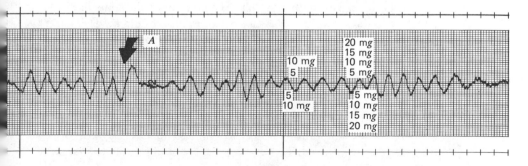

Figure 16.9. Rail ride characteristics as measured by a recording accelerometer. *A* indicates poor rail alignment with peak-to-peak horizontal acceleration exceeding 15 milli*g* where *g* is the acceleration of gravity (32 fps² or 9.76 mps²).

the fault. Interpretation of the same chart can indicate other faults such as poor rail joints and worn or flattened roller guides.

FLOODING

In some areas flooding may be a hazard that will require special design and elevator operating considerations. This is prevalent in cities along rivers which are subject to flooding or in coastal cities during extreme high tides or hurricanes.

The elevators will normally serve a ground floor but when a water level rises as indicated by a float switch in the pit, the elevators are prevented from traveling to the ground floor and the uppermost floor so that neither the car or the counterweight can enter the flooded pit area.

Secondary limit switches are provided and all pit equipment such as buffers and electrical switches are designed to withstand being submerged. Operating speed is reduced because compensating ropes or chains and traveling cables may pick up water and soak other parts of the hoistway.

Such operation should be limited to emergency use since part of the elevator protective system is reduced. If for any reason the elevator were to travel to the pit area, the buffer might be submerged and rendered ineffective.

POWER FAILURE

Buildings that have an essential function, such as hospitals, should have standby power systems so that essential elevator service is maintained in the event of utility power failure. In practice, usually two sources of outside utility power are supplied if possible and means to switch from one to another provided. The elevator system is designed to operate from two power feeders, some of the elevators on one feeder and the remainder on a second feeder run from this main building power panel to the machine rooms. If one source of power fails or if a feeder is destroyed, at least half the elevators will be available.

When a power failure does occur a number of events should be designed to take place. First, car lighting and a small circulating fan should go on in each elevator car immediately, supplied by a battery pack on the car itself. This lighting is required by the A17.1 Elevator Code and the ventilation can be provided by a fan with a permanent magnet motor which draws minimum battery current. The standby generator should automatically start and, when it is up to speed, each elevator should automatically return to a main floor, park with its doors open, and the standby power switched to the next elevator. When all elevators are returned, the emergency personnel should be able to manually choose which elevator they wish to use.

In a high-rise building it is desirable to size the standby generator to run at least two elevators in the building. A service elevator serving all floors is dedicated to the standby power system for use by emergency personnel and the

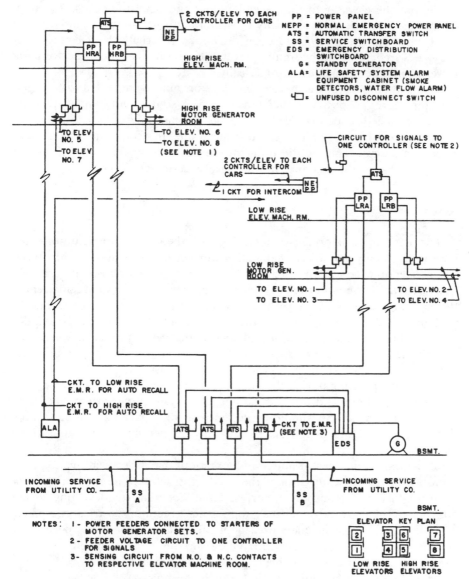

Figure 16.10. Circuitry for standby power where split elevator feeders are provided. Split feeders are provided to a group of elevators so that only one-half of the elevators will be out of service if one feeder fails (Courtesy of Jaros, Baum and Bolles).

remaining elevators evacuated on a one at a time basis until all are accounted for. Figure 16.10 shows an arrangement for standby power utilizing the split feeder approach.

Hydraulic elevators can be arranged to operate on standby power by utilizing the valving arrangement which allows the hydraulic elevator to be lowered without causing the pump to start. Standby power is used to operate the down valve, run the elevator to the lower landing, and open the doors. The same arrangement can be done manually by operating the down valves and using an emergency unlocking device to open the doors. Alternatively, a battery pack may be wired into the hydraulic elevator operating system to perform the lowering and door opening automatically.

With any standby power system, the elevators must be provided with a signal that the source of power is the standby power source and limited. Elevator controllers need to be arranged to operate so that only one or a designated minimum number of elevators can operate at one time to provide the mode of operation desired. The specifications for the elevators must include all the necessary details for the guidance of the elevator design engineer.

NOISE

The noise an elevator causes operating with the normal ambient noise in a building is usually nonobjectionable. Elevator noise may be noticed and objected to under quiet conditions such as a residential building or hospital during the quiet periods. The counterweight or elevator passing sleeping rooms adjacent to the elevator hoistway will cause disturbing noises and any vibration that may be present can be transmitted through the walls and structure. This can be avoided by providing sound and vibration isolators on the elevator brackets where they are fastened to the building structure (Figure 16.11). The best approach is to architecturally place such rooms away from the hoistways.

Elevators operating during the night create a certain amount of noise when they stop at a floor and doors open and close. Operating systems that park elevators at upper floors should be designed so such parking is done without door operation. Door operations should only happen in response to a car or landing call.

Structural borne elevator noises and vibrations may be transmitted to places remote from the hoistway through the building structure. The elevator machine should be isolated from its support and the secondary or deflecting sheave attached to the machine rather than supporting structure. Electrical equipment such as isolation transformers used with solid-state motor drives need to be isolated from the building structure to break up sound paths and guard against the transmission of objectionable structural borne frequencies.

As a high-speed elevator travels through the hoistway various noises are generated. A "puff" may occur when the elevator passes each floor as the air currents eddy at landing sills, and various pockets and structural members in the hoistway can create additional "puffs." These can be overcome by aerodynamic design and structural considerations.

Wind noises around hoistway doors where stack effect is present will create whistling noises. These can best be overcome by minimizng the stack effect or, as a last resort, weatherstripping the hoistway doors, which usually requires additional maintenance considerations.

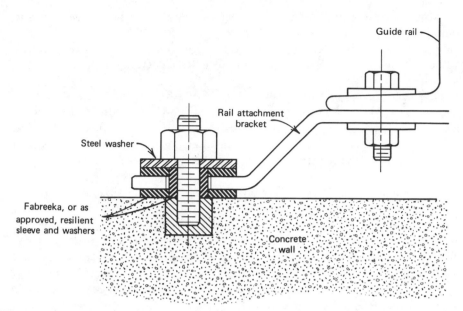

Figure 16.11. **Structure borne noise isolation arrangement of elevator rail brackets—necessary if sleeping rooms or sensitive equipment is located adjacent to an elevator hoistway
(Courtesy of Cerami Acoustical Associates).**

Each noise problem must be individually approached. Sensible initial design
can avoid the expense of later correction.

SPECIAL OPERATING CONDITIONS

In addition to the environmental considerations in designing elevators, the
habits, problems, misbehaviors, and other aspects of the people expected to use
the elevators or the special problems of the building where elevators are to be
used create unique requirements.

Security requirements, which include restricting, preventing, or apprehend-
ing people using elevators are a current prime consideration. This aspect can
also include means to monitor elevator operation both from a usage and maintenance viewpoint. Riot control is another aspect of security and the actions
that need to be taken to evacuate people from stalled elevators is both a security
and maintenance problem. A brief discussion of each of these aspects will
follow.

Security

Elevators in a building are a poor security barrier. Once people are on an
elevator they are able to get off at any floor they wish simply by either operating

the stop switch and pulling the doors open or by pulling the car door open and manipulating the interlock on the hoistway door. Security systems that electrically lock out car or landing call buttons are ineffective for this reason. The value of such a floor cutout system is simply a deterrent and the system needs to be closely supervised if used.

One such cutout system consists of a set of switches in a central location which are wired to the elevator operating system and prevent the registration of a car or landing call for particular floors. A second version consists of a card reader slot in the elevator operating panel. Inserting a magnetically coded card in the slot allows the registration of selected car buttons for certain authorized floors.

Two systems of card readers are possible. One is simply an activating device which energizes certain selected floor buttons in the elevator car. A second, more costly version, requires wiring from the car station to a central processor and back to the elevator operating system. A card inserted in the reader is checked for current status and authorized by the processor before the car call for a particular landing can be registered. Neither system can prevent a second unauthorized person from entering the elevator with the card bearer and from exiting at the floor where the card bearer gets off.

The total system requires that supervisory personnel first allow access to the elevator system before the cards can be used. It is far better to allow free access to all the floors in a building and provide the security restrictions at the entrance to the offices at each floor. A complete system would screen people entering the building and then at individual floors.

If it is desired to lock out floors, the locking of elevator hoistway doors must include magnetic bolt locks on each door panel interconnected with the elevator operating system to prevent door operation unless the bolt is withdrawn. Emergency operating procedures such as fire recall also require that such locks be opened. Failure to do so may force a fireman to use his "key," which is described as being long, red, and sharp ended, called an "axe." Safety codes also require certain doors to be always openable for emergency evacuation of stalled elevators, which will be discussed later.

Monitoring

It is desirable to monitor elevator systems for both proper operation and security.

The earliest monitoring system was the Starter's Indicator Panel found in most larger buildings when elevators were operated by attendants. The starter was able to tell at a glance where the elevators were, the waiting landing calls, and the direction the elevators were traveling. Directions were given to the individual operators by means of buzzer signals to return nonstop or to change the mode of operation.

Monitoring should be provided both for the benefit of building security and for maintenance personnel. An indicator panel in the machine room showing the position and motion of all elevators will provide maintenance personnel the means to periodically check operation. Readouts located in a building office, security room, or building fire command center are essential during emergencies and should be arranged to indicate an alarm if someone is on the elevator and calling for aid, either by means of an intercom system or the alarm bell. During a fire emergency or power failure, position information of the elevators is a necessity so emergency personnel can be effectively dispatched to the proper location. The same panel can contain standby power and other vital operating functions.

Television monitoring has been successfully used on some elevator installations. A special lens is mounted in the ceiling of the elevator which provides a view of the car interior—somewhat distorted—but recognizable with some training. If an alarm is sounded or a stop switch used, the status of the interior of the car can be viewed and an emergency action decision made. TV surveillance should be coupled with sufficient override elevator control so that, for example, if necessary, the car can be moved to a main floor without intermediate stops where emergency personnel can be waiting. Having a system similar to fire recall, except to call individual elevators rather than the group, can accomplish this. Similar decisions relating TV surveillance, monitoring, and special operations can be programmed for various situations. The security specialist, elevator design engineer, and building operating personnel need to be involved in such programming early in the building design process. Too often such considerations are made after the building is opened and the retrofit is prohibitively expensive. As a minimum, elevator traveling cable containing coaxial cable for the TV camera needs to be initially specified.

Riot Control

During a fire emergency it is desirable to return all elevators to the ground level to evacuate them and have the elevators available for emergency personnel. If a civil disturbance is taking place, it may be desirable to prevent all elevators from coming to the ground level to secure against access to the upper floors of the building.

This prevention is classed as riot control and is a relatively simple design consideration if done early in the design of the elevator operating system. Another consideration may be to provide a selective operation so individual elevators can be called to the lobby. A common standard operation is called "Car to Lobby" and consists of a key switch for each car which will call it to the lobby under normal operating circumstances, allowing it to answer car calls but not landing calls. A second, less common, variation would be individual elevator emergency recall which is related to the total recall that can take place during a fire emergency, as mentioned in the section on security.

Emergency Elevator Evacuation

As mechanical and electrical equipment, elevators are subject to unpredictable stoppage. There is no hazard if such stoppage takes place at a floor with the doors open and little hazard if the elevator is at floor level with the doors closed, since they can be opened with an emergency unlocking device and people escorted out.

The greatest hazard takes place when the car is stopped above a floor or between floors. People in the car will often force the car door open, manipulate the hoistway door lock, and, in attempting to exit, slide down and under the toe guard area of the elevator into the open hoistway. This has occurred often enough so that safety authorities are considering requiring doors to be locked except at a short distance from floor level.

Supervised evacuation is essential and can avoid many of the hazards and fatalities. If the car is in a portion of the hoistway without doors, an elevator can be brought alongside the stalled car and passengers aided through the top or side emergency exits. If the elevator is in a single blind hoistway, the A17.1 Elevator Code requires that emergency access doors be provided each 36 ft (11 m) so that a ladder may be used to access the car top and effect escape.

The admonition is that a person in a stalled elevator remain there, summon help by means of the alarm bell, intercom, or telephone, and wait for aid. It is safer to remain in the stalled elevator than try to leave unaided.

SUMMARY

The foregoing chapter simply touches upon the many aspects of elevators beyond their mechanical and electrical installation. These considerations may be of small consequence in a small building in a suburban location, but for the taller, more unique structure they may be of serious consequence. Early consultation with knowledgeable elevator consulting and design engineers is essential to avoid the hazard and cost of redesign and retrofit as well as the intangible cost of having an elevator out of service or not available when needed for an emergency.

CHAPTER SEVENTEEN

Elevator Specifying and Contracting

PREPARING TO BID

Once preliminary designs of a building have been approved, the task of preparing contract documents consisting of drawings and specifications for the various building systems is initiated.

As a current practice, the structural steel and elevator contracts are the first to be bid and awarded due to the long lead time of both. This requires that the elevator layouts be firmly established and that the architect has worked with the elevator consultant so that adequate space is available.

Space allocation is an extremely important aspect of elevator contracting especially if some innovative design is expected to be used. Outside observation-type elevators, underslung elevators, and even building elevators of a size different than current manufacturers' standards need to be planned and reviewed with knowledgeable individuals. One project had to be completely redesigned because an assumption was made that an elevator would fit; foundations were started, steel ordered, and when the elevators were bid the discovery was made that the space and location of the counterweights for the elevators could not be built as shown.

Once all the variants of space allocation both in plan and elevation are finalized, elevator specifications can proceed. Elevator bidding may take two forms, scope or final design. Scope is usually for budgetary reasons and generally proceeds from a simple outline of the project. The scope specifications are intended to be refined after working drawings are completed.

SPECIFICATIONS

Once the method of bidding is determined as either a scope or a complete basis, preparation of specifications may proceed. These must be complete if final bidding is expected or may be a simple outline if the bidding is preliminary. The areas that must be covered in the specifications are first, special conditions of the particular project, such as restrictions on the contractor, methods of payment, insurance requirements, necessary reports, and safety precautions. These are expected in construction projects and many standard forms of terms and conditions have been issued.

Secondly, the scope of the work that includes any services the elevator contractor is expected to offer is covered, such as the type and extent of post-installation maintenance, availability of elevators for temporary use by the building before actual completion, particular hazards, plus anything that may be out of the ordinary in the particular installation. Included in the scope are the owner's responsibilities. This includes construction of the hoistway, any patching and cutting, setting of inserts for rail brackets provided by the elevator contractor, and providing power for tools and for the preliminary operation of the elevator, as well as any other item that is normally the owner's responsibility or may be so on the particular job. Statements as to the procedure of approval of shop drawings and samples of finishes should be included in this section. The scope also includes an outline of the elevator equipment required. This includes capacity, speed, rise, number of entrances, type of machinery expected, signals, type of door and door operation, cab description or dollar allowance for a cab design,* plus any items considered necessary for the particular job. The foregoing is the first main section of an elevator specification.

The second main section of an elevator specification includes a description of various equipment components: machine installation, rails, pit structure, fixture description, operating panels, intercom system, cab specification or elaboration as to what the allowance does or does not include, details of entrance construction, plus any special entrance fire test or installation requirements. Any item that includes the installation of a fixture or equipment should be included in this section. If a specific operation is desired, any hardware should be specified in section 2 and the expected operation detailed in section 3.

The third section of an elevator specification should be devoted to operations both basic and special. The individual elevator or group elevator operations should be generically described and further elaborated to specify the exact features desired. For example, the A17.1 Elevator Code defines Group Automatic Operation but does not describe program and operating features such as up peak, load dispatch, load bypass, two-way traffic, down peak, parking of elevators, and other desirable operations. Unless the specifications call for such features the individual elevator contractors have a perfect right to provide what they consider their standard.

---

*An allowance for a car enclosure is a means of deferring the decision on the architectural treatment or design, just as a decision on the colors to paint the doors or other features could be deferred. Because elevator equipment takes considerable time to manufacture before actual installation begins in a building, details of visual treatment may not be firm when specifications are first being prepared. By using an allowance, the probable cost of the installation is established and the architect is allowed latitude in final design. If what is designed costs more or less than the allowance the difference is negotiated at the later date. Allowances can also be used for lobby floor entrance assemblies or for special signal fixtures.

Special operations must be spelled out in detail. Fire recall may be required by local codes and must be furnished by the elevator contractor. What may be lacking from the local code is the number and extent to which each elevator must be equipped or any additional feature that may be desired. These aspects must be distinctly specified. In computer jargon, section 2 of an elevator specification can be called "hardware" and section 3 "software."

Section 4 of an elevator specification is entitled "Execution" and should describe details of painting, electrical installation, tests, and final acceptance procedures. Any special wiring such as shielded intercommunication and telephone systems wiring, and coaxial cable for TV cameras should be included in this section. Painting specifications should include any extraordinary painting that is to be done by the elevator contractor such as the divider beams in the hoistways. Tests may include recorded accelerometer tests to be performed to qualify the ride characteristics of the elevators.

Section 5 will describe any alternatives required to the base elevator bid or added to the specifications if, for example, the base bid price is low enough to warrant extra expenditure on higher-speed equipment. An alternative may be used to call for additional equipment that may be desirable but not necessary. Alternatives may be requested should the owner decide to build part of a facility now and the rest at a future date. In any event the alternative should clearly state the intent and, if for future installation, the expected delay.

Often times it is desired to provide an illustration to supplement a word description of an operating panel, fixture, or method of installation. This can be done in a section 6 entitled "Illustrations." The paragraph in the other sections should indicate a reference to the illustration such as "see illustration, section 6." It is best not to reference illustrations by figure numbers since any change adding or subtracting paragraphs may destroy the sequence and result in a multitude of changes.

Escalator and material-handling system specifications follow the same basic outline as described for elevator specifications.

CONSTRUCTION SPECIFICATION INSTITUTE

A national organization called the Construction Specification Institute in the United States has been active in organizing specification writing and attempting to develop standards for all specification writers. Their basic specification is in three parts, Part 1. General, Part 2. Products, and Part 3. Execution. This three-part arrangement is very satisfactory for most building trades but with elevators, operation is the main goal of the installation and should be treated separately. These operations are also often affected by other building systems and close coordination or interface needs to be considered. Elevators have unique operations related to the people in the building, the electrical system, and the life safety system. The elevator specifications require the additional section describing the operation expected.

The Construction Specification Institute has set aside Division 14 as the Division used to describe Conveying Systems. The general headings assigned to elevators, escalators, and material-handling systems are as follows.

14000—Conveying System—General

14100—Dumbwaiters

14200—Electric Elevators

14700—Escalators

Section 14200 must be further broken down if parts of the elevator are separated. For example, in some localities it is an accepted practice to let separate contracts for the elevator entrances and car enclosures. For this purpose, sections 14220 for the doors and 14250 for the cabs are created.

The use of the Construction Specification Institute format will depend upon the preference of the building architect. In many areas, 14.1, 14.2, etc., are used and in others, 14A, 14B, 14C. At least Division 14 seems to prevail.

CODES AND STANDARDS

The universally accepted safety code for Vertical Transportation in the United States is in the A17.1 Code and its official title (in 1981) is The American Standard Safety Code for Elevators, Escalators, Dumbwaiters and Moving Walks. Copies may be purchased from the ASME, 345 East 47th Street, New York, New York 10017. Many localities, basic building codes, the federal government, and others use this code as the minimum basis of elevator and escalator requirements. It is continuously updated and either supplementary or new editions are issued annually. The code is administered by a main committee that meets annually. All changes are subject to public review and any question or additional interpretation will be answered by writing to the ASME at the above address.

The A17.1 code is an advisory code and must be adopted by a locality before it can be enforced as a building code. The city and state elevator inspectors are charged with code enforcement and will usually use a companion publication, the A17.2 Inspector Manual, as a guide in their elevator inspections.

Some localities have written their own codes often based on the A17.1 rules. Unfortunately, these local codes are often not updated to reflect current practices and particular attention must be paid to the local code since, even though something may be new and better, its use might not be allowed because of local requirements. Of course, the reverse may be true. Local code authorities can react to situations faster than national bodies with their extensive review process. This was especially true when fire safety rules were developed and caused the comment, "A fire is different in every political subdivision," which reflects the many different approaches local code authorities used to establish rules for elevator recall.

In Canada, the B44 Code is the counterpart to the A17.1 Elevator Code and liaison members are on both the A17.1 and B44 committees so there is a close correlation. In the Common Market countries of Europe, the CEN81 Code is being developed with the finalized versions of the elevator and escalator sections currently (1981) available. The CEN code is not as extensive as the A17.1 code since it is relatively new and European countries do not have the extent of high-rise construction found in the United States and Canada.

PERFORMANCE CRITERIA

If the building had been elevatored on the basis of performance criteria as described in Chapters Three, Four, Five, and Six, these criteria should be included as part of the elevator specifications. Lesser performance will usually require less advanced equipment and may cause the building's elevator service to suffer. Such criteria should include door time, floor-to-floor time, dwell-time, plus any other measurable quantities desired.

Bidders should be qualified before they are invited to bid; they should be able to demonstrate their competence to install equipment of the nature required. The architect or owner's representative should investigate all the installations cited as qualifying examples. If elevator service in a particular building is expected to be critical, as in a hospital or hotel, the bidder should also have adequate maintenance facilities in the immediate area. Failure to qualify on this point may result in each shutdown requiring excessive travel time for which the owner will have to pay and be without elevator service.

PROPOSAL BIDDING

The alternative to specification bidding is to outline briefly the elevator requirements and request various elevator concerns to submit a detailed proposal describing what they intend to furnish. This requires evaluation by the owner of all the criteria of a suitable installation.

Unless the invitation to bid specifically states that the lowest bid will be accepted, the owner is under no obligation to accept a low bid. This has been established by litigation and the term "lowest responsible bidder" is often used. It should be cautioned that, with proposal bidding, the items not contained in the proposal may result in unexpected expense.

NEGOTIATION

Often elevators are not bid separately but as part of the general contract for construction of a building. The general contractor then awards the elevator subcontract, seeking those who promise to meet the plans and specifications at the lowest price. If the owner does not approve of the elevator contractor who submits the lowest bid, the owner may have to add to the general contract the difference between the lowest bid and that of the preferred elevator sub-contractor.

No matter what plans and specifications are used to describe the elevator work, their scope and completeness provide only a partial guide to a good elevator installation which depends, essentially, on the reputation and the ability of the installer. For this reason it behooves the architect and owner to exercise care in choosing invited bidders.

ESTABLISHING A CONTRACT

When the bidding negotiation is complete a contract is established. This may be a letter of intent until final details of terms and conditions are agreed on, a signed proposal submitted by the elevator company, or the owner's signed contract. The work of manufacturing and installing the elevator equipment can then begin.

One of the first steps is taken by the elevator contractor in preparing and submitting a layout drawing for approval. This layout should show details of the elevator installation in the building in the space allocated coordinated with the building's plans. The architect, elevator consultant and structural engineer must check the layout with the contract documents and approve it. The general contractor must coordinate this drawing with other trades and allocate prepara-tory work that must be accomplished. Once the layout is approved the elevator contractor can get the necessary building permit from municipal authorities to proceed with the installation. Simultaneously, the owner is expected to confirm the power to be supplied by the local utility in the completed building.

Once an approved layout and power confirmation is obtained, the elevator equipment is manufactured. The elevator company's construction superintend-ent will arrange for the rail brackets or inserts to be placed while the building is rising. At the proper time rails will be delivered and installed and the elevator installation will proceed.

Meanwhile, other drawings are submitted to the architect or owner, includ-ing approvals for operating fixtures and finishes, cars and entrances, and any other items having optional treatment. By a certain time the elevator contractor will need power to start up the elevator and complete the installation. This will be well before the elevator is complete because the moving platform is often used as a scaffold to install the doors and to finish up interior hoistway equip-ment. The electrical contractor must supply power in the machine room plus wiring to the elevator controller. Additional circuits for car lighting, intercom-munication, fire safety, and other functions must also be provided in the ma-chine room.

TEMPORARY OPERATION

The general contractor or the owner may wish to use an elevator for tempo-rary service long before the building is complete. This understanding must be established at the time of bidding and the time of temporary service determined, usually a stated number of days or weeks after the hoistway and power become available. The extent of temporary operation may be such that the elevator contractor has to provide a temporary hoisting machine or the general contractor provide a temporary location for the hoisting machine at some point below the permanent machine room level.

Labor contracts have recognized that construction workers should not be required to walk up more than a certain number of floors.This requirement is met by a temporary elevator, usually the regular elevator platform with a temporary plywood cab and wooden doors on the hoistway. An operating safety test should be performed and the temporary operation licensed. To allow work to proceed on adjoining elevators the hoistway is enclosed in wire screening or otherwise protected. The temporary elevator is operated by an attendant and often stops only at every third or so floor.

The number of floors to be served by such temporary elevators must be established early because concentrated effort is usually required to get the equipment on the job and running. The number of temporary elevators is based on the number of people expected to be on a job at a particular time. Calculations can be made to determine how quickly people can be moved and elevator requirements established to meet the temporary needs.

ACCEPTANCE

Once a permanent elevator is complete it may be turned over and accepted by the owner, at which time an elevator being used for temporary operation can be completed by the elevator contractor. Elevator constructors are often the earliest on the job and the last ones to leave, starting with the foundations and finishing only when the tenants are moving in. To complete their job, elevator installers must have all the elevators in a group running so that group operations may be adjusted. With the general contractor wanting to use a car to complete construction work, tenants wanting to use one to move in, and passengers requiring the others, the final stage of completion is often drawn out.

The contract is essentially complete when the architect or owner accepts each individual elevator. By this time a regulatory authority has made a safety test and the elevator has been certified. The contractor has made a check and items of deviation noted have been cleared up. The elevator is turned over when the owner signs a final acceptance, at which point the new installation service begins.

NEW INSTALLATION SERVICE

New installation service is a maintenance service included in the original contract. The elevator installer agrees to maintain the equipment for a minimum period of three months, making necessary adjustments and seeing that the elevator is operating properly. During this period the management of the building is expected to make arrangements for continued maintenance of its equipment, either by its own staff or by a maintenance contract with the elevator company. In either event, if a major part failure occurs, its replacement is warranted by the elevator installer for a usual period of one year, based on normal use of the elevator and reasonable maintenance. For this reason it is wise to include a year's service with the initial contract.

Since elevators are completed, turned over, and accepted over a considerable period of time, an actual date for the start of new installation service for an entire installation can be controversial, especially when the initial service period is concluded and a formal maintenance contract commences. This problem is overcome by establishing an equalized final acceptance date by gives and takes. For example, if one elevator is completed 30 days before a date, one completed on a date, and the other 30 days after a date, an equalized final acceptance date could be easily established on the middle date. If one year's initial service was included in the original contract, an equalized final acceptance date would be essential to start the continuing elevator maintenance contract. In a major complex, the equalized price is based on the daily maintenance cost per unit compared to the total monetary amount in the total project included for New Installation Service. This amount would have been stated in the original contract.

ACTUAL ELEVATOR CONSTRUCTION

The foregoing sections simply highlight the process of installing an elevator in a building. The actual work depends on close scheduling with the other parts of the building so that an elevator installation proceeds smoothly and without interference with other trades. Step-by-step installation of an elevator has been thoroughly outlined in *The Elevator Erection Manual.** Designed primarily as a handbook for elevator constructors, this book covers many of the considerations necessary for installing elevators. These consist of reading elevator layouts, rigging and hoisting of equipment, fastening to steel and concrete, setting brackets and rails, erecting the car frame and platform, setting the machine and overhead sheaves, roping and electrical wiring, plus all the other facets of an elevator installation.

Coordination and timing is the key to the success of any construction project. To this end various means of scheduling are used, the most common being Gantt charts and the Critical Path Method, or CPM.

Gantt charts are simple bar graphs charted against a time base and showing the time required to deliver and complete necessary steps in an elevator installation. CPM is a more sophisticated approach and requires constant updating, usually by utilizing a computer. Scheduling specialists are used by contractors

* Available from Elevator World Magazine, PO Box 6506, Mobile, Alabama 36606.

to maintain construction schedules and the information they provide needs to be updated by inputs on actual field and material delivery conditions. Specialized texts on such procedures are available.

Each project is totally different with its complexity, size, and height determining the time required. As can be imagined, installation scheduling for a multigroup elevator project will be exceedingly complex compared to a smaller system. Similarly, an apartment house elevator is a relatively simple situation and formal schedules will probably not be prepared as they can be discussed verbally and depend on a hoistway ready date. In a commercial or industrial building, however, where considerable tie-in with other trades may be necessary, formal scheduling may be essential and should be considered before bidding, possibly with dates and delivery time included in the contract terms.

APPROVALS

One of the essential aspects of maintaining an elevator installation schedule is prompt approvals. These include layout drawings, design of fixtures, entrances, cabs, color selections, and the myriad choices that must be made to make the elevator appear and operate as desired. Although it may be difficult for an owner to decide a year or so in advance what type of metal fixture is wanted in the building, these fixtures are essential parts of the elevator and so must be ordered early. For example, the landing button must be mounted in an electrical box imbedded in the corridor wall. Because the ultimate style of the landing button fixture and its mounting box must be coordinated, the design must be firm long before the walls are built.

Similarly, hoistway doors are generally made with a factory-applied baked enamel finish. Because the doors must be installed before the locking mechanism, door color is one of the earliest approvals. This can be deferred by installing elevator entrances with a prime finish only and the final finish and color are applied when the building is complete.

With any special elevator equipment such as an outside glass cab, the manufacturing process may be six months or longer in addition to on site installation. No one will start manufacture before an approved design is received and a glass cab usually must be completely custom built and, often, must be final fitted by hand. In addition, the owner may wish to see the final product mocked up in the cab manufacturer's shop. After disassembly, packing, and shipping, six or more months may elapse after the drawings are approved. Once at the site, the elevator contractor may require a month for erection and attachment of all the operating accessories before the final adjustment of the elevator is accomplished. The time lapse may be a year or more to get a cab delivered on an elevator which, if it did not have the glass cab, may have taken only six months to complete.

DELIVERY SCHEDULING

As plans for a building become firm and actual construction begins, the need to have certain items installed at certain times becomes apparent. Some equipment, including elevators, may be critical to the overall progress of the job and require considerable study and consultation with manufacturers' representatives to determine when items are necessary.

As an example, an observation tower is to be built for an exposition expected to open at a specific date. At least two years before that date the design of the tower must be made firm. Its essential elements are the structure and the elevators, because the tower must be built before the elevators are installed and the elevators must be installed before the furnishing of, for example, a restaurant on the top can proceed. The architect, in planning the tower, must consult with structural steel and concrete contractors to determine a practical delivery and erection schedule for the tower itself. The architect must likewise consult with the elevator manufacturer to determine delivery and installation times for the somewhat special equipment the tower requires. Let us assume that the tower structural work will require a year before elevator machinery can be set on the top.

The elevator manufacturer may require a year before the design and delivery of the necessary machinery can be accomplished and, say, six months to complete its installation. No matter how many people are placed on the job, certain time requirements are irreducible because people can work only on one part at a time. The machines, for example, cannot be finally set until the rails are installed.

If the elevator contract is delayed for six months, or until 18 months before completion of the project, no time is left for contingency and the schedule is critical. If the delay is longer the elevator project must be considered for overtime and costs will skyrocket. If further delay is encountered it may be impossible to complete in the given time; no inducement will make it possible.

The first essential for any construction or elevator installation is to establish a realistic original schedule. To avoid delay during the progress of a job two courses are open. One is to provide the schedule and the necessary personnel to coordinate all trades to maintain such a schedule, which is accomplished by progress reports, follow-up, expediting, and prompt approvals by all concerned. This procedure is usually the most satisfactory and is used by the knowledgeable contractors in the field.

The second means is to establish an expected delivery date a given number of days after receipt of the contract and impose a charge on the contractor if the date is not met. If the time is reasonable and the charge is not excessive, the elevator contractor can accept, provided liability is limited to only those delays subject to control and responsibility. If the time is too short there are three choices: to refuse to participate in the job; to include, if possible, the necessary overtime to complete within the time allowed; or to estimate how many more days the job will take than the allotted time and include the cost of these liquidated damages in bidding.

The architect can rest assured that, if such a job is accepted, every delay, every inaccuracy, each interference between trades, delays due to weather and labor difficulties, and so on, will be recorded, and any delay, including time to receive approvals, will be imposed to extend the introduction of a penalty. The cost in paper work and argument may far exceed the gain expected by the penalty. As may be surmised, it is far better to schedule a job properly and make necessary allowances for contingencies than to try to substitute penalties for poor scheduling.

RELATION OF ELEVATOR WORK TO OTHER TRADES

Because an elevator, escalator, materials-handling system, or moving walkway becomes an integral part of a building when it is completed, considerable coordination is required between equipment installation and construction of the other parts of the building. Recognizing this, the general contractor should keep track of all the trades so that interference is minimized and the work flows smoothly. It also behooves the elevator installer to get equipment in place at the proper time so that the other trades will not interfere.

For example, in a steel building, the rail brackets should be set and fastened before the beams are fireproofed. If not, the elevator contractor may have to cut away concrete before the brackets can be fastened. The machine beam supports must be in place before the elevator machine beams are set. If the other trades fail to complete this task, the installation is delayed. Unless the elevator contractor gets door sills in place before the finished floors are poured, considerable cutting and patching will be necessary. Similarly, boxes for landing button and lantern fixtures must be set before walls are built. Proper power is necessary early enough so that the running elevator can be used.

When the job is ending a considerable amount of cleaning up must be done. Gypsum dust and loose masonry abounds and must be cleaned up. The hoistways can act like flues everytime an elevator hoistway door opens and draw in dirt and dust from every floor. As the air conditioning in the building is balanced, door locks must be adjusted, for the conditioned floors will be at a different air pressure from the hoistway and stack effect will tend to hold doors open.

In a large building the elevator contractor may have left the system in a state of adjustment designed to meet the initial traffic conditions. As the building becomes occupied and tenants' habits become established, additional tuning to adjust performance to traffic is necessary. This may include adjusting the load weighing that causes elevator dispatch and bypass operation, the time when peak traffic operations are established, and the traffic responsiveness of the group of elevators.

If the building is slow in being occupied these adjustments must necessarily be delayed—a condition that emphasizes one of the advantages of having the elevator manufacturer maintain the equipment on a contract basis. Adjustments that must be made will then be done at the necessary time. The alternative, with owner maintenance, is either to have the owner's mechanic do it or issue a separate order to the elevator manufacturer.

Representatives of the elevator manufacturer should call on the building management as the building is being occupied. They should give the owner instruction for maintenance, parts leaflets, aid in overcoming operational problems, and all the services necessary to make the installation a credit to the manufacturer as well as the owner. A reputable elevator company should strive to satisfy a customer so that they will recommend the installation to others in the market for elevators.

CHANGES AND ALTERATIONS

Occasionally an unpredictable development may require interrupting progress of an installation to make a change. The variety of such changes is immeasurable. A prospective tenant of the building may want an additional entrance on the elevator, floor designations may be changed to suit somebody's fancy, and additional control features may be warranted, as well as a host of other factors.

If the need for a change becomes apparent early enough it can often be made without delaying the job. If the change is late, scrapping of ready-worked material or extensive changes to installed equipment may be necessary. Needless to say, changes should be carefully investigated to limit them to only those features that involve minimum scrapping or rework.

Once equipment is installed it may be desirable to add further operating features, possibly to handle a traffic situation unanticipated at the time of planning or to meet the changed requirements of a particular tenant.

With microprocessor operation of the elevators such changes can be accomplished with minimum interference to interim building operation. The equipment is designed to be reprogrammed and elevator manufacturers with good service organizations should be able to make such program changes in a short time. Hardware changes are another matter. New fixtures or changed signs may have to be ordered and, possibly, custom built.

CONTRACT COMPLETION AND ACCEPTANCE

If the owner or the architect has engaged a consulting elevator engineer, that engineer will play a considerable role in the entire elevator design, specification, contracting, review, and acceptance procedure of the elevator installation.

The consultant has written the specifications; hence, the review and approval of shop drawings are tasks that should be performed by the consultant.

When the job is considered complete and ready to be accepted, the consultant should make thorough observations of the equipment and installation practice as an initial step. Any discrepancies in the installation of the equipment and in the interior condition of the hoistway should be noted and recorded in a job report which then becomes a punch list for the elevator and general contractors. When the elevators are operating, the consultant should be a witness to tests by local authorities for elevator licensing and may require additional testing to assure the owner that the installation meets the building requirements.

Operational tests need to be made and the performance times observed and recorded. Any interfaces with a life safety system need to be tested. For example, the smoke detectors in the elevator lobbies should be tested to ensure they provide a signal so that elevators are recalled.

The riding characteristics of the elevators need to be observed and tested. This consists of taking a number of trips and noting any unusual noises or bumps. If the elevator appears reasonably smooth, further testing is accomplished by use of a recording accelerometer to provide a visual record of the ride. A second such test should be made about a year later after the building has gone through a heating and cooling cycle, has become reasonably occupied, and, hopefully, fully compressed. The first and second readings need to be compared and corrective action taken, if necessary, to equalize them.

When solid state control has been provided on the elevator, attention may have to be paid to any complaints of electrical line pollution in the building which may affect tenants' computer equipment. Necessary corrective measures such as the installation of filters may be required. Similarly, if structure or air borne noise from the elevator equipment is apparent, corrective measures are required. Power factor correction capacitors may substantially reduce adverse electrical line pollution.

If the installation has a standby generator designed to provide emergency operation of the elevators, a thorough test of that system needs to be accomplished when all the systems are ready. Generator tests often need to be scheduled for middle of the night hours since during the test, practically all the vital buildings systems are shut down for a period of time.

In addition to the necessary physical tests the elevator consultant usually works with the owner to aid in the establishing of a maintenance contract and to help solve operational or elevator programming problems that may have been unforeseen. These may include problems due to building sway and wind as described in an earlier chapter, or material and mail handling problems if such systems were part of the elevator contract.

SUMMARY

As may be surmised elevator specifying and contracting, plus the acceptance of a completed elevator installation can be an involved process. Similar requirements will exist for escalator, materials handling and moving walk installations.

Experience is of unestimable value. If the elevator contractor has a large contract and appoints a knowledgeable project manager as a representative, a first essential step is taken. The project manager should be conversant in all aspects of the installation, be the source of information, attend the job meetings, and bear the brunt of the expected trouble. A great deal of effort is always required to create a successful installation.

CHAPTER EIGHTEEN

Economics, Maintenance, and Modernization

COST

In providing services for a building there must be two cost considerations. One is the cost of the initial construction and the second is the return from investment over the economic life of the building.

In elevatoring any building the cost-return considerations start with the initial plans. Many factors external to the building itself establish the criteria. It is very seldom that land is unlimited and that any type of building, low-rise or high-rise, can be chosen for a given site. Zoning regulations must be complied with and the restrictions of land use on a particular plot must be related to the building height. In areas where land use may be unrestricted, needs for parking and access to parking must be considered. The cost of walking long horizontal distances, in terms of time consumed, should be compared with vertical travel by elevator or escalator.

A functional analysis of the building must be undertaken. Related facilities and the communication between them must be considered. To what extent must people performing their services, such as doctors in a hospital or executives in a business, come together to confer face to face, or can they use appropriate means of electronic communication, possibly including two-way television? Do materials have to be moved from person to person or department to department or is the process one of in and out? Storage functions must be considered. For example, can materials be stored in a vertical plane to be moved into production lines when the demand arises, or must each item be processed continuously?

In formulating the initial plans the foregoing alternatives are usually considered. As with any plan, the approaches taken may represent compromises because there is often more than one way to solve any problem. In elevatoring a building the final result will generally be one of compromise between requirements of initial cost, space consumption, service quantity, and passenger-time conservation.

MULTIPLE APPROACHES TO ELEVATORING

In earlier chapters of this book various elevatoring questions and possible solutions are discussed. They include: use of high-rise and low-rise elevators rather than a group of elevators making all stops; escalators rather than elevators in low-rise buildings; separating vehicular and pedestrian traffic on elevators; providing shuttle elevators for garage levels; double-deck elevators; sky lobby arrangements; and other schemes. Their objective is either maximum-handling capacity with acceptable riding and waiting time or, if a long trip is in prospect and handling capacity secondary, minimum waiting and riding time.

ALL STOPS VERSUS LOW-RISE AND HIGH-RISE

In the foregoing chapters there are some very clear-cut examples. The selection of high- and low-rise elevators or a single group of elevators exemplifies the multiple effects of a choice on building design and operation. The two arrangements are shown in Figure 18.1. Either plan with a total of eight elevators can serve a building 18 to 22 floors of about 14000 sq ft/fl. Arrangement (a) is accomplished with a single group of eight elevators which, if they serve 22 stops, have a total of 176 hoistway entrances. All eight elevators will have the same speed; for a building of this height, about 700 fpm (3.5 mps). Arrangement (b) has two groups serving floors 1 through 12 and the high-rise group serving the first floor, skipping floors 2 through 12, and serving floors 13 through 22. These eight elevators will require only 92 entrances. The speed of the high-rise elevators will be the same as arrangement (a), 700 fpm (3.5 mps), but the low-rise can be lower speed, 500 fpm (2.5 mps), because travel distance is shorter. If we assume that the size of the cars is the same in both cases, say 3500 lb (1600 kg), the local and express arrangement will require less building space.

From the passenger's aspect the single eight-car group will have one advantage and one disadvantage. The chances of getting an elevator will be high because any passenger can take any of the eight elevators and waiting interval will be short. Once the passenger boards an elevator the ride may take longer with more interruptions, as the tendency to make stops is increased with 22 floors served.

The building management has a number of advantages. Once the peaks are over it is a relatively easy matter to detach a car for maintenance or service work. Lobbies are in the same place throughout the building. There is only one motor room location and only one group control system, although somewhat complex.

Two four-car groups give passengers both advantages and disadvantages. Although each passenger has only four rather than eight chances of getting an elevator, once the passenger boards the car the trip is swifter because there are fewer possible stops. The elevator may also be less crowded because all people in the building cannot ride all cars, some taking the high-rise and others the low-rise elevators.

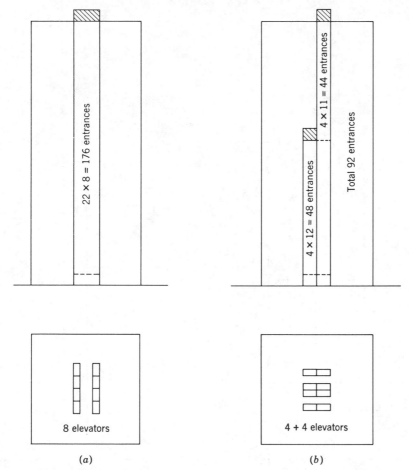

Figure 18.1. Eight elevators serving all floors versus four low-rise and four high-rise elevators. Approximate space required by the elevators and elevator lobbies. (a) 30,000 ft^2 (2800 m^2). (b) 22,600 ft^2 (2075 m^2).

The building suffers a few disadvantages. It is more difficult to take a car out for maintenance or for service repair work because the remaining group is reduced to three cars. Two groups of elevators require two group operating systems but each may be less complex than the eight-car system. Initial investment and maintenance costs are lower. Flexibility in tenant location and expansion is limited. With the eight-car group all floors of the building are immediately accessible, whereas the low-rise, high-rise arrangement imposes obvious restrictions. The local and express arrangement takes less space for hoistways, fewer openings and the space released is on upper floors where its rental value is often higher.

This comparative analysis applies in principle to larger buildings. Figure 18.2 shows two arrangements for a possible 30- to 40-story building. Based on given dimensions, the arrangement with three groups of six elevators, 18 elevators in

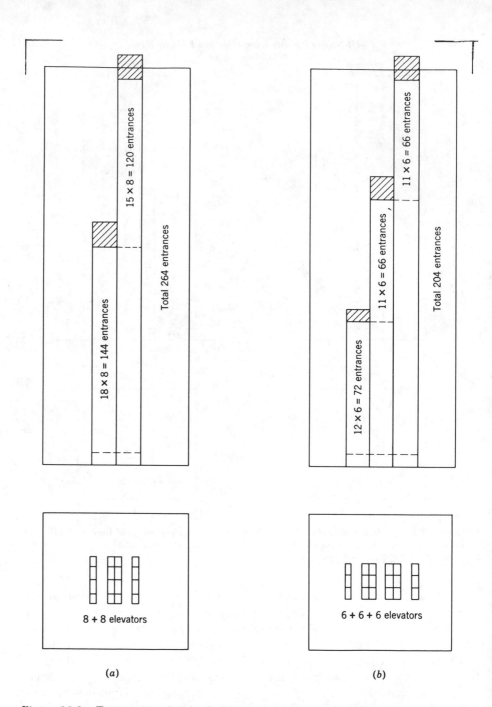

Figure 18.2. Two groups of eight elevators versus three groups of six elevators for the same building. Approximate space required by elevators and elevator lobbies. (*a*) 60,000 ft² (5580 m²). (*b*) 58,000 ft² (5400 m²).

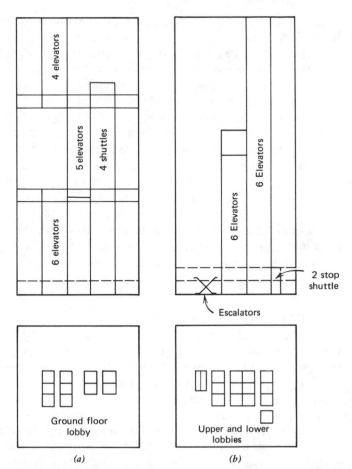

Figure 18.3. (*a*) Sky lobby arrangement with shuttle elevators to upper stops and local elevators from the two sky lobbies serving upper floors. (*b*) Double-deck elevators.

all, may be more economical from both initial cost and operating standpoints than two groups of eight cars each or a total of 16 elevators. Economy of the three-group arrangement stems from the reduced number of entrances, the reduced speed of some elevators, and the recovery of rentable space.

The comparison cannot stop there. Two other elevator schemes may be feasible for the building and need to be studied. One possible sky lobby scheme would be to create sky lobbies at floors 12 and 23. Local elevators would rise from those lobbies and the sky lobbies would be connected to the street by shuttle elevators serving floors 1, 12 and 23 only. Assuming all things being equal (theoretically), the six elevators are required for the low-rise as in the original scheme and since the express run and its time penalty of the mid-rise and high-rise is eliminated, the mid-rise locals may become five elevators and the high-rise locals four elevators. There is a rentable area recovery in the low-rise area, but also a rental penalty since additional public space is required for the sky lobby. The shuttle elevators will increase the total number of elevators in the project to at least 19 versus 18 and 16 for the prior schemes. The total cost of the installation may be considerably lower since lower-speed equipment for the lesser rises is being used. A thorough study of this scheme versus the two-group of eight or three-group of six schemes must be made. In addition the study must evaluate the recovered rentable space. This study can be compared to a scheme with only one sky lobby rather than two (Figure 18.3a).

A further study should be made of double-deck elevators for the building. Again, assuming all things being equal (theoretically) it appears that six low-rise double-deck and six high-rise double-deck elevators can provide vertical transportation for the building (Figure 18.3b). The savings in rentable area is obvious and, in general, amounts to about one-third of the area required by conventional elevators. The cost of double-deck elevators with the heavy machines and elevator structure required will be considerably higher. In addition to rentable space study, the budget costs of the various alternatives are important factors before a final decision is made.

STANDARD OR "MODEL" ELEVATORS

Elevator manufacturers are attempting to reduce the cost of elevator equipment for the lower-rise buildings by offering standardized equipment. This equipment is designed to be manufactured for stock and consists of components for standard-size hoistway dimensions, doors, door heights, cab heights, cab designs which have a number of design and finish options, operations, signals, and fixtures. By standardizing layout and ordering procedures, the manufacture and installation time is reduced and the customer can benefit from the economy of quick availability. The variations in what is being offered as "models" change constantly and the current manufacturers' literature needs to be appraised.

The highly competitive standardized elevator will require that the purchaser pay greater attention to the needs of the building. Since the equipment will often be sold by catalog rather than the personal service of the elevator company's representative, either the architect, contractor, or owner may pick the standard elevator that is appealing, with possibly minimal regard to application. The knowledgeable elevator engineer can recognize improper application and advise against it whereas a contractor or architect may buy simply on the basis of price.

For example, a particular building requires a four-car group of elevators operating with a Group Automatic system. Since a manufacturer's model is limited to two-car operation only, two two-car groups are purchased and installed. The landing buttons face the same lobby so a prospective passenger may register both landing calls and call two elevators at one time. Because of the reduced price of the two two-car model installation versus the much higher price of a nonmodel four-car group installation, a potential building elevator problem is created.

Attempts are being made to "modelize" elevators for high-rise installations of the higher-speed 500-fpm (2.5-mps) and 700-fpm (3.5-mps) gearless range. This can only meet with partial success for a number of reasons. Architects cannot be standardized, no two building plots are ever the same, zoning laws will impose design restrictions on the building, and all other building components are designed to be custom fitted. A viable means to support such a standardization program would be to standardize all elevator structural elements except platform dimensions, door size, and cab design. Flexibility in door height and design needs to be provided and a programmable operational control employed so that the elevators can be adapted to varying building project requirements.

ELEVATOR MAINTENANCE AND REPAIR

Once an elevator is installed and operating, periodic maintenance is required, and, as its use continues, replacement of wearing parts. Maintenance, which consists of ensuring the elevator is properly lubricated and operating, should take place as often as weekly in a heavily used installation, and as little as once a month in a lightly used facility. Replacement of minor wearing parts such as contacts on controllers is part of regular maintenance, and major replacements such as replacing wire ropes will need to be done at periods from 8 to 10 years, depending upon the use, condition, and design of the elevator. Failures such as the burnout of a motor, a bearing failure, or other repair items have to be corrected as needed.

Practically all elevator manufacturers offer a maintenance contract to provide maintenance care and needed replacement of their equipment. This manufacturer's maintenance is available in several forms, the most desirable and highest priced being called full maintenance (F.M.).

Full maintenance is a contract made between an elevator company and the building owner which requires the elevator company to periodically check the equipment and provide all necessary replacements and repairs. For a monthly fee, which is adjusted annually to reflect metals and labor indices, the elevator company will provide the necessary service to keep the elevators running.

Although the original manufacturers of elevators are very active in offering service on their equipment, there are many companies that specialize in service rather than installation. With over 400,000 elevators and escalators in the United States and Canada (in 1980), the market is quite extensive and with monthly maintenance prices from $50.00 to $600.00 or more per elevator, depending upon size and complexity, represents substantial dollar volume.

The company maintaining an elevator will write various forms of contracts ranging from a complete coverage for an unlimited term subject to annual price adjustment to a minimum contract which provides periodic lubrication, all other items being an additional charge. The buyer must use caution in dealing with the manufacture and understand what is *not* in the contract, rather than what is written. Variations in written words may be in the form of an elaborate list of what is included which may not mention vital parts, such as motor armatures, which, since they were not included, are excluded. Any failure of the "not mentioned" item can result in a substantial additional cost. (See section entitled Proposal for Elevator Maintenance.) Some companies, unfortunately, approach elevator maintenance as a means to get as much out of the installation until major repairs are imminent and then will cancel the contract or do such a poor job of servicing that the owner will cancel. In some instances, tactics, such as claiming that the equipment is obsolete and unserviceable and must be replaced at substantial additional cost, are used.

Switching from one maintenance company to another usually involves the new company's requiring that a contract for extensive premaintenance repairs be signed before the new maintenance is begun. In addition, the new company may want to prorate the major wearing parts of the elevators such as ropes, armatures, sheaves, and other items. Proration means that if any replacement is required during the term of proration, the owner and the elevator company share the prorated cost, which may be at the elevator company's inflated price.

There are no easy answers to good elevator maintenance but there are a few guidelines to sensible contracting. The owner should deal with a company of good reputation, well known in the area, and with a substantial financial rating. The owner should require that the maintainer check in on a regular basis with a person in charge, and the person in charge should periodically accompany the maintainer to observe the maintenance. An accurate record of elevator outages should be kept and early discussion should begin with the elevator company's supervisory personnel if frequent calls for service occur for the same problem, especially if they are after hours at premium billing rates. If major repairs are not included in the contract, the owner should reserve the right to a second opinion and the right to award the repair independently.

In general, for the smaller installation, the manufacturer should be given a preference in maintaining the equipment. For larger installations, the manufacturer should also be given a preference but the contract terms should go beyond the manufacturer's standard form. Items such as the number of hours the mechanic is to spend at the site, the billing rates for after hours calls, review process, process in the event of disputes, the replacement of an incompetent mechanic, credit for excessive elevator shutdown or excessively long times for repair and replacement, credit for elevators taken out of service, credit for light service on the elevators if the building is not fully occupied, plus a myriad of other items including insurance and indemnification should be part of the contract terms.

Elevator maintenance is billed at a monthly rate. The elevator companies count on sufficient maintenance business each year to write off the cost and to make a return on their investment in tools, manpower, inventory, training, and all the incidentals of doing business. There is no reserve built up to cover next

Figure 18.4. Before and after photographs of modernized elevator equipment. (*a*) Gearless hoisting machine with auxiliary leveling machine and brake. (*b*) Auxiliary leveling machine replaced by a new main drive machine brake. (*c*) Old and new elevator controllers. (*d*) Old and new elevator safety governors. (Photos courtesy of *Elevator World Magazine.*)

year's expenses. If one building does not require major replacement this year, it partially pays for some other building that does. This approach suggests that if there is a tax advantage to building up a reserve for future replacements for an organization, it may pay to arrange with the elevator maintenance company to just pay for labor expended and maintain an inventory of needed parts and pay for major repairs as required.

Maintenance contracts should include the costs of making safety and operational tests as may be required by governmental and insurance agencies. They do not include the cost of making any changes to the equipment either due to obsolescence or as may be required by new laws, such as the addition of a life-safety system. The rapid progress in elevator design and the changing taste in architectural treatment can lead to the need to completely modernize elevators at the age of 20 to 25 years. In fact, many buildings approaching age 50 are undergoing their second generation of elevator modernization (Figure 18.4).

PROPOSALS FOR ELEVATOR MAINTENANCE

The following two sections show examples of maintenance coverage. The italics highlight the differences.

Extent of Coverage

Under the terms and conditions of this agreement subsequently set forth we will maintain the elevator equipment described in this proposal, using skilled elevator personnel whom we directly employ and supervise. This proposal includes 24-hour emergency adjustment call-back service only at no additional charge.

We will systematically and regularly examine, adjust, lubricate as required, and, when we consider it necessary, repair or replace components including the following:

MACHINE	*Worm shaft bearings—thrust bearing—gear shaft bearings—* slack cable device—machine limit device—*regroove* drive sheave—gear oil—packing.
MOTOR	Armature—commutator—field coils—interpole coils— brushes—brush holders—rotor—stator windings—slip rings—motor bearings—motor oil.
MOTOR-GENERATOR	Armature—commutator—field coils—interpole coils— brushes—brush holders—rotor—stator windings—slip rings—motor bearings—motor oil.
BRAKE	Linings—magnet coils—linkage-bushings—contact springs— plunger—dashpot—adjusting screws.
CONTROLLER	Magnet coils—copper contacts—carbon contacts—shunts— springs—fuses—insulators—resistors—rectifiers—timing devices—solid state devices—overload heaters.
SELECTOR	Sheaves—drive chains—cables and tapes—contacts—floor bars—springs—shunts—reduction gears—selector carriage— advance motors—coils.

WIRING	Traveling control cables—hoistway control wiring.
HOISTWAY	Hoistway limit switches—car limit switches—speed governor—bearing boxes—governor tension sheave—interlocks—door rollers—buffers—*counterweight guide shoes*—bottom door guides.
CAR	Guide shoes—guide shoe linings—guide shoe rollers—levelling units—car door operator motor—belts—gears—cables—sheaves—gate switch—safety edge—photoelectric devices.
ROPES	*On a pro-rata basis*: hoist—drum counterweight—car counterweight—operating—governor—car safety rope.
SIGNALS	Pushbuttons—contacts—indicator lights—hall lanterns.

Extent of Coverage

Under the terms and conditions of this agreement subsequently set forth we will maintain the elevator equipment described in this proposal, using skilled elevator personnel whom we directly employ and supervise. This proposal includes 24-hour emergency adjustment call-back service only at no additional charge.

We will systematically and regularly examine, adjust, lubricate as required, and, when we consider it necessary, repair or replace components including the following:

MACHINE	*Worm gear—worm shaft*—worm shaft bearings—thrust bearing—gear shaft bearings—slack cable device—machine limit device—drive sheave—gear oil—packing.
MOTOR	Armature—commutator—field coils—interpole coils—brushes—brush holders—rotor—stator windings—slip rings—motor bearings—motor oil.
MOTOR-GENERATOR	Armature—commutator—field coils—interpole coils—brushes—brush holders—rotor—stator windings—slip rings—motor bearings—motor oil.
BRAKE	Linings—magnet coils—linkage—bushings—contacts—springs—plunger—dashpot—adjusting screws.
CONTROLLER	Magnet coils—copper contacts—carbon contacts—shunts—springs—fuses—insulators—resistors—rectifiers—timing devices—solid state devices—overload heaters.
SELECTOR	Sheaves—drive chains—cables and tapes—contacts—floor bars—springs—shunts—reduction gears—selector carriage—advance motors—coils.
WIRING	Traveling control cables—hoistway control wiring.
HOISTWAY	Hoistway limit switches—car limit switches—speed governor—bearing boxes—governor tension sheave—interlocks—*door hangers*—door rollers—*buffers—counterweight—guide shoes*—bottom door guides.
CAR	Guide shoes—guide shoe linings—guide shoe rollers—levelling units—car door operator motor—belts—gears—cables—sheaves—gate switch—safety edge—photoelectric devices.

ROPES   Hoist—drum counterweight—car counterweight—operat-
                         ing—governor—car safety rope.
SIGNALS   Pushbuttons—contacts—indicator lights—hall lanterns.

ELEVATOR MODERNIZATION

There are many good reasons to modernize an elevator installation and a few
wrong ones. The good reasons are that elevator efficiency and utilization may
be increased in view of continuing development in group operation, the introduction of more efficient door operation, and greater elevator reliability or the
changing nature of the building. The wrong reasons are that a new operating
system can save considerable elevator energy, or that some operating "gim-
mick" can give such a tremendous increase in operating efficiency that the
number of elevators required to serve the building can be reduced. This latter
point may be true but considerable traffic study and projection of eventual
building use need to be made first.

Condition Survey

Any consideration of an elevator modernization should begin with a thorough survey of the condition and traffic-handling ability of the existing elevator
installation. Oftentimes operating inefficiencies develop due to poor maintenance or, simply, the lack of a demand for a high standard of operation. Abuse
of the elevator system often contributes to poor service. In a hospital this may
be unsupervised use whenever elevators are held out of service while attendants
deliver material. In an office building abuse may stem from the use of a vital
passenger elevator for freight deliveries during a peak pedestrian rush hour.
Other deficiencies may be a result of excessive dwell-time (door-hold open time)
at floors, or simply misunderstanding the operating system where the elevators
have multiple programs and are locked on the wrong program because no one
ever investigated a change.

If the survey shows that the state of operation, maintenance, and traffic
handling are reasonably good, then the next considerations are the age and
expected life of the equipment plus the possible need for improved efficiency. A
plan of modernization should be developed and include considerations such as
providing a programmable operating system, life safety systems, standby generator operation, and modernization or replacement of cabs, fixtures, door assemblies, and other visual features. The condition survey should determine which
mechanical and electrical parts need to be repaired, modernized, or replaced.

Modernization Contracting

Modernization of a small installation may only require that an outline of the
work to be done be made and proposals solicited from reputable companies
which can be compared and a satisfactory contract established.

A major installation may require that bidding documents such as terms and conditions and material and performance specifications be prepared and bids solicited from qualified companies.

Such specifications should include a time schedule to show how many elevators can be out of service at one time and which ones. The specifications should also include a schedule of work that will be done by contractors other than the elevator contractor, which may include adding of ventilation to the machine room, new power feeders, standby generator, life safety system, additional floors to be served by the elevators and necessary construction work, plus many other possible items.

Job conditions must be outlined and include where material must be stored, removal of old material and receiving times of new material so that interference with building operations is avoided, safety barricades, conduct of elevator contractor's employees and facilities they can or cannot use, and their relation to building security, as well as any other aspect which may affect the relationship between the elevator contractor and the owner.

An elevator modernization specification must be more thorough than the specification for a new elevator since most building conditions exist as contrasted to a new building where they are created. Specific requirements must be made for material to be retained or removed. For example, any electrical wiring or other piping in the hoistways not needed for the modernization should be removed.

On a new building project many trades such as carpenters, bricklayers, marble workers, and others are present. On an existing building any cutting and patching, stone work, iron work, or electrical work is in addition to the elevator modernization contract and suitable arrangements must be made.

The Economy of Elevator Modernization

One of the elements of an elevator modernization is the opportunity to reduce operating expense of the building. With improved elevator service it may be possible to eliminate one or more elevators, thus releasing space for other building needs such as air conditioning ducts or new electrical feeders. This reduction in the number of elevators needs to be well established by traffic surveys and calculations of expected elevator performance, as well as by projections on building use and population. Very often, an older building has evolved from a high-density, single-tenant-type occupancy to a completely diversified occupancy with low density where individual offices have been created or where multiple firms occupy various floors as contrasted to single, large, multifloor operations.

If the building has attendant operated elevators, a great opportunity for reduction in operating expense exists. Each elevator requires at least one person to operate it and, with vacations, breaks required by union agreements, uniforms, locker space, lunch hours, and other benefits, each elevator may require the employee equivalent of two or more personnel. If 24-hr service is required, the operating cost alone can often pay the cost of the modernization in a reasonable time period.

The maintenance price of a modernized elevator may be higher than the original installation if a number of additional accessories have been added which will increase the required maintenance time and replacement parts. Such accessories may include power operated doors, landing lanterns, intercom systems, life safety features, standby generator operation, and security systems.

Existing Equipment

It is fortunate that basic elevator equipment, such as drive machines, the counterweight, the structural car frame and safety device, and guide rails seldom needs to be replaced. The motor armatures can be removed and rebuilt or this can be done on the site if necessary. All other wearing parts can be replaced or overhauled. In this way a major cost of removal and possible building renovation is avoided. This is especially true in buildings built after about 1910 when most elevator companies changed design from drum-type machines to traction-type. Even the drum-type machine is capable of a satisfactory modernization but the obsolescence and expense may be such that its replacement by a new traction machine is more cost effective. Elevator guide rails, if they are of metal and standard size, can always be reused.

ADDITIONAL CONSIDERATIONS

Escalators

The overall modernization of the vertical transportation system of an existing building may include the modernization of the escalators or the addition of escalators, or an automated mail and materials-handling system to improve the efficiency of the building.

Escalator modernization can be accomplished if the existing truss can be reused and new tracks, step chains, steps, driving machine, and architectural features can be installed in the existing truss. This is possible with some previously manufactured, currently obsolete escalators but due to age and design, may be totally impossible with others. For example, it would be impossible as well as undesirable to try to modernize an old Reno type escalator. It may be undesirable to modernize an early wooden-cleat-type escalator. The alternative is to remove the entire truss, rearrange the building structural steel, and install a completely new escalator.

One effective means of accomplishing this if building conditions permit is to completely assemble the new escalator off site and transport it to the building for moving in and installation on a long weekend. This is a common practice in Europe where most escalators are preassembled, and has been accomplished at a number of locations in the United States.

A pair of new escalators can be effectively used to create a second main lobby for a single tenant occupying the floors served by a group of elevators. With increased emphasis on security, this is an extremely viable approach since the entrance to an escalator is an excellent checkpoint. Escalators can also be effectively used to improve elevator service. Where buildings have two main entrances at different levels, causing all traffic to use a single lobby elevator entry floor can reduce elevator round-trip time by 20 or more seconds and provide a substantial improvement in elevator operating interval and handling capacity. Similarly, if a cafeteria is on a second or basement floor, eliminating elevator service to that cafeteria floor and providing escalators can also improve elevator efficiency. Consideration must be given to the needs of people unable to use escalators which can usually be accomplished by special operating features on the elevators.

Automated Materials Handling

The addition of an automated mail- and materials-handling system in an existing single-tenant building can be an extremely cost-effective and elevator traffic reducing step. A thorough survey of the number of people involved in carrying mail and supplies throughout the building must be made and the benefit of an automated materials-handling system calculated. This analysis must include the capital cost, maintenance cost, and the number of employee equivalents required or the number that can be reduced by the use of the new system. Various consolidations of mail handling, supply room, reproduction center, print shop, or other frequently used facilities must also be considered. For example, the difference in cost of having infrequently used reproduction machines at each floor versus the cost of an intensively used machine at a central location related to the cost of transportation of documents to and from that center needs to be determined.

Full-time Employee Equivalents

As mentioned previously, any job that requires an employee to be on duty full time requires more than one employee to provide the full-time coverage. This is true of any essential personnel, such as a security guard, elevator operator, or mail delivery personnel upon whom the day-to-day functioning of the business depends.

The determination of a full-time employee equivalent (FTE) is calculated in the following manner.

1. Actual working time per employee

Days per year	365
Less: Saturdays and Sundays	104

Vacation	10	
Sick time	3	
Personal time	2	
Holidays	11	
	119	119
Actual days worked		235
At 40 hr per week		× 8
Hours worked		1880

2. Efficiency: coffee breaks, wasted time, waiting for assignments. Estimated at 70% = 70% × 1880 = 1316, actual hours worked per employee.
3. If a job requires full-time attention, hours required 8 × 250 days (5 days per week less holidays) 2000 hr.
4. Therefore to fulfill the time requirements, 1.5 employees (2000 ÷ 1316) are required for each full-time job and the cost of a full-time employee equivalent is 1.5 times the cost of a single employee. Stated otherwise, it requires 1.5 employees to man a position that requires full-time manning.

For a mail-handling function, the number of hours required to perform the activity is determined by survey. The cost of the function can then be established by multiplying the hours required by the FTE. The budget cost of the automated facility plus the number of personnel hours required to service the automated facility is then determined and a study of the financing and a discounted cash flow is developed.

Figures 18.5a to d show such a study made to determine the cost effectiveness of a material-handling system requiring a capital investment of $1,200,000 but capable of replacing 7 full-time equivalent employees with an annual total payroll of $10,000 each including all benefits and indirect costs.

Vertical Transportation Safety and Liability

Elevators and escalators are judged to be the safest form of transportation in use today. Millions of passengers are carried each day and, even in the short distance of an elevator hoistway, the distance traveled can add up to a considerable number of miles daily.

As an example, based on manufacturers' statistics there were approximately 400,000 elevators in public use throughout the United States in 1980. Using this total to develop some logical but arbitrary statistics, consider the following.

1. The urban population of the United States is about 150,000,000 people. If 25% use elevators and they will generally use them at least four trips per day, in and out of a building at the beginning and end of either their

RETURN ON INVESTMENT ANALYSIS
MATERIALS HANDLING SYSTEMS

	1984 $(000)	1985 $(000)	1986 $(000)	1987 $(000)	1988 $(000)	1989 $(000)	1990 $(000)	1991 $(000)	1992 $(000)	1993 $(000)
A. Manual Delivery of Mail, Reproduction and Office Supplies by Cart to Entire Building										
Capital Cost	10									
Direct Labor With Fringes (41%) (9%/Annum Escalation of 7 Persons)	115	125	137	148	162	177	193	210	229	250
Total	(125)	(125)	(137)	(148)	(162)	(177)	(193)	(210)	(229)	(250)
B. Robot Vehicle With Elevator and Special Guide Path										
Capital Cost	383									
Maintenance Costs	40	44	48	52	56	62	67	73	80	87
Total	(423)	(44)	(48)	(52)	(56)	(62)	(67)	(73)	(80)	(87)
Annual Cash Flow Savings	(298)	81	89	96	106	115	126	137	149	163
DCF @ 14%	1.00	0.88	0.77	0.67	0.59	0.52	0.46	0.40	0.35	0.31
Net DCF	(298)	71	69	64	63	60	58	55	52	51
Cumulative DCF	(298)	(227)	(158)	(94)	(31)	29	87	142	194	245

(a)

Figure 18.5. Return on investment analysis of various automated material-handling systems vs manual delivery of mail and office material and supplies. (a) Robot vehicle vs manual:

RETURN ON INVESTMENT ANALYSIS
MATERIALS HANDLING SYSTEMS

	1984 $(000)	1985 $(000)	1986 $(000)	1987 $(000)	1988 $(000)	1989 $(000)	1990 $(000)	1991 $(000)	1992 $(000)	1993 $(000)
A. Manual Delivery of Mail, Reproduction and Office Supplies by Cart to Entire Building										
Capital Cost	10									
Direct Labor With Fringes (41%) (9%/Annum Escalation of 7 Persons)	115	125	137	148	162	177	193	210	229	250
Total	(125)	(125)	(137)	(148)	(162)	(177)	(193)	(210)	(229)	(250)
C. Robot Vehicle With SVC, Tote Boxes and Standard Guide Path										
Capital Cost	351									
Maintenance Costs	31	34	37	40	44	48	52	57	62	67
Total	(382)	(34)	(37)	(40)	(44)	(48)	(52)	(57)	(62)	(67)
Annual Cash Flow Savings	(257)	91	100	108	118	129	141	153	167	183
DCF @ 14%	1.00	0.88	0.77	0.67	0.59	0.52	0.46	0.40	0.35	0.31
Net DCF	(257)	81	77	72	70	67	65	61	59	57
Cumulative DCF		(176)	(99)	(27)	43	110	175	236	295	352

(b)

Figure 18.5. (*b*) Robot vehicle and tote box conveyor vs manual:

RETURN ON INVESTMENT ANALYSIS
MATERIALS HANDLING SYSTEMS

	1984 $(000)	1985 $(000)	1986 $(000)	1987 $(000)	1988 $(000)	1989 $(000)	1990 $(000)	1991 $(000)	1992 $(000)	1993 $(000)
A. Manual Delivery of Mail, Reproduction and Office Supplies by Cart to Entire Building										
Capital Cost	10									
Direct Labor With Fringes (41%) (9%/Annum Escalation of 7 Persons)	115	125	137	148	162	177	193	210	229	250
Total	(125)	(125)	(137)	(148)	(162)	(177)	(193)	(210)	(229)	(250)
D. Robot Vehicle With Carts and Lift Standard Guide Path										
Capital Cost	344									
Maintenance Costs	31	34	37	40	44	48	52	57	62	67
Total	(375)	(34)	(37)	(40)	(44)	(48)	(52)	(57)	(62)	(67)
Annual Cash Flow Savings	(250)	91	100	108	118	129	141	153	167	183
DCF @ 14%	1.00	0.88	0.77	0.67	0.59	0.52	0.46	0.40	0.35	0.31
Net DCF	(250)	81	77	72	70	67	65	61	59	57
Cumulative DCF		(169)	(92)	(20)	50	117	182	243	302	359

(c)

Figure 18.5.   (c) Robot vehicle and carts vs manual:

	1984 $(000)	1985 $(000)	1986 $(000)	1987 $(000)	1988 $(000)	1989 $(000)	1990 $(000)	1991 $(000)	1992 $(000)	1993 $(000)
A. Manual Delivery of Mail, Reproduction and Office Supplies by Cart to Entire Building										
Capital Cost	10									
Direct Labor With Fringes (41%) (9%/Annum Escalation of 7 Persons)	115	125	137	148	162	177	193	210	229	250
Total	(125)	(125)	(137)	(148)	(162)	(177)	(193)	(210)	(229)	(250)
E. Telelift System										
Capital Cost	1210									
Maintenance Costs	61	66	72	78	85	93	101	111	121	131
Total	(1271)	(66)	(72)	(78)	(85)	(93)	(101)	(111)	(121)	(131)
Annual Cash Flow Savings	(1146)	59	65	70	77	84	92	99	108	119
DCF @ 14%	1.00	0.88	0.77	0.67	0.59	0.52	0.46	0.40	0.35	0.31
Net DCF Cumulative DCF	(1146)	52 (1094)	50 (1044)	47 (997)	45 (952)	44 (908)	42 (866)	40 (826)	38 (788)	37 (751)

(d)

Figure 18.5. (d) Telelift system vs manual.

residential or working day and up and down once at a lunch hour or for some other purpose, it can be stated that elevators carry about 150,000,000 people per day. For 250 working days each year that is a total of 37.5 billion passenger trips per year.

2. If it is considered that each elevator trip is an average of 100 ft (30 m), the grand total is 3750 billion ft per year or 71 million mi per year (125 million km per year). The distance is insignificant compared to automobile and airline travel but—to quote the elevator operator's answer to an irate patron who asked, "Where have you been?"—"Where can you go on an elevator?"

Elevators in tall buildings individually travel 25 to 30 thousand mi (48,000 km) per year. The safety record for elevators is excellent. There are fatalities, but practically none can be attributed to a falling elevator. The fatality usually occurs when people try to get out of a stalled elevator by manipulation of the doors and door locks and slip into the open space between the bottom of the elevator and the landing, or when a person who, in an older building, leaves an elevator at a floor with the doors closed and returns, opens the door and steps into an open hoistway because someone else has moved the elevator. Rules for effective locking of hoistway doors have been developed and legislation is needed to require their retrofitting to existing elevators. Such locks can prevent people from leaving stalled elevators unless help is present. New rules requiring elevator doors to be openable only from outside the elevator or when the elevator is within a safe distance of a landing are under active consideration.

In 1979, 26 fatal accidents occurred on elevators and none on escalators. A total of 607 accidents of all types were recorded on the approximately 400,000 elevators in use, mostly due to people being struck by doors and gates. A total of 801 accidents were recorded on the approximately 20,000 escalators in use, most due to passengers falling.

Specialist elevator consultants, insurance companies, and elevator companies themselves are concerned and thoroughly investigate all elevator and escalator accidents. Since many such accidents result in litigation, actual causes or the results of investigations are seldom available until after a legal settlement is made. Legal agencies will classify such information and use the results of previous cases in current cases. An interesting and informative compilation of unusual accidents can be found in the book *Court Cases of Consequences*, based on a series of articles written by John Miller, and published by the *Elevator World Magazine*.

Building owners are subject to liability since they offer elevators as public transportation. Their insurance carriers will write such policies and provide elevator inspection services to ensure the elevators they cover are in good condition. The elevator maintenance company is also a party to the safe condition of the elevators and the original manufacturer has considerable concern. The best approach to any liability is to avoid it by the assurance that everything has been done to make the installation as safe as possible.

SUMMARY

Once elevators and escalators are installed they must be maintained as close to the level of the original installation as possible. Proper passenger service and high reliability can then be assured and both the owner and passengers benefit by trouble-free, safe vertical transportation service.

Like any mechanical equipment, new designs, improvements, changing public taste, architecture, and, simply, age will cause obsolescence. At this point the elevators and escalators can be modernized and the improvements that have taken place over the years can be incorporated in the renewed equipment.

Buildings have economic lives of 60 or more years and the basic parts of an elevator such as machines, structure, and guide rails can last that long. Replacing controls, door operators, fixtures, and parts which are subject to intense wear, as well as door panels and entire cabs, can provide an elevator which is essentially new.

Although elevators and escalators are an extremely safe form of transportation, proper attention to their maintenance, operation, and timely repair can minimize the possibility of accidents and maintain their excellent safety record and as well as reduce the cost of liability insurance.

CHAPTER NINETEEN

Traffic Studies and Performance Evaluation

GOOD ELEVATOR SERVICE

There may be two buidings side by side. One has poorly operating elevators with long floor-to-floor time, poor operation, and a group automatic operating system with only half the features working, but the tenants are not complaining. In the other building, floor-to-floor time is excellent, the door operation is superb and all the group operating features are doing what they are supposed to do; however, there are always complaints of long waits and obvious backing up of traffic in the lobby.

The reason for the differences may be obvious or may be hidden. The first building may have more than enough elevators for the population so poor operation is tolerated since ample or excess service exists. The second building may be densely populated and the elevators are called upon to meet or exceed their potential handling capacity. The second building may also have a number of tenants who require their employees to be at work at strict given concurrent times, whereas the first building may have the same type of tenants who have starting times that are separated by a half-hour or so.

EVALUATING ELEVATOR SERVICE

There are no hard and fast rules to conclusively state that elevator service is good or bad. Judgment is usually made in a negative way—if no one is complaining, elevator service is good. If there are a great number of complaints by rank and file employees, they are generally ignored, but if the people who sign the leases complain, something is usually done.

Complaints can start from a number of events. Backup in the lobby at starting time is a common complaint resulting in the inability of employees to get to their stations at starting time and having lateness blamed on the elevator system. Excessive waiting time at upper floors is also a common complaint and people who complain usually state the wait is 10 min or more where in actuality it may be 2 or 3 min. The judgment has to be related to the state of mind of the people waiting—if they are anxious, the wait always seems longer. The evening rush to get home seldom elicits complaints, unless elevators fail to stop at certain floors. People will pack elevators when they stop or will even use stairways to get out of the building at quitting time when they are on their own time.

Traffic survey methods have been developed to evaluate and both qualify and quantify the common complaints. This chapter will review the methods.

FEAR OF ELEVATORS

A number of people have a fear of elevators. Stories have been told of people who will walk 10 or more floors to get to work in order to avoid riding elevators. Claustrophobia can cause excessive anxiety for some elevator passengers and complete panic may arise if such a person is ever in a stalled elevator.

Psychiatrists and psychologists have formed special group therapy sessions and have developed treatment for people with such fears. Much of the therapy is provided by educating people about elevators, how they work, how safe they are, what would take place if they were ever stalled, plus association of elevators with benefits.

As an example, one therapy session starts with a slide show and lecture by a qualified elevator person showing safety features and elevator design. The group may be taken on an elevator ride with people who can reassure them. In one instance, a gourmet dinner was served on an elevator to a person who, reportedly, enjoyed the meal so much he forgot where he was and lost his association of an elevator with fear.

Elevator companies can do a great deal to reduce people's fears. Quietness of operation and smoothness of ride are important factors. There is nothing more disconcerting than an elevator that rattles, bangs, and squeaks as it travels and nothing more harrowing if it strikes a switch or a loose item in the hoistway at high speed with a loud noise. Car position indicators with lights that move as the elevator travels rather than jump ahead to the floor where the car is programmed to stop can be reassuring. Impressionable people claim that the rapidly moving lights, especially in the down direction, indicate that the elevator is falling.

Excessive preopening of landing doors is also disconcerting. People anxious to get off or get on are unnerved when the doors open a foot or two before the elevator is level and while the elevator is leveling to a landing.

Architectural features, good signage and signals can also contribute to the feeling of well being and safety on and around elevators. Car interiors should be tastefully decorated and well lighted. Signals should give their messages clear and without being misunderstood. Landing lanterns should indicate which direction the car is going. Lanterns that light both up and down to indicate that an elevator is free to travel up or down depending upon the next call are totally confusing to the average person. In a double-deck elevator, the doors should open each time the car stops whether a person wants service at that floor or not. Having a person stand in a stopped elevator with the doors closed for a period of 10 or more sec is excessively long.

TRAFFIC STUDIES

There are a number of different traffic studies that can be made of elevators in a building. The simplest and most revealing is a lobby count of traffic on and off the elevators. Good observations in the lobby during the morning, noon, and evening peaks will give considerable information about the way the elevators are operating, people's attitude, interval, bunching, car loading, and door operation, as well as elevator-handling capacity.

A lobby traffic count is performed by having one observer for every three or four elevators who will record the number of people leaving and entering each elevator as it arrives. This is done for 5-min periods with other events observed such as people backing up in the lobby, the bunching of elevators, i.e., a number of cars arriving or leaving together within a short time, interferences with operation, and the carrying of packages, freight, or carts on elevators, as well as anything unusual.

A sample form used for a lobby count is shown in Figure 19.1 and has been marked up to show the record taken on a typical installation for three 5-min periods during a morning incoming traffic period in section A. Section B shows the count during a noontime two-way traffic period, and section C shows an upper-floor lobby count where traffic may enter or leave from up or down traveling elevators as indicated by the small arrows.

In actual practice, earlier building research would determine the time of the various traffic periods. For example, the observation and recording during the incoming peak may be from 8:30 a.m. to 9:15 a.m., noontime traffic from 11:15 a.m. to 1:30 p.m., and the evening peak from 4:45 p.m. to 5:15 p.m., depending upon the expected practice at the building.

Traffic counts should also be augmented by inconspicuous observations at the building at other times. When the counters are present many abuses are often discontinued for that day. The next day, elevators may be taken out of service during the peak by the maintenance worker, the lobby attendant may hold an elevator for a favored tenant, the mailman may get exclusive use of an elevator, plus other elevator abuses may occur which would not happen when the elevators are under observation. Employees are often suspicious of the counters and tend to avoid practices that may otherwise be a normal action, such as holding an elevator for a slow-moving friend.

Results of a Lobby Count

Referring to Figure 19.1, the results of the study during various periods can be analyzed as follows:

Assuming that the period from 9:00 a.m. to 9:15 a.m. included the heaviest traffic observed, the peak traffic was approximately 75 people in 5 min and the elevators operated at an average loading interval of 49 sec. The interval must be averaged over a period of time since it is difficult for any counter to cut off counting at the exact end of 5 min to start the next line. Average interval is calculated by dividing the total number of trips into the total time period in seconds. The peak is approximately 75 people since one carload more or less will have an effect.

The bracket shows that three elevators left very close together in time, indicating "bunching." Similarly, two elevators arrived close together in time, also indicative of "bunching." Frequent bunching requires further investigation of dispatching time adjustment.

If the elevators are of 3500-lb (1600-kg) capacity, the full car was the one with 16 people. If it was observed that that elevator did not leave promptly upon reaching full load, the load dispatch switch is incorrectly set and requires adjustment.

A calculation of the elevator system based on handling 75 people up and 10 people down, and using an average car loading of 10 people up and 2 people down can be made to determine the potential interval of the elevators and compare it to the observed interval. If the actual attendance of the building is known for that day, the 75 people up can be used to establish a percentage handling capacity. If the attendance is relatively low compared with normal attendance, this percentage can be adjusted upward to predict what can be expected on a day, such as payday, when most people will be in the office.

The two-way traffic study, if based on the same population as the up peak, indicates that a substantially higher percentage of the building population uses the elevators during noontime. This may indicate that the morning starting times may be staggered, and should be confirmed by questions and by computing the total arrival count over a long period compared to the attendance. The difference in interval between up-peak and two-way traffic especially with the greater handling capacity indicates there may be other deficiencies during the up-peak operation or there may be considerable interfloor traffic during the up-peak with the possibility that elevators are not returning to the lobby. This will require further investigation and additional traffic study, to be described later.

The section C tabulation on Figure 19.1 shows the results of observations taken at an upper floor such as the transfer floor between groups of elevators, at a lobby floor where there is a basement served by the elevators, or at an upper-floor cafeteria floor. These results must be compared to counts at other floors and may form the basis of providing additional operating circuits to ensure additional elevators at that floor or, if a lobby, to provide separate

UP ELEVATOR TRAFFIC

NUMBER OF PASSENGERS ENTERING AT 1ST FLOOR (BY CAR LOADS)

NO. OF ELEVS.	HOUR	MINUTES	Number of passengers entering (by car loads)	NO. PASS. ENTERING	
4	9	00 – 05	6 10 11 14 10 6 16 2	75	} A
		05 – 10	4 8 10 3 2	27	
		10 – 15	2 10 10 7 14 16	59	
		15 – 20			
4	12	20 – 25	8 4 5 7 8 9 6 2 10 12	72	} B
		25 – 30	2 8 8 3 6 7 10 12	56	
		30 – 35	10 12 5 8 7 2 4 2 0	50	
		35 – 40			
4		40 – 45	3↑ 1↑ 4↑ 5↑ 2↑ 3↑ 1→	15↑ 4↓	} C
		45 – 50	8↑ 9↑ 2↑ 0→ 0↑ 1↑ 3↑	21↑ 2↓	
		50 – 55	6↓ 10↑ 2↑ 10↑ 2→ 0↑ 1↑	23↑ 8↓	
		55 – 60			
TOTALS					

DOWN ELEVATOR TRAFFIC

NUMBER OF PASSENGERS LEAVING ELEVATOR CAR

NO. OF ELEVS.	HOUR	MINUTES	Number of passengers leaving elevator car	NO. PASS. LEAVING	TOTAL NO. PASS.	
4	9	00 – 05	0 1 3 2 0 1 2	10	85	} A
		05 – 10	2 3 0 1 0 0	6	33	
		10 – 15	4 2 1 2 4 2	15	74	
		15 – 20				
4	12	20 – 25	3 2 0 0 1 2 2 8	25	99	} B
		25 – 30	10 11 12 10 0 2 9 3	62	118	
		30 – 35	6 5 7 2 3 1 0 8	24	74	
		35 – 40				
4		40 – 45	0↑ 1↓ 2↑ 0↑ 1↓ 6↑ 6↓	8↑ 8↓	23↑ 12↓	} C
		45 – 50	0↑ 0↓ 0↑ 1↓ 2↑ 2↑ 6↓	8↑ 3↓	29↑ 5↓	
		50 – 55	3↓ 0↑ 2↑ 3↑ 0↑ 0↓	3↑ 6↓	26↑ 14↓	
		55 – 60				
TOTALS						

Figure 19.1. Tabulation of pedestrians on and off elevators at a lobby floor.

447

service to the floor below, if the additional traffic-handling capacity is required on the main elevators.

On-car Traffic Counts

Where elevator traffic is between floors or it is suspected that a considerable interfloor traffic exists, it is necessary to ride the elevators to survey the traffic. People are counted as they enter and leave the elevators at various floors and the interactions of the elevators in the group are coordinated by time.

Figure 19.2 shows a form used for an in-car traffic test. It has been marked to show some typical situations and some of the notations used to show what is happening. As with the lobby count, any unusual occurrences should be noted for information and, perhaps, additional investigations.

A person is assigned to each car or, on a preliminary basis, sample readings are taken to determine if a thorough study must be made and what time it should take place. Each counter should have a watch, preferably one showing hours, minutes, and seconds, and all watches must be synchronized among the counters.

Referring to Figure 19.2 the people are counted on and off the elevator in each direction and the time the elevator leaves a terminal, reverses, or waits is recorded. All stops are recorded, zeros indicating that no one got on or off. If the elevator reverses before reaching a terminal, that is shown by an arrow changing from an up to down trip.

By comparing the survey taken on a number of elevators over a period of time, it is possible to determine the extent of interfloor traffic. Determination can also be made if the elevators are operating on a zoned basis in accordance with the group automatic operation specifications and if they are being held or parked for extensive periods of time. With modern operating systems, elevators may seldom go to the terminal floors unless there is a demand for them to do so. This operation favors interfloor traffic, but can be detrimental if frequent demand exists at the main lobby as would be so during an up-peak period. Solving this problem may require clock control to force elevators to the lobby.

Using the on-car traffic test in conjunction with a lobby count will reveal additional facts about the operation of the elevators in the building and may suggest corrective measures if they are required.

On-car traffic tests are especially important in hospitals and hotels where a number of important functions may be located above a main floor. In hospitals, considerable interfloor traffic can take place and there are frequent movements of carts and equipment between floors. This is especially true on hospital service elevators where an in-car traffic test is necessary to determine the number of patient transfers on beds or stretchers and even the movement of empty stretchers.

In a hotel, the room service elevators must be surveyed by an on-car count since the kitchen is usually at a different level than houskeeping and frequent cart and food movement may occur between these levels and upper floors.

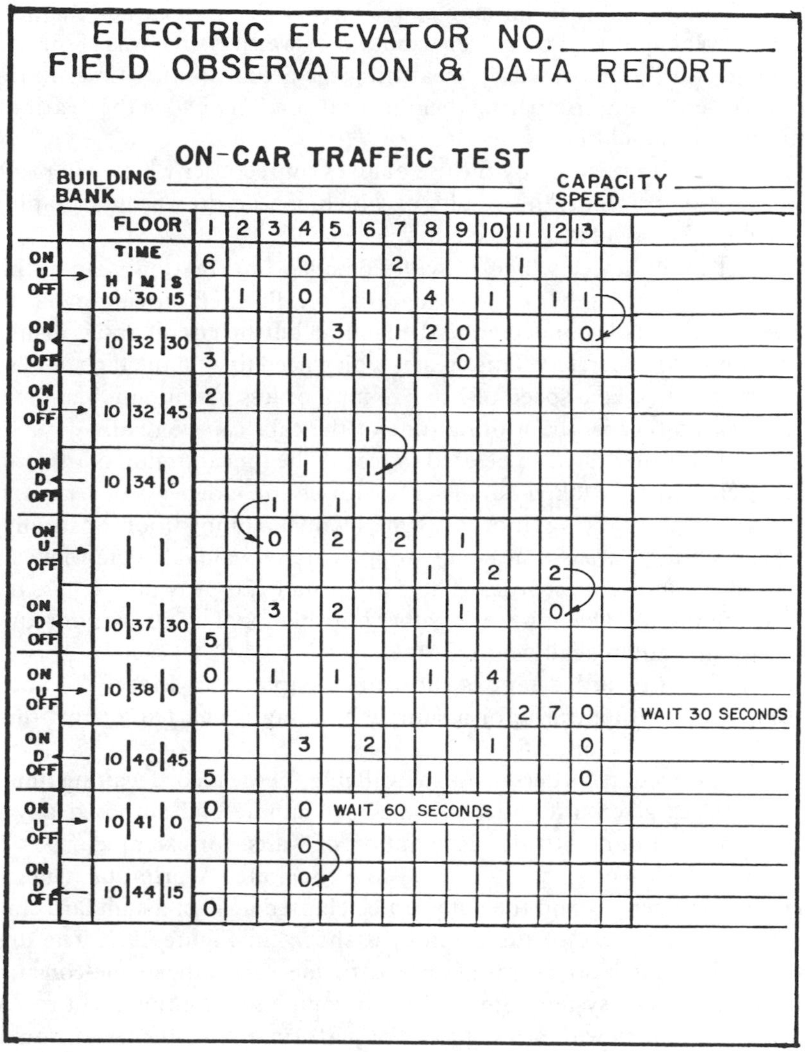

ELECTRIC ELEVATOR NO. _____

FIELD OBSERVATION & DATA REPORT

ON-CAR TRAFFIC TEST

BUILDING _____   CAPACITY _____
BANK _____   SPEED _____

	FLOOR	1	2	3	4	5	6	7	8	9	10	11	12	13	
ON U OFF	TIME H M S 10 30 15	6			0			2				1			
			1		0		1		4		1		1	1	
ON D OFF	10 32 30				3		1	2	0				0		
		3			1		1	1	0						
ON U OFF	10 32 45	2													
					1		1								
ON D OFF	10 34 0				1		1								
				1		1									
ON U OFF	‖			0		2		2	1						
									1		2		2		
ON D OFF	10 37 30			3		2			1				0		
		5							1						
ON U OFF	10 38 0	0		1		1			1	4					
												2	7	0	WAIT 30 SECONDS
ON D OFF	10 40 45				3	2				1			0		
		5							1				0		
ON U OFF	10 41 0	0			0		WAIT 60 SECONDS								
					0										
ON D OFF	10 44 15				0										
		0													

Figure 19.2.   Tabulation of pedestrian movement on and off an elevator observed from within the elevator.

449

Another important aspect of elevator traffic-handling ability is the length of time people wait until their landing call is answered. In an earlier chapter the relation between interval and the distribution of waiting time was shown. This distribution can be confirmed by measuring the actual waiting times during a busy traffic period and comparing them to a calculation of what the distribution of waiting time should be.

Waiting time is measured by the use of an event recorder which will produce a paper tape or printout, the results of which can be drawn on a graph and compared to the calculated expectation.

The most exact device which provides detailed information is an event recorder which has a number of pens marking a calibrated moving paper tape. Each pen is electrically connected to the landing button circuit and is displaced when the button is operated. This creates a displaced line on the tape, and since the tape moves at a fixed speed of 1 in. (25 mm) or less per minute, the time the pen is displaced (hence, the waiting time of the call) can be read.

The event recorder can also be used to record the time duration of other data. How long elevators are taken out of service for use on independent service, how long they park at a parking floor, the time spent at a lobby floor, or simply the time spent running versus time spent stopped are all valuable data that can be determined by an event recorder. There are usually 20 pens on a single meter and all of them should be utilized to gather information if all 20 are not required to measure the landing call waiting time.

A sample of a recorded tape is shown in Chapter 5 as Figure 5.4 and the distribution of waiting times for a sample building is charted against the expected standard in Figure 5.5.

Digital type event recorders are also available. These record waiting times by recording pulses given a number of times per minute and automatically sort them by various floors into the distribution of waits, for example, floor 2 up between 0 and 10 sec, 10 to 20 sec, 20 to 30 sec, etc. A printout is made at predetermined intervals and the data can be charted against a standard similar to Figure 5.5. A sample of the printout is shown in Figure 19.3. The digital event recorders can also be programmed to measure other time-consuming events in the elevator system. Some elevator companies use this event recorder to provide customers with a record of the performance of their elevators and utilize the results as a means to sell improved systems or service.

With the increased application of microprocessor programmable controllers, it is expected that such event recording means will be a built-in feature and available to the building owner to periodically check on elevator service or when complaints occur. Even if the irate passenger is shown that the maximum recorded wait was only 2 min, it is, at least, a defense against the irate passenger's claim of a 10-min wait. Even 2 min maximum is too long unless it is a very infrequent occurrence.

If the chart of the distribution of waiting times shows a high percentage of long wait calls, it is an indication that the elevators are operating bunched and require corrective dispatching or operating measures. When this is determined, the engineers of the maintaining company need to be questioned as to what corrective measures can or need to be taken with the equipment in that particular building.

For example, the Westinghouse Mark IV or V system has long wait timers; if a call waits beyond a preestablished time, a priority system is set up to travel an available car to that call to minimize excessive waiting times. With an Otis system, the number of calls and the waiting time of each in a zone are electrically totaled and if the total time exceeds a predetermined maximum, corrective measures are taken. Each elevator company's control designers have different approaches to this common elevator operating problem.

16:00 (Recording time period) (c)

(a)	(b)	0	10	20	30	40	50	60	90	120	150	210	300
FLR	TC	10	20	30	40	50	60	90	120	150	210	300	+
12D	0037	0014	010	005	006	002	000	00	00	00	00	00	00
11U	0015	0006	006	001	001	001	000	00	00	00	00	00	00
11D	0029	0016	008	003	001	000	001	00	00	00	00	00	00
10U	0042	0023	013	003	001	000	002	00	00	00	00	00	00
10D	0025	0012	008	004	000	001	000	00	00	00	00	00	00
9U	0019	0012	004	003	000	000	000	00	00	00	00	00	00
9D	0021	0012	008	000	001	000	000	00	00	00	00	00	00
8U	0014	0009	002	001	001	001	000	00	00	00	00	00	00
8D	0003	0000	002	001	000	000	000	00	00	00	00	00	00
7U	0000	0000	000	000	000	000	000	00	00	00	00	00	00
7D	0049	0034	012	002	000	001	000	00	00	00	00	00	00
6U	0000	0000	000	000	000	000	000	00	00	00	00	00	00
6D	0009	0005	002	002	000	000	000	00	00	00	00	00	00
5U	0016	0007	006	002	001	000	000	00	00	00	00	00	00
5D	0043	0034	006	003	000	000	000	00	00	00	00	00	00
4U	0012	0009	002	001	000	000	000	00	00	00	00	00	00
4D	0015	0010	004	000	000	001	000	00	00	00	00	00	00
3U	0027	0019	006	001	000	001	000	00	00	00	00	00	00
3D	0038	0024	009	003	001	001	000	00	00	00	00	00	00
2U	0064	0047	011	004	001	000	001	00	00	00	00	00	00

16:01 (d)

		0	10	20	30	40	50	60	90	120	150	210	300
FLR	TC	10	20	30	40	50	60	90	120	150	210	300	+
RT	00474	00290	00119	0038	0014	0009	0004	000	000	000	000	000	000

1. Floor designations.
2. TC = total calls during the time period indicated.
3. Headings $\dfrac{0 \quad 10 \quad 20 \quad 30}{10 \quad 20 \quad 30 \quad 40}$ represent the number of calls waiting that length of time in seconds.
4. Summary of total calls and the distribution of waiting times for the time period.

Figure 19.3. Computer printout of landing call waiting times (Comput-O-Chek^{TM} by Otis Elevator Co.).

PERFORMANCE EVALUATION

One of the easiest of traffic tests can be performed any time and only requires a stop watch. An observer rides the car and records the time it takes doors to close, car to start, floor-to-floor travel, door-open time, and the amount of time spent at a floor. This is done in both directions and, preferably, with the elevator empty, partially full, and close to a full people load condition.

The results are compared to design criteria which are used in calculating performance as described in the early chapters of this book. If any of the operating times exceeds the design normals for the mechanical and electrical features of the elevator, readjustment to bring it close to the expected time is required. If the dwell-times are excessive and are not different for car button stops and landing button stops, or if the time the door stays open after reversing while closing is significant, adjustment also needs to be made to match those times with the observed traffic movements in the building.

Figure 19.4 shows a sample form that can be used in in-car performance evaluation. The various times are developed either by direct readings or by the difference between events. For example, to establish the time it takes the car to start from the time the doors are fully closed would require a measurement of from start of door close to door fully closed and a second measurement of start of door close to car start, the difference being the time it takes to get the car started after the doors are fully closed.

If the doors are arranged to preopen, that is, start to open before a car is level to a floor, this preopening time can be recorded by using a chalk to mark the car sill at the position of the doors when the car is level. On the next stop, the time between the doors passing the chalk mark and full open can be measured and preopening time established. If such preopening time is excessive, the results will be obvious by observing the car from the landing side as it comes to a stop at a floor.

Since doors are "checked" or slowed prior to their full open or full close position, there is an argument that claims door open and close time should be measured from a three-quarter open or close position. Checking is a normal part of door operation and the door operation should be designed to provide good checking so the door neither slams closed nor bangs open. For that reason, the full travel should be measured. The tables of door times in the early chapters include the total travel and should be used for comparison.

The total performance of an elevator system can be recorded and presented in graphic form. By connecting recording meters which produce a tape on a time base to each elevator, a profile of elevator action showing travel up, stopping at a floor, continued travel and stops, reversal, and downward travel and stops can be recorded. This tape can be matched with similar tapes of other elevators all on the same time base to develop a composite of all the various trips a group of elevators make over a given period of time.

A sample of such a recording is shown in Figure 19.5. This enables the consultant to analyze the interaction of the various elevators and suggest corrective actions that may be taken if deficiencies are apparent.

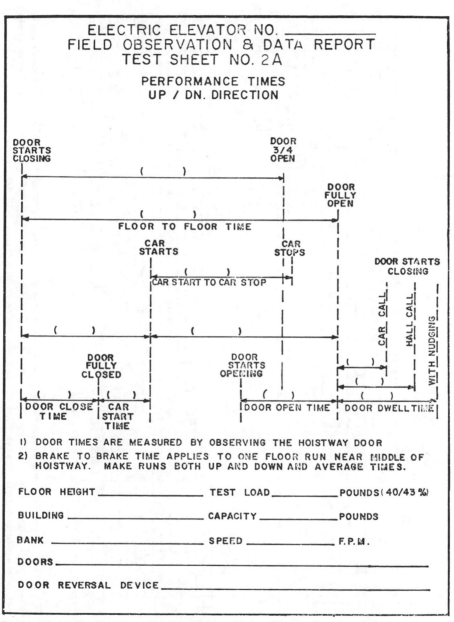

Figure 19.4. Performance time chart for an elevator floor-to-floor travel showing the various elements of a one-floor elevator trip (Courtesy Jaros, Baum and Bolles).

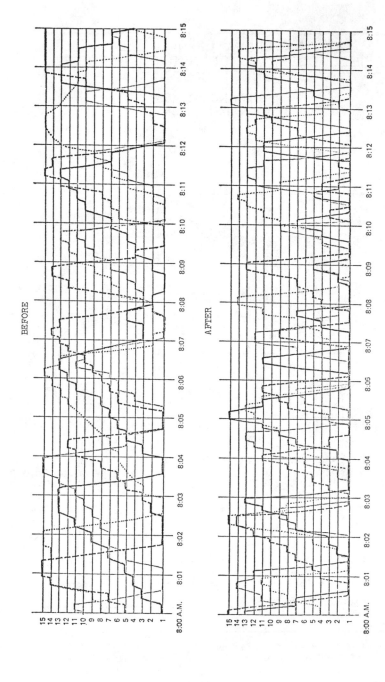

BEFORE

AFTER

Figure 19.5. Taped time recording of elevator travel up and down showing time spent on floor-to-floor travel and time spent stopped at the various floors. "Before" chart shows serious elevator "bunching." "After" chart shows performance after corrective measures have been taken (Courtesy of Performance Profiles Inc.).

A series of recordings taken before and after corrective actions are made can be valuable in providing a record of improvement and a basis for periodic future checking to establish that proper operation is maintained. The recording also provides a positive means to establish if elevator capacity is efficiently used or is being wasted, especially if the recordings are made in conjunction with a survey of landing call waiting times for a concurrent period of time.

EVALUATING RESULTS

Once the various traffic studies and tests are performed, the results must be evaluated and recommendations for corrective measures taken. These are usually developed into report form and presented so that the building operator is aware of the extent of the problem and the various corrective measures that are available to be taken.

In the discussions of the various types of tests, suggestions are made as to the meaning of some of the observations. One of the first steps is to establish what the elevators are capable of doing, then compare what they are actually doing with their potential.

The traffic study must therefore go beyond the actual elevators and into the occupancy and management of the building, and perhaps, compare that building with others in the same neighborhood with similar functions. This is akin to almost starting with a planned rather than existing project. In a new project, the population density and arrival rate would have to be established before sufficient elevators to handle that demand could be established. In an existing building, the density and arrival rate or demand can be observed. If there is access to the record, the original elevator design of the building can be compared to what was furnished or what was intended to be furnished.

The next question would be, Are the elevators operating as they should and performing as designed or capable of performing? Answers that would follow would include the number of people they are required to serve and the elevators capability of serving those people.

In preparing a traffic study report it will soon be discovered that the report must be constructed backward with the last pages established first to provide the foundation of the summary and recommendation.

A typical outline of the sections of a traffic study report is as follows:

1. Title.
2. Table of contents.
3. Executive summary which is a concise single-page summary of the conclusions and recommendations.
4. Recommendations in detailed terms.
5. Building data including population.
6. Evaluation of performance or traffic studies.
7. Elevator data including observed and calculated performance.

8. Survey data including observations.
9. Appendixes containing detailed survey sheets, lobby counts, compilation and charts of waiting time, plus other data necessary to back up the evaluation done in section 5.

The recommendations are the most important section of the report. They must include specific directions for the owner to take and should cover any management as well as elevator hardware and software corrective measures. The recommendations should also be the basis of forming an outline of detailed specifications of any mechanical or electrical elevator work which may have to be contracted. Budget data should also be included if appropriate.

CORRECTIVE MEASURES

Based on the assumptions that the elevators are operating properly, performance time is up to standard, and no physical plant problems exist, some of the common moves that can be recommended to solve elevator traffic problems may be as follows.

If the building has more than one entry floor and the elevator system serves both, one of the floors should be discontinued and alternate transportation such as escalators or shuttle elevators provided, if at all feasible. Alternatively, operating systems that can share the elevators between the two lobbies and create equalized dispatching may be a consideration.

Similarly, if the elevators serve an upper-floor cafeteria which is open during peak incoming traffic periods, service to that floor may have to be restricted during the times that there is no car at the main lobby by temporarily discontinuing landing call service until a car is at the lobby available for loading. If the cafeteria is on a basement or second floor, separate elevator or escalator service needs to be considered.

If the elevator system serves odd floors such as basements or penthouse, consideration should be given to eliminating such floors and providing alternative service. The use of a shuttle elevator for basement or garage service is especially beneficial if it results in basement or garage service being taken off the main elevators.

Elimination of overlapping floors between two groups of elevators or changing such floors to transfer floors with up landing call and down car call service is beneficial. People on the overlapping floors are then forced to use only one group of elevators which can relieve the other group of unnecessary stops. Reference should be made to the earlier discussion on transfer floors.

The elimination of freight service and mail cart abuses can do much to relieve a critical elevator situation. Freight service hours must be enforced to occur before or after peaks. An automated mail-handling system can be considered to eliminate carts on elevators. A cart can be equivalent to two or three "people spaces." This abuse is especially prevalent in hotels with minimum service elevators where luggage and valet carts are often used by employees to move only one piece of luggage.

In hospitals, as suggested earlier in this book, means need to be developed to measure a full elevator when it is occupied by a stretcher rather than people—area load sensing rather than weight load sensing. An elevator with a stretcher which is immediately sent to its destination is obviously more efficient than having the elevator stop at intermediate floors where it is impossible for people to get on. The use of special key switches for such service is effective only to the extent that abuses are not tolerated.

Each elevator traffic problem may be somewhat unique but more often a similar type problem has been previously solved. Communication with the elevator companies may often lead to a similar situation and solution. There are also cases where the only solution was to chop through floors and add elevators. This has happened where the nature of the building has been changed, such as an apartment house converted to an office, or where proper elevator advice was not heeded in the design, a mistake was made, or the builder "took a chance." The value of good elevator study and competent elevator consulting work in the initial design cannot be overemphasized as the best means to avoid future elevator problems.

Sometimes a solution may have nothing to do with the elevators. Staggering of working or lunch hours is one such solution. Moving a tenant to another floor of a building so that a group of elevators is solely for a single tenant is another example. Ingenious solutions have also been found. Consider a situation where constant complaints about long waits for elevator service abounded in a building. The solution—put mirrors at each floor around the landing buttons. The actual waiting times were the same, but the complaints ceased as a result of the perceived waiting time.

CHOOSING AN ELEVATOR CONSULTANT

Competent help to solve (or avoid) elevator problems can be found in many areas. The elevator manufacturers usually are more than willing to help. The confidence the elevator companies inspire may be simply a result of asking the proper questions or their willingness to perform traffic studies as outlined previously in this chapter. Anyone who has a maintenance contract with an elevator company is a favored customer and should be helped if such help is requested.

If certain difficulties persist, there may be a point of frustration reached with an elevator company and the owner may want an unbiased opinion. There are elevator consultants available with experience and knowledge who are capable of performing all the aspects of elevator work. Many are specialists in certain areas such as maintenance repair, new installations, or legal aspects or construction practices.

Prior to engaging elevator consultants, their credentials and extent of experience or specialization should be investigated. A person who has only experience in maintenance should not be engaged to design a new building or vice versa. Previous clients of the consultant should be contacted to determine the consultant's performance on past projects. Elevator consultants are not licensed, nor are there any qualifying criteria subject to professional and governmental review.

A good elevator consultant will appraise the problem and provide a proposal for the time to make the necessary tests and develop the information required. The proposal should be for a specific job or include all the aspects of a total job broken down into phases. For example, the development of an elevator system for a new building would be done in four phases. Phase I would be the necessary traffic analysis and elevator design. Phase II would include Design Development, specifications and review of bid drawings and documents. Phase III would be aid in contract negotiations and review of shop drawings and Phase IV would be checking the completed installation for contract compliance and performance.

Elevator consulting work should seldom be performed on a per hourly basis unless a great number of unknowns are apparent and a preliminary investigation is necessary to establish the parameters of the job. Once the extent and object of a job is established, a fixed fee proposal should be developed.

SUMMARY

The foregoing outlined some of the many varied traffic studies that can be performed on elevators, and developed suggestions to appraise the results of such studies.

The important reason for an elevator study is to develop suggested corrective measures which not only include items mentioned in this chapter, but also lessons presented in all the chapters of this book. If the answer cannot be found, the better alternative of rebuilding a building and adding elevators is all that is left.

CHAPTER TWENTY

The Changing Modes
of Horizontal and
Vertical Transportation

TRADITION

The vertical elevator propelled by a traction machine and equipped with a counterweight has been developed over the past 80 years and is established as a safe, reliable, and relatively standard product. There are many firms engaged in its production and safety codes have been written to ensure its proper installation. People have accepted elevators based on the traction principle as the standard of good elevator service.

The hydraulic-type elevator enjoys somewhat the same reputation. The appearance to the public is the same as a traction elevator since the same doors, cab, controls, and signals are common to both.

Both the traction and hydraulic elevators are developed as very rigid, permanent installations in a building, seldom moved from one building to another. In addition, it is difficult to install most traction elevators and the direct-plunger-type hydraulic elevator in an existing building, since very stringent pit, overhead, hoistway, and structural requirements must be met. It is also difficult to adapt either type to some extreme vertical lifting requirements since certain physical limitations are present which require additional expensive considerations to meet the requirements. For example, using a traction elevator to serve a 1000-ft (300-m) high television antenna tower requires roping which will not be affected by extreme wind.

Other approaches to elevator drives have been developed to overcome some of the necessary restrictions of conventional traction and hydraulic elevators. In other ways, conventional hydraulic and traction elevator machineries have been adapted to other than pure conventional enclosed elevator applications to meet special lifting requirements. The hydraulic elevator has been adapted to lifting stages in theaters, the traction elevator machine to moving vehicles horizontally, and a class of new propelling devices consisting of racks and pinions, screw lifts, and scissor jacks has been adapted to elevator lifting means. In addition, elevators, both hydraulic and traction, are capable of unusual approaches to transporting personnel.

An entire new field is opening up which borrows a great deal of elevator people-moving discipline and activity. The vast experience that the elevator industry has in opening, closing, and protecting people from moving doors during transfer is being adapted to the horizontal people-moving field. Even the escalator and moving walk technology is being adapted to develop a new vehicle called the accelerating moving walk which can transport people from a walk, accelerate them to 10 mph (4.5 mps), and then gradually slow them back down to a walk.

These aspects will be briefly discussed in this concluding chapter and some of the current applications introduced in Chapter Fourteen expanded to introduce innovations. During the next decade it is expected that some will be more extensively developed and will broaden the entire realm of personnel transportation.

RACK AND PINION ELEVATORS

Rack and pinion elevators are elevators in which the driving means is carried on the car platform and provides motive power by having a motor drive a pinion gear operating on a stationary rack connected to either a self-supporting or supported tower. Guide shoes on the platform act on rails on the tower to maintain the pinion in alignment with the rack. A safety device which consists of a governor operated brake and separate pinion operating on the same rack is provided. If the car overspeeds, the brake applies and the safety pinion locks the car on the rack.

Power to drive the machine is carried to the car by means of a traveling cable and the operating device and controller are mounted on the car. In some applications, the electric driving motor is replaced by a gasoline powered motor—an outboard motor if you wish—operating through a clutch and reversing gear. If the elevator runs out of gas or if the electric power fails, manipulation of the motor brake can be used to lower the car at a controlled speed.

The lack of ropes on the elevator and the possible use of a gasoline powered motor make a rack and pinion elevator ideal for very tall TV towers, utility shafts in mines or underground power stations, maintenance elevators for smoke stacks and powerhouses, or any other place where periodic use of an elevator is necessary.

Since the rack and pinion elevator can be tilted to follow the contours of a structure, it is ideal where inclines or a combination of inclined and straight travel is required. Figure 20.1 shows an example of such an arrangement.

Figure 20.1. Rack and pinion elevator showing a combination of vertical and inclined travel.

The rack and pinion elevator is designed to operate from a self-supporting structure and its application as a temporary elevator on a construction site is quite extensive. By adding a counterweight and a multiplicity of drive machines, loads as high as 6000 lb (2800 kg) at a speed of 300 fpm (1.5 mps) have been obtained. For the normal personnel lift work the standard load and speed is usually 650 lb (300 kg) at 125 fpm (0.62 mps).

SCREW LIFTS

There is nothing new about screw lifts—they have been around since the days of the ancient Greeks. What may be new is their refinement into high-speed devices and adaptation to passenger elevator application. Industrial screw lifts have been used to lift extremely high loads at very low speeds usually measured in inches per minute (millimeters per second). Recent developments have made screw jacks available which can move moderate loads of up to 2000 lb (1000 kg) at 100 fpm (0.5 mps). Such equipment can be used to move a small elevator by operating with the screw in tension and being driven by a motor on the top. As the screw is turned, the elevator travels up or down in its guides (Figure 20.2). By using the screw in tension, the bending that could occur if the screw were used to push from below is avoided.

Large size screw jacks are used to lift platforms and can be exceedingly valuable to raise a section of floor to form a platform since a screw jack has a natural tendency to lock the load when the turning power is off. Unlike a

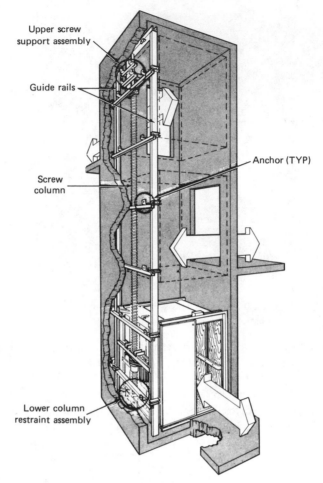

Upper screw
support assembly

Guide rails

Anchor (TYP)

Screw
column

Lower column
restraint assembly

Figure 20.2.   A screw lift elevator.

hydraulic lift which may creep as the oil cools or if a slight leak develops, a screw jack raised platform can be left in place indefinitely. If more than one screw jack is required to lift the load, the screw jacks have a further advantage of being self-equalizing if driven through gearing by a single-drive motor.

Such screw jack arrangements are used to lift the bottoms of swimming pools to change depth from diving to wading, lift building walls to close off entrances during nighttime hours, and to move heavy vault doors down and out of the way when a vault is opened.

SCISSOR JACKS

If one visualizes two X's placed one on top of the other so the touching points are connected and moved apart or closer together by a hydraulic piston

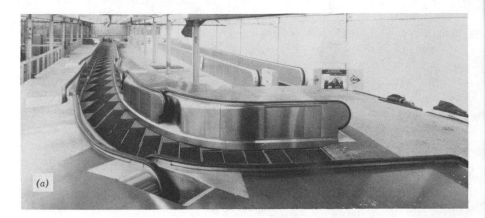

Figure 20.5.   An accelerating moving walkway. (*a*) Prototype model showing step tread and handrail arrangement. (*b*) Diagrammatic view of changing step configuration as people are accelerated. (Photo courtesy of *Elevator World Magazine*.)

portation terminals. As more people use such facilities, the larger, more spread out the facilities become. Walking is accomplished from about 120 to 240 fpm (0.6 to 1.2 mps). A moving walkway at 160 fpm (0.8 mps) and a person walking on a moving walkway can travel at 400 fpm (2 mps). In some terminals the center is 2500 ft (800 m) from the loading areas which can be a tiring 10-min walk. In other facilities parking is 5000 ft (1600 m) away which requires bus transportation and vehicle and driver expense. An accelerating moving walk at 1000 fpm (5 mps) can reduce a 5000-ft (1600-m) travel distance to 5 min. Short, single-speed moving walkways from that point can make parking or transportation loading quickly accessible.

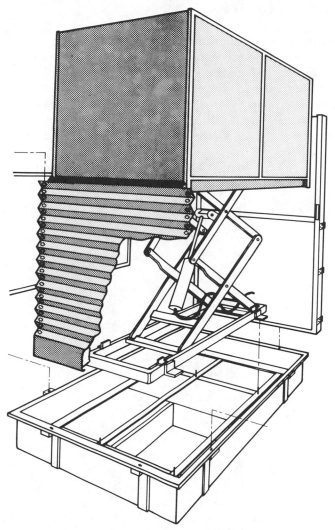

Figure 20.3.   A scissor lift.

or screw mechanism, the principle of the scissor lift is apparent (Figure 20.3). Imagine the X's 10 or 20 ft (3 or 6 m) high and made of heavy steel with a platform on the top. Such devices are used to form elevators and to move such items as trailer trucks or other large loads. One such application is on shipboard where advantage is taken of the extremely low profile when the scissor lift is collapsed. Other applications consist of multiple scissor lifts mounted on a wagon to raise small platforms 50 or more feet (15 or more meters) to service light bulbs in a high hall.

OTHER LIFTING DEVICES

Development work and application is progressing on a new adaptation of the old roped hydraulic elevator using roller chains or ropes and telescoping hydraulic pistons. This lifting arrangement is frequently used in high-rise forklift trucks and is being adapted for elevator use both in North America and abroad. The arrangement offers a hydraulic elevator which can be classed as "holeless" since the piston is mounted at the side of the elevator, does not have to be sunk into the ground, and only needs to travel half the rise of the elevator if a two-to-one chain or rope arrangement is used (Figure 20.4).

Telescoping pistons are also being extensively used to avoid the expense of drilling and to provide higher-rise hydraulic elevators. A piston and cylinder are located on each side of the elevator and act in unison to lift the car. Special equalizing means are used to ensure that the load is shared by each.

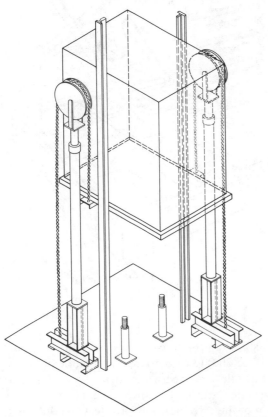

Figure 20.4. Hydraulic elevator using a combination of cylinder and chain. One foot of piston travel moves the elevator platform 2 ft. A safety device is required for an elevator of this type similar to those used on cabled elevators. The advantage is that no hole needs to be drilled for the cylinder.

Extremely heavy loads are lifted by the use of jacking arrangements although such an approach can hardly be called an elevator. Hydraulically actuated jacks are arranged so that one set of jaws grabs a rod while another set of jaws locks on a rod and both are moved by hydraulic cylinders. As each set reaches its extreme travel, the action interchanges so that the device actually climbs the rod. Such a device has been used to lift bridges and one platform application was used to raise a solar furnace weighing some 60 tons (54 t) to the top of a tower for testing in the center of a field of reflective mirrors.

There are certain to be other approaches developed as the need for lifting continues. An on-going study approached by many scientists and researchers is the quest for antigravity. Perhaps in the future the entire field of vertical transportation will be revolutionized if antigravity can ever be firmly established and controlled.

ACCELERATING MOVING WALKS

In 1980 three systems of providing a horizontal moving walkway that can allow people to enter at a walk, accelerate them to a high speed, and decelerate them back to walking speed have been prototyped. Continuing development work is proceeding and, perhaps, within the next decade such a device will be in daily use by the public (Figure 20.5).

The development is fraught with problems both mechanical and human. Steps that can start at a slow speed and be changed in configuration so they move along at high speed have been developed. One approach is to have combs that spread apart, another is to turn the step gradually so it moves at a faster speed, and a third approach is to have a series of small diameter rollers moving at increasing speeds. They all work for people's feet but the handrail synchronization has been difficult to attain.

One approach to a synchronized handrail is the use of grabs which spread apart as they are speeded up. Another is the use of multiple handrails, each operating at different speeds so a person moves their hand from one to another. Additional approaches need to be developed.

A people problem remains. If people were not social and would space themselves along the fast portion of the moving walk, they would have plenty of room to exit when the walkway is slowed. Since people are social, they tend to get close to the person in front and, as the walkway slows, the space between each decreases and a bumping can develop. This problem may prove insurmountable.

An advisory guidance code has been established by the A17.1 Elevator Code Committee which sets up some preliminary safety considerations for the future design of accelerating moving walks. An active program is underway to develop a working model for public demonstration and use, and if successful, may introduce a new era in short-range personnel transportation.

One may ask why there is a need for such a development. The need comes from the increasing population and congestion in the world especially in trans-

Figure 20.6. Escalator with an intermediate flat section (Courtesy Schindler).

Similarly, although not being actively considered, escalators or moving inclined walkways which have the capability of allowing a person to walk on, be
accelerated to a higher vertical speed, and then be slowed to allow exiting at a
walk would be valuable in high-rise (or deep) transit stations. This would be an
expected outcome of a perfected accelerated moving walk.

A recently installed escalator in Europe rises at an incline, flattens, and then
rises again (Figure 20.6). Visualize the flat area as a long distance when people
can be moved at high speed and imagination can establish possible application,
such as a pedestrian walk across a wide expressway.

PEOPLE MOVERS

Solving the problem of moving people from place to place in a safe reliable
way is one of the goals an accelerated moving walkway program is attempting

to meet. It has been accomplished in other ways and is in active public use in many airports, in amusement parks, and between buildings in a medical center. This approach is the people mover where passengers are encapsulated in an automated moving vehicle traveling on its own dedicated guideway.

Two major systems are in use. One is an automated "trolley car" which is a self-propelled vehicle operating on rubber tires and traveling on a dedicated

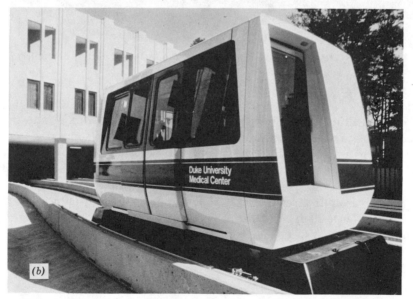

Figure 20.7.   (*a*) High-capacity, horizontal automated personnel rapid transit vehicle (Courtesy Westinghouse Electric). (*b*) Air cushioned, linear induction propelled, automated personnel rapid transit vehicle (Courtesy Otis Elevator Co.).

guideway (Figure 20.7*a*). Motive power is picked up from a trolley wire raceway in the guideway and stopping and starting is controlled by controller logic located on the vehicle. At the terminals, automatic doors with safety protective edges open and close and are timed to permit passenger transfer. Operation can be programmed to be on schedule, i.e., vehicles traveling at time periods designed to maintain a headway timing between vehicles, or "on call," i.e., operating when someone indicates a service requirement by operating a call button at a station. The vehicles are large and designed to accelerate and decelerate at about 0.1 *g* (3 fps^2 or 0.9 mps^2). Emergency stopping is limited to about 0.3 *g* (10 fps^2 or 3 mps^2) even though that rate is rather severe. The passenger capacity is based on vehicle design and the usual vehicle is designed for a comfortable load of about 50 people in 150 ft^2 (13.5 m^2) with a maximum load based on 2.3 ft^2 (0.2 m^2) per person.

A smaller people mover designed close to elevator capacities of 15 or 20 people employs an air cushion suspension and is propelled by linear induction motors (Figure 20.7*b*). This approach allows the lateral movement of the vehicles so that loading and unloading can be done off the main guideway and other vehicles can pass. The elevator-type door is also employed and the system can be programmed for on call or scheduled operation. The advantages of the lateral docking are to use a multiplicity of vehicles and provide increased service when traffic demands increase. The air cushion and linear induction motor drive also provide an easy way to switch from one track to another and to negotiate curves without the wheel or tire squeal that occurs with wheeled vehicles. The disadvantage is that a relatively smooth and flat guideway needs to be provided.

This second level access is being extensively applied in colder cities in North America even though the automated vehicles are not, as yet, provided. Downtown areas in at least three major cities have a great number of buildings connected by "skywalks" which are serving to make the second floors major retail areas and protect patrons from extreme weather. The street levels are reserved for bus and auto transport and have narrow sidewalks. From an elevator standpoint, the skywalks also create a second elevator lobby which can add to elevator round-trip time. Escalators between the street and the skywalk level are therefore recommended.

CABLE-DRIVEN PEOPLE MOVERS

In North America the quest is for an automated people mover to avoid the cost of operating attendants, whereas in other parts of the world cable cars, tramways, and funiculars are common and are all operated by attendants stationed in a stationary control cabin. There are some notable installations in the United States which are used primarily as tourist atractions, and at least one as a commuter transportation means in New York City (Figure 20.8*a*).

Medical Center System

(b)

Figure 20.8.   (a) Suspended cable car (tramway), Roosevelt Island, N.Y., N.Y. (Courtesy VSL Corp.); (b) Cable-driven people mover—horizontal elevator (author's design).

Ropeways, as cable cars are often called, are a product of a highly refined technology and are in a specialized field of their own. The vehicle is suspended on a wire rope designed to support the traveling load and propelled by driving ropes connected to a traction drive. The cars are interconnected so that the down traveling cars or the ones traveling in one direction counterbalance the up traveling cars or the cars traveling in the opposite direction. At the terminals the cars may be freed from the load bearing rope system for loading and unloading or may be captive so that when one stops, all stop.

A system wherein the car is suspended from rails similar to a monorail with a lower guide and propelled by cables driven by elevator-type traction machines has been proposed by the writer and it is hoped that it will be developed (Figure 20.8*b*). The system can be described as a horizontal elevator and has all the attributes of an elevator system. It would be ideal for short-range transportation in airport terminals to serve satellite terminals, and in shopping centers between parking areas and the mall. It could also be used to bridge busy thoroughfares, and could even be used in industrial applications to bring workers, for example, from a mine field entry point to the head of the mine.

WATER TOWERS

Most towns have a water tower to supply local needs but few, if any, have taken advantage of the possibilities such towers present. Perhaps water does not enjoy the same importance in the northern hemisphere as it does in the more

Figure 20.9. Water tower with public space developed on the top of the water storage area (Courtesy Schindler).

arid parts of the world since water towers are usually no more than tanks on stilts. In the Mideast, where water is valuable and well respected, the water tower is a prominent and important feature in any area, so much so that water towers have been developed into centers and main attractions. The top of the water tower is leveled off and the area used for recreation, restaurants, libraries, civic offices, or a gathering place for the citizens (Figure 20.9). This is accomplished by a simple elevator traveling through the tower to serve the upper level. Perhaps as water shortages persist, water towers in North America may become symbols of civic pride and be more intensively developed.

TRANSIT STATIONS

Most developers of transit systems have visions of hoards of people entering and exiting each station as packed trains arrive and depart. It often follows that the only way to serve those people is by escalators so that the design of the station may be compromised to accommodate the escalators even though well-designed elevator systems can accomplish the same people-moving capability.

As a 10-car train arrives, people are discharged over a horizontal distance of about 600 ft (180 m). To effectively use an escalator system, all these people must walk to a central location. The natural tendency is for people to want to board the train at a point that will be near the escalator at their destination where they can enter it when they arrive.

The escalator operates at a 30° incline so the boarding point is 1.7 times the rise of the escalator from either end. With a deep station, the train station may be a block or two away from the street-level entry or exit point.

Consider elevators spaced along the platform. People are near an elevator no matter where they get off the train and the vertical elevator exits right above the station.

The argument continues that people must wait for an elevator and an elevator ride is too long. A train load of people must also wait for an escalator and a station that is 200 ft (60 m) deep will require almost a 3-min ride by escalator.

An elevator system which can serve such a deep station and carry almost as many people as an escalator with a much faster travel time is practical and is shown diagrammatically in Figure 20.10. This system was proposed by the writer to be used in a proposal for a subway system that could be economically bored out of solid rock below a city, as opposed to conventional subway systems which are built by cut and cover. The elevator system increases the utilization of one hoistway approximately three times over that of a conventional elevator and provides proportionally greater elevator-handling capacity for a given area of hoistway. This was an important consideration since the hoistway must also be bored vertically out of rock. Figure 20.10c shows the time cycle usage of the vertical shaftway. An elevator cab is traveling up or down while the cabs at the floors are loading or unloading.

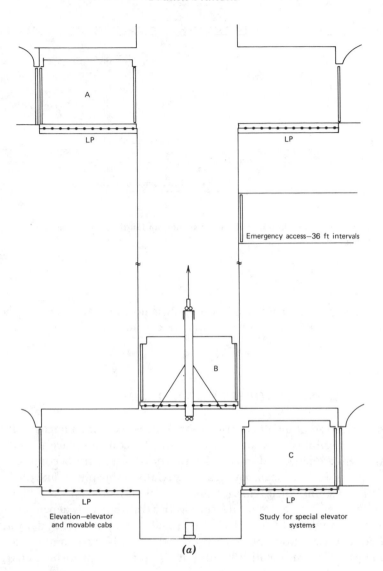

Figure 20.10. (*a*) **Elevation sketch of a high-capacity elevator system with three elevator cabs and a single hoistway.**

The concept of the entire elevator cab transferring on and off the elevator can also lead to consideration of a completely integrated horizontal and vertical transportation system. Visualize tenants on a floor of a future super apartment house entering the elevator of tomorrow, digitally coding their destination, traveling up or down, being transferred horizontally across town, traveling up or down, and arriving at the desired office building floor a few minutes later.

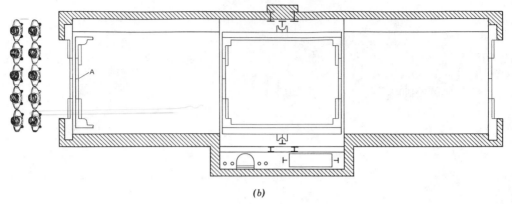

(b)

Figure 20.10. (*b*) **Plan sketch of the top and bottom landing floor arrangement and passenger loading area.**

The idea may sound far-fetched but it really is not. Instead of being delayed in an expressway traffic jam, our commuter can now be delayed in a lateral-elevator traffic jam.

AUTOMATED SELF-PROPELLED VEHICLES

The application of an all electric automobile is receiving increased attention and it is a possibility that such vehicles will be commonplace in the future. Similarly self-propelled automatic electric transporters are becoming increasingly common in European countries, especially in hospitals, and their introduction into use in North America may be only a matter of time.

As with the electric automobile, the electric transporter is dependent upon a battery that can take sufficient charge to provide for a full cycle of daily activity. Unlike an electric automobile, the transporter can be programmed to find a charging point and automatically hook itself up for a replenishing charge and then continue the chores that await it.

Such transporters are designed to follow a guidepath laid out with either metallic strips or with infrared-light-reflective paint. They can also be programmed to stop and automatically pick up a waiting cart and proceed to an indicated destination. Such destination may involve an elevator trip, which, by the logic built into the system, is also automatically taken—the cart indicates a call, waits, automatically boards, rides and exits the elevator, and continues on its way.

The expansion and development of the concept and the hardware leads to limitless opportunities. Material, supplies, and ultimately people, can be auto-

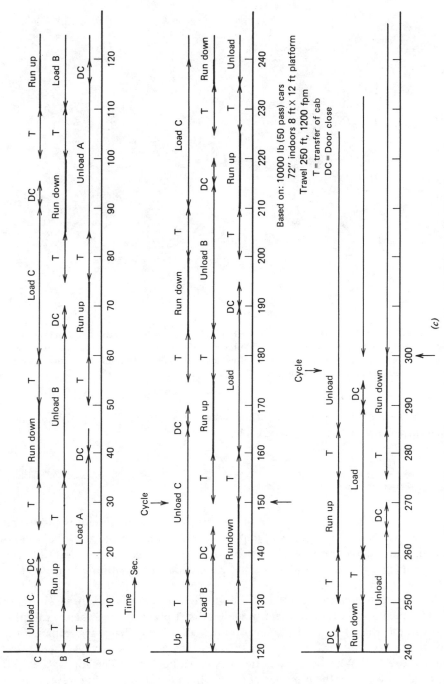

Based on: 10000 lb (50 pass) cars
72'' indoors 8 ft × 12 ft platform
Travel 250 ft, 1200 fpm
T = transfer of cab
DC = Door close

Figure 20.10. (*c*) Diagram of elevator travel and loading-unloading time cycle.

Figure 20.11. Self-propelled, wire-guided vehicle programmed to transport people in an airport. The vehicle follows a wire imbedded in the floor and makes programmed slowdowns to pick up and discharge people. Sensors slow down and stop the vehicle if an obstruction is encountered and restart is attempted.

matically transported within a building and, eventually, from place to place, including elevator trips (Figure 20.11).

CONCLUSION

There is no permanent conclusion to this book. By the time it is published changes will have taken place which may change some of the approaches and concepts that have been developed and reported herein. This is certainly the history between 1967, the date of the publication of the first edition of this book, and today.

The student of vertical transportation technology must keep abreast of these developments and evaluate them in view of past practices and the overall philosophy of what is trying to be accomplished. The challenge is to change the seven famous last words of any organization from "We've always done it that way before" to "What new and better way can it be done?" The approaches and methods outlined in the previous chapters are a development and represent an increased discipline from those in the first edition. In the future the assurance can be had that the methods and approaches will be somewhat different and greatly expanded. The single development of the microprocessor and the ability to try completely different elevator operating strategies quickly are certain to lead to new discoveries and development in people-serving techniques which will assuredly be reflected in both *Vertical and Horizontal Transportation.*

Appendix: Literature on Elevators and Escalators

Although the available literature on elevators and escalators is very limited, there are a number of interesting publications available for those who wish to pursue or increase their knowledge of the many aspects of vertical transportation. A listing of those volumes familiar to the author will be given below.

An additional source is the promotional publications and magazine article reprints available from the various manufacturers and can usually be obtained by a written request. The Otis Elevator Company publishes two excellent booklets designed for lay people. They are entitled *Tell Me About Elevators* and *Tell Me About Escalators*. Single copies can be obtained free of charge by writing to The Otis Elevator Company, Public Relations Department, Farmington, CT. 06032.

Other publications are either obtainable from public libraries or the publisher. A brief description of their content is as follows.

Elevator World Magazine, P.O. Box 6506, Mobile, Alabama 36606, William Sturgeon, Editor and Publisher. This is a monthly publication published continuously since 1953. It contains articles and picture stories of unique, timely, or technical aspects of elevators and escalators as well as articles about the people and companies in the vertical transportation business. Once a year, an Elevator World Annual is prepared and describes, in detail, some aspect of the elevator industry or activity. There are a number of past issues of annuals which report the elevator business and practice in various sectors of the world such as Europe, South America, or The Far East. Other annuals include studies of elevator çodes and elevator consulting, safety, and modernization. Back copies are generally available and may be ordered from the publisher.

In addition to the magazine, *Elevator World* compiles and publishes special studies including *In Search Of the Past*, which gives a brief history of elevator development and *Court Cases of Consequence*, which is a compilation describing various legal actions and their outcome involving elevators. *Elevator World* also maintains an extensive literature and photo library in their Mobile headquarters and welcomes visitors at any time.

William "Bill" Sturgeon is the Founder, Editor, and Publisher of *Elevator World* and its publication has been his pride since he gave up his own repair and service company. He is most helpful in finding information and I value him as a good friend.

Electric Elevators, by Fred A. Annett, McGraw-Hill, New York, 1960. This book was one of the first efforts to compile a volume on the many aspects of elevators, especially from a technical standpoint. Mr. Annett was an editor of

Power Magazine during the 1950s and 1960s and the first edition of *Electric Elevators* was a compilation of various articles published in that magazine. Later editions of the book were updated with data and information gained as a result of Mr. Annett's visits and discussions with the engineers in various companies. The book is presently out of print but copies can be found in libraries. If one can be bought it represents an excellent commentary on the state of the art at the time of the edition's publication. Since elevators seldom wear out, much of the equipment described in *Electric Elevators* is still in every day use.

Vertical Transportation, by Rodney R. Adler, American Elsevier Publishing Company, New York 1970. Rod Adler was a publicity writer for the Otis Elevator Company during the 1960s and early 1970s. He did me the service of reviewing the first edition of *Vertical Transportation* during its preparation. Adler's book is an overview designed for lay people who are involved in understanding elevators and escalators and their importance to a building. It is not a technical publication but rather an excellent description of the many aspects and considerations that would be of interest to people involved in rental, financial, advertising, or building committees by helping them to understanding what the architects and engineers are talking about.

Pedestrian—Planning and Design, by Dr. John J. Fruin, Metropolitan Association of Urban Designers and Environmental Planners, 1971. Currently out of print but republication is planned. This volume is considered a vital reference for anyone engaged in the design and layout of any facility where people must be served in any quantity such as terminals, lobbies of buildings, exhibits, stadiums, or museums. This work is the only source known to the author wherein the area where people can walk or assemble is clearly defined in terms of how much space must be allocated related to pedestrian traffic volume. It is a source of information on how many people (and at what rate) can use stairs, corridors, revolving doors, sidewalks, subway turnstiles, gates, doors, and so on. It is an ideal companion work to *Vertical Transportation* if all the aspects of pedestrian accomodation in any facility are considered.

The National Elevator Industries Incorporated Installation Manual (NEII Installation Manual), originally published by NEII and now republished by Elevator World Magazine. This book could almost be called a "do it yourself" guide to installing elevators. It gives descriptive detail of the various steps required to install an elevator in a building, starting from the setting of rail brackets to the finished, operating elevator. Construction managers and architects would find the book instructive, as would anyone engaged in the vertical transportation field who may not have had the opportunity to observe installation progress at a construction site.

The book is fully illustrated with photographs and diagrams showing each of the major steps. It is also a guide to the unique elevator language where words such as rails, shaft, DBG, and others have highly specialized meanings. In addition, a dictionary of elevators and escalator terms is available as follows.

NEIEP (National Elevator Industry Educational Program) *Elevator Terms— An Illustrated Glossary*, available from National Elevator Industries Incorporated, 600 Third Ave. New York, N.Y. 10016. This small volume is a first attempt to put all of the unique terms used in elevator and escalator work together in a single source. It is quite informative and valuable to anyone who has extensive dealings with elevator or escalator documentation.

FEM—Section 7—Terminology for Lifts, Escalators and Passenger Conveyors, 1981. For the person who deals with vertical transportation on an international level, this extensive volume of elevator and escalator terms is available in eight languages—English, Dutch, German, Spanish, French, Italian, Danish, and Portuguese. Copies are available from *Elevator World Magazine*.

Elevators in The Eiffel Tower, Smithsonian Institute Bulletin No. 288, published by and available from the Smithsonian Institute, Washington, D.C. Although this booklet is primarily a historical and technical description of the elevator systems in the Eiffel Tower, it is an excellent history of the development of vertical transportation and provides an interesting report of some of the international negotiations that went on while the Eiffel Tower was being built. Readers will find that the original Eiffel Tower elevators were both double decked and designed to travel in an inclined, parabolic path with passenger seating arranged to articulate and maintain people's bottoms parallel with ground level. The first part of the booklet provides an in-depth study of the development of elevators both in the United States and abroad. The later section gives a commentary on international competition at the time the Eiffel Tower was being built, which is especially interesting in view of the recent (1982) modernization of the elevators.

The American Standard Safety Code For Elevators, Escalators, Dumbwaiters and Moving Walks, A17.1 and *The Inspector's Manual, A17.2*, American Society of Mechanical Engineers (ASME), 345 East 47th St., New York, N.Y. 10017. The A17.1 elevator code has been in continuous publication since 1922 and has served in an advisory capacity as a model for elevator safety standards throughout the world. It is continuously updated—the latest edition published in 1981 and the next edition planned for 1984. Each edition includes revisions to reflect the state of the art or clarification of previous rules. The code is supplemented by a tabulation of official interpretations which are consensus opinions based on inquiries to the code committee on how the various rules should be applied in specific situations. New interpretations can be obtained by writing ASME if a question arises about a rule and its application. A standard inquiry format appears in the front section of the code.

The code is developed by a main committee that meets about every six weeks. Actions are delegated to subcommittees who investigate and make recommendations for changes. These are discussed, voted upon by the main committee, and balloted, as well as published for public review. Upon resolution of negative comments, the altered or new rule or interpretation is included in the next edition.

The Inspector's Manual A17.2 follows the latest edition of the code and provides ways and means to judge if a rule is properly followed or applied. In addition to the inspector's manual, various check charts are available to aid in the inspection of elevators and escalators.

People interested in serving on the various code committees should contact ASME. Service is voluntary and people donate their time and skill to this rewarding work. I am currently Vice Chairman of the Main Committee and have enjoyed code work since 1970.

Reference to the code is a must for anyone concerned with vertical transportation. Various communities have enacted the A17.1 code or a modified version of it into local ordinances and these should be carefully consulted when working in that area. Major code documents are published in other countries and are similar to the A17.1 code. In Europe and the United Kingdom, the CEN 81 code is applied. In Japan, the Japanese Standard Code, which is almost a direct translation of A17.1, is used. In Australia, the Australian Standard Code includes many features of the A17.1 code plus more. These are all examples of national codes and are usually available from the standards group or elevator industry representatives in that country.

Also available from ASME are compilations of the various interpretations of the A17.1 code and a *Code Handbook* which describes the application of various rules of the A17.1 code with narrative and illustrations.

There are a number of additional volumes that may be of specialized interest to a reader. The following is a listing of those known to the author.

Aerial Ropeways and Funicular Railways, by Zbigniew Schneigert, Pergamon Press, New York, 1966.

Lift Traffic Analysis, Design and Control, by G. C. Barney and S. M. dos Santos, Peter Pergrinus, Ltd. United Kingdom, 1977.

Electric Lifts, by R. S. Phillips, Pitman, United Kingdom, 1973.

Monograph on Planning and Design of Tall Buildings, Volume SC, Tall Buildings Systems and Concepts, American Society of Civil Engineers, 1981. Chapter SC4 entitled "Vertical and Horizontal Transportation" is an overview on the relation of transportation systems to tall building design. It was prepared by Committee 2A, Richard T. Baum, Chairman, Rudiger Thoma, Vice Chairman, and William S. Lewis, Editor.

Also in that same volume is an appendix which lists a High Rise Building Base including tabulations of 100 tallest buildings in the world, the tallest buildings in major cities, a glossary of building terms, and an extensive bibliography on building construction.

The Soviets have also written a book on elevators which describes their standard elevator systems primarily designed for residential buildings. Its title is *Safe Operation of Elevators* and it is written in Russian. An excellent book for people who can read Czechoslovakian is *Lifts and Escalators*, by Lubomir Janovsky and Josef Dolezal. This volume is elaborately illustrated with many fine mechanical drawings of elevator equipment.

Index of Tables and Charts

Index of Examples

Subject Index